Library of
Davidson College

PERGAMON INTERNATIONAL LIBRARY
of Science, Technology, Engineering and Social Studies
*The 1000-volume original paperback library in aid of education,
industrial training and the enjoyment of leisure*
Publisher: Robert Maxwell, M.C.

ENERGY
A GLOBAL OUTLOOK

The Case for Effective International Co-operation

THE PERGAMON TEXTBOOK
INSPECTION COPY SERVICE

An inspection copy of any book published in the Pergamon International Library will gladly be sent to academic staff without obligation for their consideration for course adoption or recommendation. Copies may be retained for a period of 60 days from receipt and returned if not suitable. When a particular title is adopted or recommended for adoption for class use and the recommendation results in a sale of 12 or more copies, the inspection copy may be retained with our compliments. The Publishers will be pleased to receive suggestions for revised editions and new titles to be published in this important International Library.

Other Pergamon Titles of Interest

BACH *et al.*
Renewable Energy Prospects

DE MONTBRIAL
Energy: The Countdown

EGGERS-LURA
Solar Energy for Domestic Heating and Cooling

EGGERS-LURA
Solar Energy in Developing Countries

GABOR
Beyond the Age of Waste, 2nd Edition

GARDEL
Energy: Economy and Prospective (also available in French)

GRENON
Future Coal Supply for the World Energy Balance

GRENON
Methods and Models for Assessing Energy Resources

GRENON
The Nuclear Apple and the Solar Orange

KLITZ
North Sea Oil

MESSEL
Energy for Survival

SECRETARIAT FOR FUTURES STUDIES
Solar Versus Nuclear: Choosing Energy Futures

SIMEONS
Coal: Its Role in Tomorrow's Technology

STARR
Current Issues in Energy

UNECE
Coal: 1985 and Beyond

Pergamon Related Journals

ENERGY

ENERGY CONVERSION AND MANAGEMENT

INTERNATIONAL JOURNAL OF HYDROGEN ENERGY

JOURNAL OF HEAT RECOVERY SYSTEMS

PROGRESS IN ENERGY AND COMBUSTION SCIENCE

SOLAR ENERGY

ABDULHADY HASSAN TAHER, Ph.D.
Governor, General Petroleum and Mineral Organization (PETROMIN)
Riyadh, Saudi Arabia

Photograph: Dick Massey
Saudi Research and Marketing

ENERGY
A GLOBAL OUTLOOK

The Case for Effective International Co-operation

ABDULHADY HASSAN TAHER, Ph.D.

Governor, General Petroleum and Mineral Organization (PETROMIN)
Riyadh, Saudi Arabia

PERGAMON PRESS

OXFORD · NEW YORK · TORONTO · SYDNEY · PARIS · FRANKFURT

U.K.	Pergamon Press Ltd., Headington Hill Hall, Oxford OX3 0BW, England
U.S.A.	Pergamon Press Inc., Maxwell House, Fairview Park, Elmsford, New York 10523, U.S.A.
CANADA	Pergamon of Canada, Suite 104, 150 Consumers Road, Willowdale, Ontario M2J 1P9, Canada
AUSTRALIA	Pergamon Press (Aust.) Pty. Ltd., P.O. Box 544, Potts Point, N.S.W. 2011, Australia
FRANCE	Pergamon Press SARL, 24 rue des Ecoles, 75240 Paris, Cedex 05, France
FEDERAL REPUBLIC OF GERMANY	Pergamon Press GmbH, 6242 Kronberg-Taunus, Hammerweg 6, Federal Republic of Germany

Copyright © 1982 A. H. Taher

All Rights Reserved. No part of this publication may be reproduced, stored in a retrieval system or transmitted in any form or by any means: electronic, electrostatic, magnetic tape, mechanical, photocopying, recording or otherwise, without permission in writing from the publishers.

First edition 1982

British Library Cataloguing in Publication Data
Taher, Abdulhady Hassan
Energy: a global outlook. - (Pergamon international library).
1. Petroleum industry and trade
2. Energy policy
I. Title   II. Series
333.8'2   HD9560.6   80-41616
ISBN 0 08 027292 4 (hardcover)
ISBN 0 08 027293 2 (flexicover)

In order to make this volume available as economically and as rapidly as possible the author's typescript has been reproduced in its original form. This method unfortunately has its typographical limitations but it is hoped that they in no way distract the reader.

Printed in Great Britain by A. Wheaton & Co. Ltd., Exeter

بِسْمِ ٱللَّهِ ٱلرَّحْمَٰنِ ٱلرَّحِيمِ

ٱقْرَأْ بِٱسْمِ رَبِّكَ ٱلَّذِى خَلَقَ ۝ خَلَقَ ٱلْإِنسَٰنَ مِنْ عَلَقٍ ۝ ٱقْرَأْ وَرَبُّكَ ٱلْأَكْرَمُ ۝ ٱلَّذِى عَلَّمَ بِٱلْقَلَمِ ۝ عَلَّمَ ٱلْإِنسَٰنَ مَا لَمْ يَعْلَمْ ۝

IN THE NAME OF GOD, THE MOST GRACIOUS, THE DISPENSER OF GRACE:

(1) READ in the name of thy Sustainer, who has created-(2) created man out of a germ-cell! (3) Read-for thy Sustainer is the Most Bountiful One (4) who has taught [man] the use of the pen- (5) taught man what he did not know!

The Word of Almighty God is the truth.

CONTENTS

Foreword — xv

Part I An Economic and Political Evaluation

Introduction — 1

Chapter 1 - Historical Evolution of the International Oil Industry — 4

Chapter 2 - Global Energy Supply and Demand Balance — 7

Chapter 3 - A Historical Review of OPEC's Creation and Actions — 11

Chapter 4 - The Rationale for OPEC — 18

Chapter 5 - The Organization of Arab Petroleum Exporting Countries — 33

Chapter 6 - Historical Review and Rationale of the IEA's Creation, Policies and Action — 41

Chapter 7 - Impact of Structural Changes on the International Energy Industries — 51

Chapter 8 - Petromin, Saudi Arabian Oil Policies and Industrialization through Joint Ventures — 55

Chapter 9 - North-South An International Energy Dialogue — 67

Chapter 10 - Structural Changes and New Strategies — 73

Chapter 11 - Towards an International Energy Development Programme — 81

Chapter 12 - Epilogue — 86

Part II Energy Scenarios, Historical and Regional Analysis

Chapter 13	Global Statistical Review of Primary Energy	91
Chapter 14	Energy Scenarios for 1985 and 1990	104
Chapter 15	Energy Scenarios for the Year 2000	115
Chapter 16	The US Energy Situation	123
Chapter 17	The West European Energy Situation	139
Chapter 18	The Japanese Energy Situation	147
Chapter 19	The OPEC Developing Countries' Energy Situation	152
Chapter 20	The Non-OPEC Developing Countries' Energy Situation	159
Chapter 21	The USSR Energy Situation	170
Chapter 22	The East European Energy Situation	176
Chapter 23	The Energy Situation of the People's Republic of China	180

Part III Statistical Data and Appendices

Statistical Tables — 184

Historical Analyses	nos 1 to 34	184
Projections	35 to 55	219
Comparison of Forecasts	56 to 59	240

Appendix I	OPEC Resolutions, Agreements and Declarations	247
Appendix II	International Energy Agency	304
Appendix III	Report of the Conference on International Economic Co-operation	318
Appendix IV	Energy: Definition, Glossary, Explanatory Notes, Costs, Supply Lead Times, Units and Conversion Factors	341
Bibliography		374
Index		381

Contents ix

Illustrations

Facing page

Figure 1	Actual and Prospective Energy Consumption Mix of OPEC Member Countries, and Actual and Prospective Energy Consumption Mix of Non-OPEC Developing Countries	8
Figure 2	Actual and Prospective Energy Consumption Mix of the Industrialized Countries, and Actual and Prospective Energy Consumption Mix of the Centrally Planned Economies	9
Figure 3	The Relationship Between GNP per Capita and Oil Reserves to Production Ratio among OPEC Member Countries	24
Figure 4	World Oil Trade, Exports	55
Figure 5	Official and Spot Market Prices of Arabian Light Crude	74
Figure 6	World Proved Oil Reserves, Production and New Reserves Added	82
Figure 7	Energy Consumption Per Capita	83
Figure 8	World Economic Growth (GDP)	100
Figure 9	Energy Scenario Projections for 1985 and 1990	106
Figure 10	Energy Scenario Projections for the Year 2000	117
Figure 11	Actual and Prospective Energy Consumption Mix of the USA	136
Figure 12	Actual and Prospective Energy Consumption Mix of Western Europe	145
Figure 13	Actual and Prospective Energy Consumption Mix of Japan	151
Figure 14	Actual and Prospective Energy Consumption Mix of OPEC Member Countries	156
Figure 15	Actual and Prospective Energy Consumption Mix of Non-OPEC Developing Countries	168
Figure 16	Actual and Prospective Energy Consumption Mix of the USSR	174
Figure 17	Actual and Prospective Energy Consumption Mix of Eastern Europe	178
Figure 18	Actual and Prospective Energy Consumption Mix of China	182

Energy: A Global Outloook

Tables

Table 1	World Primary Energy Consumption - Year 1979	184
Table 2	Composition of World Primary Energy Consumption Percentage Breakdown (1979)	185
Table 3	World Primary Energy Consumption in 1979, Percentage Shares and Growth Rates	186
Table 4	Production and Consumption of Oil and Consumption of Other Forms of Primary Energy $\pm$ Average Annual Rates of Change (Percentage)	187
Table 5	Estimates of Total World Ultimately Recoverable Reserves of Crude Oil for Conventional Sources and NGLS	188
Table 6	Estimates of World Ultimately Recoverable Reserves of Natural Gas	190
Table 7	Estimated Proved Reserves of Oil and Natural Gas	191
Table 8	Relationship between Proved Recoverable Reserves and Estimated Additional Resources of Oil and Natural Gas by Geo-Political Groupings	192
Table 9	World Solid Fossil Fuel Resources and Reserves	193
Table 10	World Energy Resources Proven Remaining Recoverable Reserves	194
Table 11	The USA Indigenous Energy Resources	195
Table 12	US Primary Energy Historical Supply/Demand Trends	196
Table 13	Western Europe Indigenous Energy Resources	197
Table 14	West European Countries Indigenous Resources of Oil and Natural Gas	198
Table 15	West European Estimated Oil Production	199
Table 16	OECD European Countries Primary Energy Historical Supply/Demand Trends	200
Table 17	Japan Indigenous Energy Resources	201
Table 18	Japanese Primary Energy Historical Supply/Demand Trends	202
Table 19	OPEC Developing Countries Indigenous Energy Resources	203

Contents xi

Table 20	OPEC Developing Countries Indigenous Resources of Oil and Natural Gas	204
Table 21	OPEC - Crude Oil Production and Output Capacity	205
Table 22	OPEC Developing Countries Oil - Historical Supply/Demand Trends	206
Table 23	Non-OPEC Developing Countries Indigenous Resources of Oil and Natural Gas	207
Table 24	Non-OPEC Developing Countries Reserves and Resources of Solid Fossil Fuels	208
Table 25	Non-OPEC Developing Countries Uranium Resources	209
Table 26	Non-OPEC Developing Countries Hydro Resources - Installed and Installable Capacity	210
Table 27	Non-OPEC Developing Countries Indigenous Energy Resources	211
Table 28	Non-OPEC Developing Countries Oil - Historical Supply/Demand Trends	212
Table 29	The USSR Indigenous Energy Resources	213
Table 30	USSR Primary Energy Historical Supply/Demand Trends	214
Table 31	Eastern Europe Indigenous Energy Resources	215
Table 32	Eastern Europe's Primary Energy Historical Supply/ Demand Trends	216
Table 33	People's Republic of China Indigenous Energy Resources	217
Table 34	People's Republic of China's Primary Energy Historical Supply/Demand Trends	218
Table 35	Actual and Prospective Oil Balances Production, Consumption and Net Trade Implications	219
Table 36	Actual and Prospective Balances for Oil Shale, Tar Sands and Synthetics Production, Consumption and Net Trade Implications	220
Table 37	Actual and Prospective Natural Gas Balances Production, Consumption and Net Trade Implications	221
Table 38	Actual and Prospective Solid Fuels Balances Including Biomass Production, Consumption and Net Trade Implications	222

Table 39	Actual and Prospective Supply and Demand for Nuclear Energy	223
Table 40	Actual and Prospective Supply and Demand for Hydroelectric and Geothermal Power Including Solar and Other Renewable Forms of Energy	224
Table 41	Actual and Prospective Total Primary Energy Balances Production, Consumption and Net Trade Implications	225
Table 42	Actual and Prospective Energy Consumption Mix of the World	226
Table 43	Actual and Prospective Energy Consumption Mix of All Non-Communist Countries	227
Table 44	Actual and Prospective Energy Consumption Mix of the Industrialized Countries	228
Table 45	Actual and Prospective Energy Consumption Mix of the Centrally Planned Economies	229
Table 46	Actual and Prospective Energy Consumption Mix of the USA	230
Table 47	Actual and Prospective Energy Consumption Mix of Western Europe	231
Table 48	Actual and Prospective Energy Consumption Mix of Japan	232
Table 49	Actual and Prospective Energy Consumption Mix of Canada, Australia, New Zealand	233
Table 50	Actual and Prospective Energy Consumption Mix of OPEC Developing Countries	234
Table 51	Actual and Prospective Energy Consumption Mix of Non-OPEC Developing Countries	235
Table 52	Actual and Prospective Energy Consumption Mix of the USSR	236
Table 53	Actual and Prospective Energy Consumption Mix of Eastern Europe	237
Table 54	Actual and Prospective Energy Consumption Mix of China	238
Table 55	World Economic and Energy Consumption Growth	239
Table 56	A Comparison of Projections for Oil Production	240
Table 57	A Comparison of Projections for Oil and Energy Consumption	242

Table 58	A Comparison of Projections for World Oil Prices	245
Table 59	An Analysis of Changes in Oil and Energy Projections over Time	246

FOREWORD

Traditionally the world has remained heavily dependent on a single resource for the bulk of its energy needs. First it was wood - until the early 1800s; then it was coal - until the early 1900s. Since then it has been oil. The events of the past decade have brought about a timely realization that the world's energy needs have reached such enormous proportions that it simply cannot afford to rely any longer on any single non-renewable resource. To avert the danger of future global energy shortages in the medium to long term, the world needs to diversify its energy sources, to lessen its current heavy dependence on the fast depleting oil resource and develop other conventional, as well as non-conventional, renewable and non-renewable sources of energy.

This work addresses itself to significant historical, current and future issues related to the international oil industry and the global energy situation. It is based on the analysis of the typical oil and energy related problems faced by certain countries and groups of countries, and by the world as a whole. I have taken into consideration the divergent viewpoints and interests of the oil producing developing countries, the developing countries which do not produce oil and the industrialized countries. The objective is to emphasize the urgent need for international co-operation to cope with the situation, and to identify areas of such co-operation.

One may enquire, why a new book on energy? In recent years there has been a flood of books on this subject. Many of these books have been a result of research that took place over several years. Such books, research papers, and articles have followed a highly objective, one might even say scientific, approach, and have produced data of considerable enlightenment on the complex subject of energy. The transnational major oil companies have also contributed a fair share in producing various scenarios of world energy problems and, in particular, of world oil problems. It appeals to certain other writers to speak of world energy problems in terms of an energy or an oil crisis. So they also have written a number of books and many articles about the subject.

It is obvious that all over the world questions about oil and about other sources of energy keep arising almost every day. So research institutes, university professors, journalists, government officials, international organizations and oil companies try to answer them. In many cases their answers to these questions or 'problems' emphasize one or more 'particular' solutions, but in some cases a more 'comprehensive' set of solutions have been provided. In a few cases, even military options have been considered.

Before I lose track of my first question, I would like to emphasize that this book relies heavily on the most up to date available data about energy. That is to say, it relies on an analysis of the performance of the energy sector in various countries in recent years and of the impact of changes in the energy sector on the world economy. There is no magic about the solutions proposed here. They imply reasonable discipline through the adoption of appropriate fiscal policies; the recognition of the importance of investment for both conservation and substitution in the energy sector; and a genuine willingness to try and understand the preoccupations and problems facing other countries, whether they be exporters or importers of oil or other forms of energy.

Reference has been made to other books and studies which have been published. So, to explain the reason for a new book, I suggest that the main difference between this book and many others in the same field is that this one is an attempt to examine the energy problem on a global basis, within a resource-oriented as well as a political framework. It is an attempt to evaluate the various energy problems within a framework of international relations, not only between national governments but also between them and transnational oil or 'energy' companies. It takes into consideration the global energy resource base and its potential, and approaches these problems in an integrated and 'aggregative' manner. I think, it is only through such an approach that one can gain a real appreciation of the magnitude of the problems involved, economic, political or otherwise, and hence develop a more comprehensive assessment of the complexity and significance of the solutions. In this way, decisions to adopt certain solutions may be made within a more rational atmosphere.

In order to make it easier for the reader to have a quick access to the main conclusions of this study, it is divided into three main parts that complement each other. Part I embodies the economic and political evaluation of the energy problems and the potential solutions. Part II comprises the various energy scenarios, historical and regional analyses on which the evaluation has been made, supported by the statistical data and appendices in Part III. So, for those who want to examine the underlying data and assumptions, Parts II and III should be read first. Otherwise, Part I contains the principal conclusions of the study.

This work incorporates many lectures and speeches which I have delivered over the fifteen years since I published my first book 'Income Determination in the International Oil Industry'. My initial impulse was to publish these speeches in their original form. However, reflecting further on the subject, I was convinced that to keep pace with a highly dynamic international economic and political environment and changes in the oil market, I must re-evaluate and integrate the various ideas and views contained in these speeches.

During the Conference on International Economic Co-operation, held in Paris from December 1975 to June 1977, I had the privilege of listening to a variety of views and of appreciating the economic and energy problems facing different countries. Before this Conference the emphasis in my thinking was orientated towards the financial aspects of the international oil industry and towards those oil issues related to Saudi Arabia in particular. I must admit, the Conference enlarged this perspective and I started to think more about the global aspects of the energy and economic problems.

This was the time when I started to conceive the idea of writing this book in its present form. I have tried to re-evaluate all the main ideas and views contained in my previous speeches within the perspective of the present and projected global energy situation.

This was not a simple assignment. Inevitably, it involved research assistance. During the completion of the first draft of this book this research assistance was provided by Mr Mohammad Jamil, who has been serving Petromin's Market Research Department as an adviser since December 1968, and who also served as my research adviser during my tenure as the co-chairman of the Energy Commission in the Conference on International Economic Co-operation.

By the time I completed the first draft, I recognized the need for revisions of various energy scenarios and projections contained in it. The rapid succession of new developments, the availability of more up to date statistics and the clearer emergence of new trends since 1973 made this essential.

At this stage I requested Mr Michael Matthews, who has been serving Petronal as an adviser since 1978 and who has been associated with the petroleum industry since 1959, to join Mr Jamil to provide the additional research assistance required, to re-examine the first draft, to update the statistics and the energy scenarios, and to participate in the final editing of the book.

I wish to extend my sincere sense of appreciation to both Mr Jamil and Mr Matthews for their valuable assistance which helped me to prepare this book in its final form. Neither Mr Jamil and Mr Matthews, nor Petromin, are responsible for any of the views presented in the book. This responsibility is entirely mine.

Jeddah, Saudi Arabia
April, 1981

Abdulhady H Taher

Part I
An Economic and Political Evaluation

INTRODUCTION

When oil supplies were relatively short after the 1956 Suez war, exploration and production of oil and oil derivatives picked up tremendously. Oil became more and more abundant as the Gulf, Libyan, Algerian, Nigerian and other sources of supply increased over the years. Hence, in the sixties, the world began to believe in the wishful dream of endless cheap oil supplies. The industrialized world consumed ever increasing quantities of oil at unbelievably low and falling real prices, and people in most parts of the world got accustomed to the abundant and cheap supplies of oil.

In 1971, I was invited by the Japanese Institute of Energy to exchange views with them on oil questions related to Japan's future oil demand. During the course of the discussions I reviewed with them their estimates of Japan's anticipated demand for oil in the seventies, quantified in the hundreds of millions of tons of crude. I asked them a very simple question: from where would they get such huge quantities of oil? The reply was equally simple: from the Middle East, of course. This was a highly knowledgeable group of oil specialists. Hence I suppose that their simple reply did not really represent their inner scepticism about the validity of that statement. The very fact that they wanted me to look at the estimates of their future oil demand was in itself an indication of their doubts about the availability of such large quantities of crude. In fact, as late as 1975, when we met in Paris in the Conference on International Economic Co-operation (CIEC), to discuss, among other questions, oil availability to the industrialized and developing countries, one of the delegates questioned the limitations of future availability of crude oil. He made the comment that the oil producing countries were playing mathematical games with oil supplies in order to increase oil prices.

Nevertheless, it is probable that future availability of crude oil can hardly meet demand, even at mid-1981 prices, unless serious steps are taken by various countries to develop additional oil supplies as well as alternative sources of energy, and to improve the efficiency of oil use. This, in fact, is the problem in simple terms. It is possible to talk about a hypothetical gap between the absolute maximum potential worldwide availability of crude oil and the potential worldwide demand for such oil required to enable the world economy to grow without an energy constraint.

In the market, at any one point in time, a gap of the type postulated above never exists, since demand and supply actually are always in balance. What may happen in such situations is that those who cannot afford the equilibrium price will have to forego acquiring additional supplies. If there is an oil crisis or the danger of one happening, it is meant in this sense. This potential gap is, in a way, the bomb waiting to explode, either in terms of acute shortage of world oil supplies, or in very high equilibrium prices, or most probably, some combination of both these things happening simultaneously. It is postulated in this book that through individual country by country actions together with some form of international or regional co-operation, such an explosive situation can be averted.

It might be relevant at this stage to mention that the international oil industry, the most important energy production and trading industry, has grown and prospered in a multi-national environment. Its survival, as I said many years ago in my first book about this industry, is a function of its ability to deal with environmental changes, especially those related to its external environment. Such an environment might be described in terms of its owners, those working in the industry, the consumers of oil products, the general public, the other organizations in the same industry, the government or governments which are related to the organization, the national and world economy, and finally, technology with all its ramifications. One may add, today at least, one other factor that may be considered within the environment of the international petroleum industry, namely, access to hydrocarbon reserves and supplies.

So, in order to examine the global aspects of the energy problem, one needs to deal first with the international oil industry, since what is called an energy crisis has a great deal to do with this industry. The great contribution made by this industry to increasing world oil reserves and supplies cannot be ignored, especially that of the major transnational oil companies and the oil industries of countries like the USSR, Romania and China.

To some, it may seem that a very simple solution could be to ask the industry to accelerate its efforts to find and produce more oil in order to avoid a future energy crisis. Unfortunately, the situation is not that simple, in spite of the fact that the world oil industry is doing the best it can under the circumstances. In the first place the circumstances are very complex. In some ways, they inhibit the effectiveness of the industry at certain times and in certain countries, due to political, economic and institutional constraints. I have examined these constraints and have tried to arrive at certain guidelines as to possible modifications of them in order to reduce, as far as is reasonably possible, their inhibiting effects.

Nevertheless, owing to the decreasing probability of finding large hydrocarbon reserves with the present technology, a situation quite different from that which prevailed in the fifties and sixties has arisen. This means that alternative sources of energy such as hydro as well as nuclear, coal and solar must be tapped whenever possible in order to conserve oil for uses in which these alternative sources cannot compete, either for technological or economic reasons.

Accordingly, an examination of such alternatives seems to be in order so that we can assess future oil supplies in a more rational manner. Hence, the energy scenario projections become relevant to the study and evaluation of various solutions, be they proposals related purely to investment or to international co-operation, or to institutional changes that may have to be established in order to achieve the desired goals.

In this process, one is bound to analyse not only the resource base available, but also the various factors within the external environment. Furthermore, the introduction of the accelerated development of alternative sources of energy on a worldwide basis becomes an integral part of the analysis.

Necessarily, one must give serious consideration to governments. They have a major role to play in the attempt to solve world energy problems. I stated in my book 'Income Determination in the International Petroleum Industry', that there are five significant factors in the international petroleum organization's relationship to governments, namely, the three governments of the "host producing countries, host consuming and home countries, in addition to groups of host governments whether producing or consuming".

This was a statement made in 1964. World conditions have changed, and one should now also highlight the role of the developing countries.

To sum up, in order to gain an objective appreciation of the global energy problems, I propose to examine the energy supply and demand picture of various countries or groups of countries. This is basically done in Parts II and III where up to date statistical information is compiled from various sources around the world. The results of the statistical analysis of the relevant information for various countries will be used from time to time in the review of the energy problems and their possible solutions in Part I, as well as in the analysis of international relations involving energy problems. Other chapters will deal with the questions of relations between transnational oil companies and producing, as well as consuming, countries, be they classified as industrialized or developing. Understanding the energy policies of some countries that have a major effect on the global energy scene forms one section of this book. Understanding OPEC and IEA comprises another section. The role of national oil companies as an extension of the oil policies of the oil producing as well as the oil consuming countries forms a third major section of this book. A central issue relevant to the role of the national oil companies, to OPEC and to the IEA, is the examination of oil prices and supplies.

Finally I propose to analyse the world energy situation in global terms and see if I may come up with some useful guidelines to contribute to the solution of the world energy problems.

1
HISTORICAL EVOLUTION OF THE INTERNATIONAL OIL INDUSTRY

In order to examine the global aspects of the energy problem it is necessary to deal first with the international oil industry, since the so-called 'energy crisis' has a great deal to do with this industry.

EARLY HISTORY

The oil industry came into being in the year 1859 when 'Colonel' Edwin L Drake successfully drilled the first commercial oil well at Titusville, Pennsylvania in the United States of America.(1)

After establishing a domestic oil industry, American private oil companies were the first to expand their operations into the international markets. By the early 1920s, these companies had become the main suppliers of oil to countries all around the globe and were well established as sales organizations handling their own oil in foreign countries. They also started the search for foreign oil reserves to supplement their home-based oil reserves to meet the growing international demand for oil.(2)

The British, who were at the height of their imperial power at the beginning of this century, had little oil either in the United Kingdom or in their empire and commonwealth countries. Not wanting to remain dependent for oil supplies on the United States of America, they also started to search for oil in the Middle East which, at the time, was conveniently in their sphere of influence.(2)

TRANSNATIONAL OIL COMPANIES

The above explains why the seven predominant major oil companies who virtually controlled the international oil industry from its inception until quite recently all had the USA or Britain as their home bases.(2) These companies, namely, Standard Oil of New Jersey (now Exxon), Standard Oil of California, Socony-Vacuum (now Mobil), Texaco, Gulf, Anglo-Persian (now BP) and Royal Dutch/Shell (60% Dutch, 40% British) are known as 'The Seven Sisters' or 'The Transnationals'. The latter name signifies the fact that although these companies are mainly owned by stockholders in a single or a few industrialized countries, their operations are internationally based. The transnational character of these oil companies was also accentuated by the fact that the main oil producing/exporting areas - the Middle East, North and West Africa, Venezuela and Indonesia - had relatively small domestic oil markets because of their lack of economic development, whereas the major oil consuming developed countries of Western Europe and Japan had hardly any oil of their own until the recent oil discoveries in the North Sea. The USA, which was traditionally self-sufficient in oil, has also become a substantial importer of oil since 1970, and the largest in the world since 1973. Japan's indigenous oil production remains negligible in relation to the country's oil consumption.

The rapid rise of the transnationals to their commanding position in the international oil industry could be attributed to two major factors. The first and foremost was undoubtedly the economic, political, military and naval power of their home countries, whereas most of the oil producing countries whose resources they exploited were relatively weak, underdeveloped countries. The second factor was the strength of their integrated operations. Having control over large market shares in the major consuming countries and having control over all phases of the oil industry, namely, exploration, production, transportation, refining and marketing, the transnationals had all the economic advantages to survive and prosper in an industry characteristically involving high risk and being capital intensive in all its phases.(2)

OTHER INTEGRATED OIL COMPANIES

As a competitive response to burgeoning oil demand and high profits, many other oil companies of intermediate size with integrated operations, but without an international position like the transnationals, also became actively established in the international oil industry. These include the US based independents, and companies based in France, Italy and Japan as well as a few other companies in other countries. The influence of these companies on the structure and performance of the international oil industry became quite significant during and after the late 1950s.(3)

NATIONAL OIL COMPANIES

The gradual evolution of national oil companies can be traced back to the early 1920s. The first state oil corporation of a developing country was founded in Argentina in 1922, and Mexico nationalized her oil industry in 1938. Among the industrialized countries Austria can be cited as one example where a national oil organization has been operating successfully since the mid-1950s.(4) By 1977, the total number of national oil companies operating in the developing countries had reached eighty.

During the post World War II period of prosperity in the Western industrialized countries most of the oil producing developing countries were subsisting under conditions of extreme under-development. Their natural energy resources were exploited by the transnational oil companies under long term concession agreements. At the time when these concession agreements were signed, most of the host countries neither had the means to study and realize their full long term significance, nor had they the legal capacity or the political strength to protect their national interests in all respects. With the increase in their awareness of the importance of oil as an exceptionally versatile source of energy and a precious commodity in international trade, the oil exporters began to feel the need for organizations which could bring about effective national control over their petroleum resources so that these could be exploited to realize their national development aspirations.

The establishment of the national oil companies fulfilled this objective partially. It is only during the past decade or so that the trend towards the creation of national oil and gas agencies has received an overwhelming momentum. The creation of the Organization of the Petroleum Exporting Countries (OPEC) in 1960 and the Organization of Arab Petroleum Exporting Countries (OAPEC) in 1968 has also played a pivotal role in the power struggle over oil production and price policies. New national oil companies with more or less similar objectives have also been established in Norway, Canada and the UK during the past decade.

As a reaction to what was seen in some quarters as confrontation between some of the OECD countries and OPEC, the International Energy Agency (IEA) was created in 1974.

During the time preceding the creation of the IEA, the OECD countries, including those belonging to the EEC, did not pay enough attention to the global aspects of their oil industries and energy policies. Also the countries of origin of the transnationals failed to give adequate attention to the growing struggle between these companies and the oil producers. There was no incentive to do so, since these countries were reaping some of the benefits enjoyed by the transnationals as a result of their powerful position. The creation of the IEA was conceived as heralding the demise of OPEC. However, the IEA has evolved ever since as a paradoxical supporter of some of OPEC's policies, as we shall see later on.

In order to gain further appreciation of the energy problem, it seems advisable that this historical background should be supplemented with a comprehensive set of energy facts and figures. Let us first consider the current thinking about the global energy supply and demand balance.

References

(1) Sell, George, F. Inst. Pet. The Petroleum Industry. Oxford University Press, London, 1963.

(2) Frankel, Paul H. The Rationale of National Oil Companies. Keynote Address at UN Symposium, Vienna, 7-15 March 1978.

(3) Hardesty, C. Howard Jr. The Role of the Independents in the International Oil Industry. Speech at OPEC Seminar, Vienna, 10-12 October 1977.

(4) Baum, Vladimir. Introduction. Speech at UN Symposium, Vienna, 7-15 March 1978.

2
GLOBAL ENERGY SUPPLY AND DEMAND BALANCE

HISTORICAL PATTERN OF ENERGY CONSUMPTION

During 1979, total primary energy consumed in the world is estimated to have been 6.9 billion tons oil equivalent - 4.8 billion tons oil equivalent (69 per cent) in the non-Communist world and 2.1 billion tons oil equivalent (31 per cent) in the Communist bloc countries including the USSR, Eastern Europe and China (see Tables 1 and 2 in Part III). The percentage distribution is shown for 1979 and compared with 1973 and 1967 in Table 3 to provide some historical perspective. This is complemented by a presentation of compound average annual growth rates in Tables 3 and 4 which shows the sharp changes for oil and gas between 1973 and 1979 compared with the preceding six year period.(1) Reliable estimates for 1980 are not available as this book goes to press. Furthermore, energy demand in 1980 for non-Communist countries as a whole is likely to have been lower than in 1979 mainly because growth of economic activity in the industrialized countries was below the medium term trend rate in 1980. Thus it is considered appropriate to examine the data for 1979 and make comparisons of trends up to that year.

NON-COMMUNIST WORLD

In the non-Communist world oil remains the predominant source of energy. In 1979 oil's share in total primary energy consumption was 51.9 per cent. Corresponding shares of other forms of energy were: natural gas 18.0 per cent, coal 19.9 per cent, hydroelectric power* 7.3 per cent and nuclear* energy 2.9 per cent (see Table 3).

Consumption of energy in general and of oil in particular has tended to be concentrated in the industrialized, developed countries. In 1979 the developed countries consumed about 81 per cent of total primary energy and 79 per cent of the oil. The developing countries' share of oil consumption was only 21 per cent, whereas they produced around 71 per cent of total oil and accounted for 73 per cent of the population.

Between 1973 and 1979 the industrialized countries did manage to reduce their dependence on oil by 2.5 percentage points, to 50.7 per cent of total primary energy demand. But in the four years after the trough of the recession in 1975, the reduction

---

* on a high factor, or fossil fuel input basis

was only 0.5 percentage points. In 1978 their use of oil actually represented a higher proportion of their primary energy demand than it had done in 1975. This reflected a decline in the real oil price over this period.

International trade in energy, at present, is concentrated in crude oil and petroleum products. International movements of coal and natural gas are small compared with consumption of such locally produced energy, and electricity movements internationally are insignificant. In fact, the one important energy trade movement is the export of petroleum by developing countries to the developed industrialized countries.

Oil is not only a source of energy, but has many non-energy uses. At present there are no satisfactory substitutes for oil based lubricants. Feedstock for petrochemicals, fertilizers and pharmaceuticals are other significant non-energy applications for petroleum. Until the year 1973, prices of oil were exceptionally below their replacement values. The versatility of oil's use, combined with the availability of sufficient quantities for all consumers at exceptionally low prices, resulted in very rapid growth in worldwide oil consumption, and a fast-growing shift from dependence on coal as a traditionally predominant source of energy to increasing dependence on oil. Development of other sources of energy remained comparatively uncompetitive and slow. During the period 1967 to 1973 the consumption of primary energy in the non-Communist world was growing at an average annual growth rate of 5.4 per cent. The corresponding growth rate for oil was 7.5 per cent. During the same period, the share of oil in total energy consumption rose from 47.8 per cent to 54.0 per cent, while the share of coal declined from 26.0 per cent to 19.2 per cent. The share of natural gas increased from 18.5 per cent to 19.1 per cent. The share of nuclear energy increased by less than one percentage point and the share of hydroelectricity declined.

The upward readjustment of oil prices in 1973-74 was intended to place some check on the unbridled oil demand growth, and to shift some of the burden of supplying the world's growing energy needs away from oil to other more plentiful energy resources. During the period from 1973 to 1979 the average annual growth rate of primary energy consumption was reduced to 1.7 per cent only, and that of oil consumption to only 1.0 per cent. In consequence, the oil share of primary energy demand fell back to 51.9 per cent. Over the six years to 1973 oil's share of primary energy consumption rose 6.2 percentage points. Over the six subsequent years it fell by only 2.1 percentage points.

COMMUNIST BLOC COUNTRIES

In the Communist bloc countries energy consumption is heavily concentrated on what is produced locally. Coal predominates, accounting for 48.4 per cent of primary energy consumption in 1979, though having declined from 59 per cent ten years previously. Oil consumption amounted to only 633 million tons or little more than one quarter of that in the non-Communist world, though the Communist countries accounted for 55 per cent of the net growth in world oil consumption from 1973 to 1979. Even so, oil represented only 30.0 per cent of primary energy consumption in the Communist bloc countries in 1979, up 1.5 percentage points from six years previously. Corresponding shares of other forms of energy in 1979 were: natural gas 18.0 per cent, hydroelectricity 2.8 per cent, and nuclear energy about 0.8 per cent (see Table 3).

ENERGY RESERVES

Estimates for the size of the world's ultimate and remaining recoverable reserves of energy resources are shown in Tables 5 and 10. Table 10 shows a distribution of oil and gas reserves in 1979 by categories of countries.

Figure 1

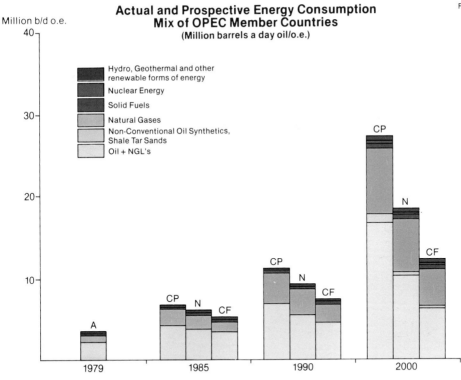

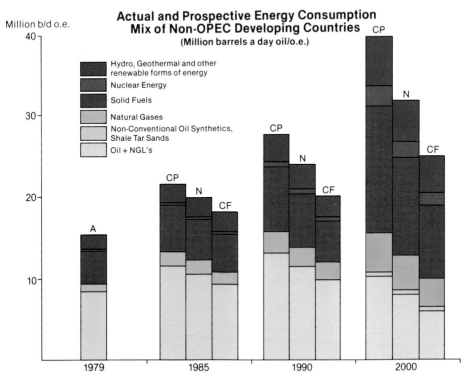

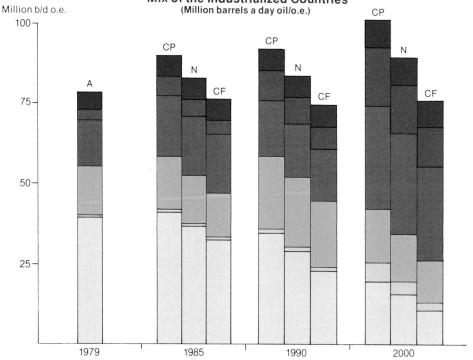

Actual and Prospective Energy Consumption Mix of the Industrialized Countries
(Million barrels a day oil/o.e.)

Figure 2

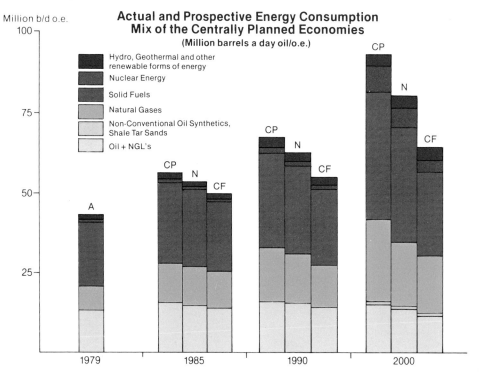

Actual and Prospective Energy Consumption Mix of the Centrally Planned Economies
(Million barrels a day oil/o.e.)

Legend:
- Hydro, Geothermal and other renewable forms of energy
- Nuclear Energy
- Solid Fuels
- Natural Gases
- Non-Conventional Oil Synthetics, Shale Tar Sands
- Oil + NGL's

Estimated coal reserves are sufficient for nearly 200 years at current rate of coal usage.

On the other hand, the remaining non-renewable proven recoverable reserves of oil and natural gas liquids in the world are depleting at a very rapid pace. These reserves have been declining in recent years. Rapid depletion has occurred to sustain a high level of consumption, and the finding rate has fallen below the level of demand. There are now serious concerns about the life expectancy of the remaining proven reserves if the present trends continue. Consequently, there is urgent need to search for and discover new petroleum reserves which could provide enough lead time for research, development and production of alternative non-renewable, as well as renewable, sources of energy to meet the medium to long term international energy requirements without economic disruption, especially in some of the developing countries.

One of the main reasons why this problem did not emerge in the late sixties and early seventies is that although oil demand was rising rapidly to very high levels, there were also very large finds of new oil reserves. For some years now, however, oil consumption has exceeded the rate of finding new oil reserves, in spite of the much lower average rate of growth of oil demand since 1973. It is now widely accepted that the probability of finding significant numbers of new giant oil fields in the future is low.

A further important fact is that a high proportion of oil discovered over the last 120 years has been located in a relatively few giant fields. The rate of finding and proving new oil reserves may well fall from the level of 14 to 15 billion barrels in recent years to less than 10 and perhaps as little as 5 billion barrels annually by the year 2000. Even though there is some scope for allowing oil reserves to production ratios to fall in many producing areas, it would be unwise to allow these to decline too drastically, otherwise the transition away from dependence on oil could become unmanageable. Improvements in the recovery factor for oil from existing reserves have been achieved with the development of sophisticated secondary and tertiary recovery techniques. Further increases in the recovery factor will be achieved, but these hold promise of simply alleviating the problem of the transition and extending its duration for some years. No medium term solution to the problem of the transition can be relied upon from this quarter, though some commentators are optimistic about significant improvements being made in the recovery factor in the long term.

1973 A TURNING POINT

The year 1973 can be seen as something of a watershed in the history of the oil industry. It marked a turning point, in that, up to that year, the industry had been dominated by the exponential growth in demand for oil which was doubling each decade. Since then, the initiative has been with the production end of the business.

Several changes were taking place in the oil industry over the period 1970 to 1973. These culminated in two large increases in the price of oil exported by members of the Organization of the Petroleum Exporting Countries during a period of less than three months.

There is a case for claiming that OPEC was the catalyst which brought about sudden change in the international oil industry by imposing these unilateral price increases and by other initiatives. There can be little doubt that OPEC actions accelerated the transition from a demand to a supply orientated industry. However, the underlying cause of change was the pressure of exponential rates of demand growth pressing on a finite reserve of a unique and depleting non-renewable resource, petroleum. A change would have occurred anyway, in due course of time, even if OPEC had never been formed. Market forces would have led to such change, most probably with greater shocks.

History may yet record that the timeliness and extent of OPEC moves led to a less cataclysmic change in the world's energy mix over the long term than if the initiatives of the 1970s had never been taken. If that should prove to be the verdict, a consequential smoother evolution of political, economic, social and technological change might be seen as having contributed to the survival of our system. A continuation of increasingly heavy reliance on cheap oil to satisfy the world's energy needs throughout the 1970s might actually have jeopardized mankind's future. Catastrophic change and a major discontinuity would have become increasingly probable, the longer the pre-1974 trends continued. The long lead times required to develop alternative forms of energy - even more painfully evident in 1980 than in 1973 - continue to make a smooth transition quite difficult enough seven years after the first large oil price increases.

The events which have occurred since late 1978 have had a further dramatic effect in markedly increasing oil prices. This was not consistent with the growth in world oil production in 1979 when there was a negligible change in consumption. The recent changes in the world oil balance are discussed elsewhere in this book, notably in Chapter 13 of Part II.

No group of countries is likely to secure satisfactory economic growth over the next twenty years without adequate resolution of oil and energy problems. This will entail planning and achieving the most appropriate energy mix over an extended period of time. It will be necessary to take into consideration the long lead times needed to develop alternative sources of energy such as nuclear as well as other energy related constraints discussed throughout this book. Energy sector difficulties are likely to beset almost all countries including OPEC members. The latter are quite alive to the potential problems which will surely affect them if they allow domestic demand for oil to grow more rapidly than their resources will allow.

Figure 1 shows comparative energy scenario projections for OPEC member countries as a group and for other developing countries. Figure 2 shows similar projections for the industrialized countries and for the Communist bloc. The discussion of these projections forms the subject matter of Part II of this book. Tables of statistics are included in Part III.

Reference

(1)   BP Statistical Review of The World Oil Industry 1979 and earlier issues. The British Petroleum Company Limited.

3
A HISTORICAL REVIEW OF OPEC'S CREATION AND ACTIONS

Traditionally, the international oil industry was described as an oligopoly with a competitive fringe. At one time the seven transnational oil companies, having completely integrated networks, were controlling over 73 per cent of the world oil market. The competitive fringe was composed of some large and many small independent oil companies with varying degrees of partially integrated networks. These were competing for the remaining 27 per cent of the market.

Over the past two decades this structure has gone through irreversible changes as a result of many factors, among which the most important was the creation of the Organization of the Petroleum Exporting Countries (OPEC) and its historic actions. According to one estimate, the transnational oil companies' share, under secure contracts, of non-Communist crude exports was as low as 35 per cent by early 1980. This would represent only about two thirds of their own downstream requirements. The following is a historical review of the events leading to the creation of OPEC and subsequent actions taken by this organization.

CREATION OF OPEC

As early as the mid-1940s a general realization of the adverse effects of the unfavourable terms of the traditional concession agreements on their national development had started to emerge in the oil exporting developing countries. It was during this period that some of these countries started to initiate various measures aimed at co-ordination and harmonization of their attitudes and policies towards the concessionaire operating companies.

In order to achieve these objectives, the Arab League, ever since it was founded in 1945, had felt the need for creating a petroleum association of Arab countries. However, it was realized that without the participation of such non-Arab large exporters of petroleum as Iran and Venezuela, such an organization was unlikely to act effectively under the circumstances prevailing at that time.(1)

In 1947, at the start of negotiations between Iran and its concessionaire, the Anglo-Iranian Oil Company (now BP), Iran and Venezuela started consultations for the purpose of co-ordinating their oil policies.(1)

During 1949 and 1951 oil missions from Venezuela, Iran, Iraq, Kuwait and Saudi Arabia exchanged visits, to explore avenues for regular and closer exchanges of views. The main objective was to prevent the companies from using the differences in benefits to these countries to play off one exporting country against another.(1)

In most of the original Middle Eastern concession agreements, payments to the host governments were set at a fixed royalty per ton of oil produced and exported -- generally four gold shillings per ton, which at the time was equivalent to around $1.65 per ton or 22 cents a barrel. In 1948, the government in Venezuela raised its tax rate to provide for 50-50 division of profits between it and the companies. This formula was extended to Saudi Arabia, Kuwait and Iraq during 1951-1952. This resulted in a higher level of unit oil revenue, which rose eventually in the mid-fifties to something in the region of 80 cents a barrel from the flat royalty rate of 22 cents a barrel in 1950.(2)

During April 1959 the Economic Council of the Arab League sponsored the first Arab Petroleum Congress in Cairo, which was also attended by Venezuela and Iran as observers. Informal consultations between the delegates of Saudi Arabia, Venezuela, Iran, Kuwait, the former United Arab Republic (Egypt and Syria) and the Arab League Petroleum Department resulted in the drawing up of a document of understanding. This document published only in 1961 throws much historical light upon subsequent developments in Venezuelan-Middle Eastern relations, the establishment of OPEC and the unfolding of its policies.(2) The Congress also adopted a resolution calling on oil companies to consult the oil exporting governments before making any price alterations.

In February 1959 the transnational oil companies decided to reduce prices by 5 to 25 cents per barrel for Venezuelan crudes and by 18 cents per barrel for Middle Eastern crude oils. In August 1960, the oil companies further reduced posted prices of Middle East crudes by between 10 and 14 cents per barrel.(2) At that time these represented significant percentage reductions. The two successive cuts were decided without any prior consultations with the governments concerned. The only reason offered for these price cuts was the state of the market, a market largely monopolized by these companies. These actions were clearly in the interest of the companies and their home countries and did not take into consideration the national interests of the oil exporting developing countries.

All these events finally culminated in the historic meeting held in Baghdad, on 10-14 September 1960, where representatives from Iran, Iraq, Kuwait, Saudi Arabia and Venezuela laid the foundation of OPEC as a permanent inter-government organization with an international status. In accordance with Article 102 of the United Nations Charter, the agreement creating OPEC was duly registered with the Secretariat of the United Nations on November 6, 1962.

At first OPEC headquarters was based in Geneva. In 1965 it moved to its present location in Vienna.

OPEC's membership, which started with five countries in September 1960, has now reached thirteen countries, comprising Algeria, Ecuador, Gabon, Indonesia, Iran, Iraq, Kuwait, Libya, Nigeria, Qatar, Saudi Arabia, United Arab Emirates and Venezuela.

OPEC'S FIRST DECADE

OPEC's role during the first decade of its existence was mostly as a deterrent against unwarranted oil price cuts by companies. The availability of potential surplus production capacity to the oil companies, during this period, had produced a buyers' market which acted as a limiting factor on OPEC's capacity to act effectively against the traditional company practices.

At the time OPEC was founded, the general reaction of the transnational oil companies and major industrialized oil consuming countries was that the Organization would not last for very long. Their conviction was based on the theory that OPEC members would never be able to work out common policies because of their divergent interests. They also maintained that creation of OPEC would generally be detrimental to the long term growth and development of the international petroleum industry. By the end of the first decade of its existence, both of these contentions were proved to be wrong.

The exercise of OPEC members' sovereign right over their own petroleum resources was effected first through various actions including those related to oil prices, tax rates, royalty expensing, and later through implementation of policies related to government participation in the ownership of the concession holding companies.

Evolution of Petroleum Policy in Member Countries

Resolution XVI.90, entitled 'Declaratory Statement of Petroleum Policy in Member Countries' adopted by OPEC's sixteenth conference held in Vienna during 24-25 June 1968 (Appendix I.2) constitutes the first and one of the most comprehensive documents defining OPEC's collective policy concerning matters related to the inalienable right of all member countries to exercise permanent sovereignty over their hydrocarbon and other natural resources. This resolution became the basis of very significant subsequent developments.

Similarly, Resolution XXI.120 adopted at the twenty-first OPEC conference held in Caracas in December 1970 and Resolutions XXII.131 and XXII.132 adopted at Tehran during 3-4 February 1971 (Appendices I.3, I.4 respectively) constitute documents of fundamental importance concerning subsequent developments related to oil prices, price differentials, royalty and tax rates.

It can be said that during the first decade of its existence, OPEC succeeded in defining its objectives and formulating basic principles to gain effective control over the exploitation and development of its natural resources and initiating the process of change and reversal in the power relationship between the member governments and the transnational oil companies.

OPEC'S ACHIEVEMENTS DURING THE SEVENTIES

By the early 1970s the buyers' market had started to turn into a sellers' market. The consciousness of rapidly rising oil consumption and the long term implications of this trend for their predominant natural resource, namely petroleum, finally served to reinforce OPEC members' initiative in exercising their sovereignty over the disposal of their primary, precious and fast wasting asset.

The Tehran and Related Agreements

By 1970 the evolving OPEC versus transnational oil companies relationship had reached a new turning point. OPEC began to adopt entirely new and more assertive approaches in dealing with the problems of prices and negotiations with the companies. There were numerous elements which provided strength to the negotiating base of OPEC members. This was the time when demand for OPEC oil was growing at faster rates, reflecting the high economic growth in the OECD countries and the entry of the USA into world markets as an important importer of increasing quantities of OPEC crude. The tightening supply situation worsened when Libya decided to cut back by up to 30 per cent(3) the production rates of many independent companies operating there. This followed the failure of successive negotiations with companies to settle Libya's long standing demands. An accidental explosion of the pipeline transporting part of the Saudi oil production to the East Mediterranean terminal of Sidon further contributed towards the growing shortages of crude oil, particularly of short-haul supplies. All these developments resulted in a sudden reversal in market price movements. For the first time, realized market prices for all short-haul crudes went substantially beyond posted prices.

Libya found herself in an exceptionally better bargaining position. Unlike the Middle East, where the transnational oil companies held a virtual monopoly situation, there were over 20 independents operating in Libya. Many of them, such as Occidental, had no access to any other crude oil elsewhere. Under the circumstances, the companies accepted, one after the other, the Libyan demands of correcting the price upward by 30 US cents per barrel, increasing the tax ratio from 50 per cent to 55 per cent and, for the first time, the principle of annual price increases (2 per cent annually through 1975).(3)

Under the momentum of Libyan success, OPEC, at its twenty-first meeting in Caracas in December 1970, adopted Resolution XXI.120, whereby it was decided that the OPEC members would enter into collective negotiations with the companies on the bases of specific demands laid down in the Resolution: namely, the tax ratios were to be amended upward in line with the Libyan settlement; posted prices were to increase substantially in line with recent market increases; all discounts and price rebates enjoyed by the companies, including those of royalty expensing, were to be completely deleted; and the system of price differentials among the various crude gravities (degrees API) to be revised and some other amendments made. It was also decided that collective negotiations would be conducted on a regional basis by forming negotiating ministerial committees.

On behalf of the OPEC members bordering the Gulf, negotiations with companies were conducted by a ministerial committee composed of Saudi Arabia, Iran and Iraq.

The difficult negotiations continued for more than one month. When they were interrupted, the historic Extraordinary Meeting of OPEC was convened in Tehran during 3-4 February 1971 (Resolutions XXII.131 and XXII.132), where the members declared their governments' readiness to enforce OPEC's demands on the prices by legislation, if necessary, should negotiations with oil companies fail. They also declared their governments' readiness to impose an oil embargo on any operating oil company that did not abide by such legislation.

The oil companies, in response, resorted to strengthening their bargaining position by obtaining the waiver of the Sherman Anti-Trust Law, with the consent of the US Government.

However, in spite of stiff company resistance, and various political pressures applied by certain governments of the oil consuming industrialized countries, OPEC succeeded in concluding with the companies the historic Tehran Agreement in mid-February 1971 (Appendix I.5). This agreement was followed by a number of related agreements concluded subsequently: namely, Tripoli Agreement concluded in April 1971 for the Libyan oil; the Lagos Agreement for the Nigerian oil; the East Mediterranean Agreements, concluded with Iraq concerning Kirkuk oil exported from Tripoli/Banias; and the agreement concerning Saudi Arabian oil exported from Sidon.

The US dollar is the currency denominating the price of oil, as well as the currency of oil payments. The Tehran Pricing System also included currency protection agreements, concerning the price adjustments to offset variations in the exchange rates of the US dollar vis-a-vis the major world currencies. This agreement was further supplemented by the Geneva Agreements I and II concluded in January 1972 and June 1973, following the first and the second dollar devaluations respectively.

All these agreements, creating a new pricing system, were intended, at the time, to last five years. However they all collapsed within two years under the fast moving events in the international petroleum industry.

Events Leading to OPEC Becoming the Sole Oil Price Administrator

The joint OPEC/companies' price administration in accordance with the Tehran and related agreements proved to have inherent weaknesses that made the system incapable of coping with new market developments. By the middle of 1973 realized market prices had already risen to such heights as to surpass the posted prices at that time. The enormous size of the companies' windfall profits led the governments of the oil producing countries, partners of the Tehran Agreement, to demand the re-opening of the negotiations of the Agreement in the light of changing circumstances (Resolution XXXV.160 of 15-16 September 1973; Appendix I.6). Negotiations in the autumn of 1973 in Vienna soon proved abortive, when the companies refused OPEC governments' demands for the adjustment of the posted price to a level that would generate a net government take per barrel to reflect the market conditions, and to syphon back the extra windfall profits of the companies. In the Vienna preliminary session of the negotiations the companies proved to be intransigent and broke off negotiations on the pretext that such a price increase would need wider political consultations, because of its very important impact on the economic activities of the industrial countries. On the refusal of the companies to continue negotiations, OPEC member countries bordering the Gulf decided to meet in Kuwait on 16 October 1973, where they announced (Appendix I.7) their historic decision on the pricing of OPEC oil by the governments, independently of the companies, and in application of the principle of the right of a state to sovereignty over its natural resources. In that meeting posted prices were increased by 70 per cent, taking the price to $5.119 per barrel for Arabian Light. That increase was decided in the light of effective market realizations by some of the member countries' national oil companies, who were already in the markets, and whose crude sales showed that the market price of OPEC oil was well above the posted prices.(3) Since then OPEC members have become the sole price administrators for their own crudes, after deliberations at OPEC Ministerial meetings.

On the following day, the Arab oil exporting countries also met in Kuwait at the headquarters of the Organization of Arab Petroleum Exporting Countries (OAPEC), to take the political decision to impose the oil embargo (Appendix I.8), following the military and political support given by the United States of America to Israel in the October War, which was still going on at that stage. The oil shortages which resulted from the succession of cuts in Arab oil production created an entirely new situation in the market during November and December of that year, whereby realized prices were in certain cases as much as three times the posted prices.

Amidst those market conditions, the OPEC Conference was held in Tehran in December 1973. An increase of 127.6 per cent was decided, bringing the posted price of Arabian Light up to $11.651 per barrel.(3)

Participation

The concept of 'Governmental Participation' was defined by OPEC members collectively for the first time in OPEC Resolution XVI.90 dated 24-25 June 1968. It was entitled 'Declaratory Statement of Petroleum Policy in Member Countries' and read as follows:

> "Where provision for Governmental Participation in ownership of the concession-holding company under any of the present petroleum contracts has not been made, the Government may acquire a reasonable participation, on the grounds of the principle of changing circumstances.
>
> If such provision has actually been made but avoided by the operators concerned, the rate provided for shall serve as a minimum basis for the participation to be acquired."

A decision towards adoption of effective measures to be taken for implementation of participation was taken in OPEC's Resolution XXIV.135 dated July 1971. Saudi Arabia was assigned to conduct these negotiations on behalf of the countries bordering the Gulf.

Following the above decisions, the actual process of negotiations with the transnational companies was extremely laborious and time consuming. It was the first time that the OPEC member negotiators concerned had had to deal with all the complex and intricate issues which constitute the international petroleum industry. However, determination and perseverance ultimately resulted in the General Participation Agreement (Appendix I.9) of 5 October 1972. This Agreement provided the general legal framework within which individual 'implementing' agreements could be concluded with each Arab Gulf country.

Iran had decided previously to drop out of the Gulf group on the grounds that participation in the case of Iran was not relevant, as it had already nationalized its oil industry in the fifties. Iran was in fact negotiating a separate deal with the operating companies which would yield, as a minimum, the benefits that the other Gulf producers would obtain from their concessionaires.

Indonesia was operating its oil industry on contractual bases, so participation was irrelevant.

In June 1972, Iraq nationalized the US and Dutch and later the Gulbenkian interests in the Basrah Petroleum Company.

Later developments included Libya's successful achievement of 51 per cent participation and Nigeria's arranging 35 per cent. By 1974, Kuwait, Qatar and Saudi Arabia had secured 60 per cent participation. Many ideas present in the original participation Agreement, such as Bridging oil and Phase-in oil and other buy back arrangements proved to be only short lived. Further progress led to an agreement for one hundred per cent take-over of Aramco which has already been tabled with the Saudi Arabian Government for approval.

The transnationals' initial apprehensions about participation proved to be too pessimistic. Over the past few years they have continued to obtain much of OPEC members' crude oil export volumes. In Chapters 7 and 8 I discuss in greater detail the modes of co-operation between oil exporting countries and the foreign oil companies.

SOLEMN DECLARATION OF SOVEREIGNS AND HEADS OF STATE OF OPEC MEMBERS (Algiers 4-6 March 1975)

The Solemn Declaration issued by the Sovereigns and Heads of State of the OPEC members following their conference in Algiers on 4-6 March 1975 (Appendix I.10) is a document of profound historic significance. The basic statements of this Declaration fall within the context of the decisions taken at the VIth Special Session of the UN General Assembly on problems of raw materials and development.

The document presents a comprehensive view of the world economic crisis and proposes measures to safeguard the legitimate rights and interests of their peoples, in the context of international solidarity and co-operation and the need for a new international economic order.

This document played a fundamental role in defining the negotiating position of the developing countries during the Conference on International Economic Co-operation held during 1975-1977 in Paris.

Agreement to Establish the OPEC Special Fund

On 28 January 1976, OPEC decided to establish a new facility for the provision of additional finance to other developing countries under the name of the OPEC Special Fund (Appendix I.11). The objective was to consolidate OPEC's assistance to other developing countries, in addition to the existing bilateral and multilateral channels through which they have individually extended financial co-operation to such other countries.

OPEC's Long Term Strategy Committee

OPEC's Long Term Strategy Committee is composed of the ministers of Saudi Arabia, Iran, Iraq, Venezuela, Algeria and Kuwait. The Committee finalized its report and recommendations during February 1980 after almost 20 months of work. The recommendations were later presented at OPEC's Extraordinary Conference in Taif during May 1980. Except for some important reservations on the report's proposed long term pricing formula registered by Iran, Algeria and Libya, the recommendations were generally accepted during the Taif Conference.

The report was referred to a meeting of OPEC ministers of Foreign Affairs, Finance and Oil which prepared the ground for the projected Summit Conference of OPEC Heads of State which was due to be convened in Baghdad in early November 1980. This was postponed as a result of the Iran-Iraq conflict.

References

(1) Kubbah, Abdul Amir. _OPEC: Past and Present._ Petro-Economic Research Centre, Vienna, 1974.

(2) Seymour, Ian. _OPEC: Instrument of Change._ Macmillan Press, 1980, for OPEC.

(3) Al-Chalabi, Dr. Fadhil J. _OPEC and the International Oil Industry: A Changing Structure._ Oxford University Press, 1980.

4
THE RATIONALE FOR OPEC

There has been a general tendency, especially in the industrialized countries, to present the actions and policies of the Organization of the Petroleum Exporting Countries (OPEC) outside the context of the real historical perspective and blame the Organization for all the energy and economic woes faced by the world today. It might be expedient for some politicians and certain news media associated with groups having a vested interest, to make the general public believe that the rising oil costs are the real cause of all their worldly problems, but, in the long term, to evade the reality would be extremely counter-productive. The general public's reactions and attitudes towards specific issues play a very vital role in the process of political decision-making and policy planning in all democratic societies. It would be unrealistic to expect necessary support and co-operation on difficult issues which affect people's customary life styles from a public which has long been fed on notions that are far from reality. One recent useful contribution to the public debate was an article under the heading 'OPEC, The American Scapegoat'.(1)

As I have discussed in Chapter 3, OPEC was created in the year 1960 in direct response to the price policies of the transnational oil companies and to protect the legitimate interests of its member countries.

The trends set by OPEC are being followed elsewhere. It may be noted that the recent establishment of Statoil in Norway and the British National Oil Corporation in the UK was motivated mainly by a desire to gain effective state participation in the rapidly expanding indigenous petroleum industry.

To understand the rationale of OPEC's creation, policies and actions to date, a study of the historical perspective is imperative.

DEVELOPMENT OF ADVERSE TRENDS IN THE DEVELOPED COUNTRIES

During the post-war years prior to 1970, the average annual growth of the gross national product (GNP) recorded in the advanced industrialized nations of the Organisation for Economic Co-operation and Development (OECD) was about 5 per cent in real terms. From the perspective of the developed countries this period is generally described as the golden years, a prolonged period of unprecedentedly high per capita growth of GNP.

We can identify at least four principal factors which contributed towards this consistent high rate of growth. First, in the immediate post-war period the United States of America used its large current account surpluses to finance, through the Marshall Plan, the deficits of those countries which had been damaged by the war. Secondly, the governments concerned accepted the principle of free trade, in some contrast to the pre-war practice. The third factor was the Bretton Woods Agreement. By maintaining a regime of fixed parities, it reduced the uncertainties that generally inhibit the making of economic decisions and imposed a degree of financial discipline that probably also helped keep inflation rates down. The fourth factor was exceptionally cheap oil. In fact the real price of oil fell by half between 1950 and 1970. There is no doubt that the abundant and exceptionally cheap availability of oil was an important factor in making it possible for the European countries and Japan to stage a remarkable recovery and unprecedented expansion in their economies within a generation after the Second World War.

During this period of prosperity there also evolved life styles in the industrial societies which promoted wasteful and inefficient uses of a precious energy resource like oil and discouraged the use and development of more abundant energy resources like coal and other alternative renewable and non-renewable energy resources. These trends were inherently inconsistent with the worldwide energy resource base and were bound to lead to increasing energy supply instability in the long term.

The Bretton Woods system imposed a certain inflexibility on United States policy in as much as the USA was the only country in the system which could never change its exchange rate. This meant that America was liable to run a permanent deficit and to keep her rate of inflation low. The high rate of world inflation of the early 1970s had little to do with oil. The important reasons were the impact of the Vietnam War on the US economy, the exceptional increase in raw material prices because a large number of countries achieved unusually high growth rates at the same time, a series of bad harvests and a big increase in the rate of growth of the money supply, associated in part with the growth in budget deficits in many industrialized countries. Failure to take timely action finally led to the breakdown of the Bretton Woods currency system. The resulting currency instability itself added to inflation. So inflation was already high and rising by the middle of 1973, before OPEC decided to increase the price of oil.

PLIGHT OF OIL PRODUCING DEVELOPING COUNTRIES PRIOR TO 1970

For the developing countries the situation during the post-war period prior to 1970 was quite different. In most of these countries the petroleum resources were under the complete control of the transnational companies. The legal instrument for exercising this control was the 'concession agreement'. Under the traditional concession agreements these transnational companies obtained almost limitless privileges, whereas there was virtually no role for the producing countries to play, except to wait for the periodic royalty payments to be made by the concessionaires. A minor office in some government department or ministry was more than adequate to cope with this limited passive role of keeping payment records and filing routine reports.

Almost by definition, the transnational companies could not be expected to take into account the specific national development needs of every country whose petroleum resources they exploited. The sole criterion which determined their policies was global cost minimization on the supply side of their activities, which was quite understandable. For obvious reasons, conservation of petroleum resources did not receive enough attention. Natural gas was indiscriminately flared and wasted. Refining and petrochemical industries were concentrated in consuming countries exclusively on economic criteria, depriving the host producing countries of the benefits of industrialization. For decades, oil prices were maintained below the cost of alternatives. This was not only detrimental to the host producing countries' economic interests, but also encouraged wasteful uses of depletable resources and generated potentially dangerous long term trends in the growth of the energy mix.

Consequently, for several decades, oil exporting countries remained deprived of the full benefits that their vast petroleum resources could have contributed towards their overall development requirements. The governments in these countries were, therefore, fully convinced that mere supervision of concessionaire oil companies was not enough. In order to achieve their national development aspirations, what was needed was effective national control over their petroleum resources. The basic elements of such control assumed the national governments' exclusive right to determine the policies governing all phases of the oil and gas industry, including exploration, production, oil and gas processing, transportation, marketing and price determination, as well as direct national participation in organizational and operational activities. It was in this perspective that the governments of the oil producing countries found it necessary to create OPEC.

It is only since their historic decisions of 1973 that the developing countries of OPEC have been able to acquire effective decision-making control over policies related to their respective petroleum resources.

REASONS FOR AND IMPACT OF OPEC'S OIL PRICE ACTIONS OF 1973

Until 1970 the price of oil was close to the cost of its production. This did not reflect the cost of alternative energy forms, though this was clearly necessary to ensure an orderly transition away from increasing relative and absolute dependence on a depleting finite natural resource. The 1973-74 price corrections, therefore, had to be of a somewhat drastic character, being the first decisive attempt to correct the long established historical distortions. It was an attempt to begin the process of removing the long standing disparity between the price of oil compared with the prices of other forms of energy, and to signal the need for conservation and efficient use of oil resources, as well as to pave the way for a concerted drive to lessen the exceptional dependence on the fast depleting oil resources by the development and exploitation of other more abundant and renewable alternative sources of energy.

In the wake of the 1973-74 oil price increases, there was much talk of approaching doom for Western civilization and the imminent collapse of the international monetary system. However, all the doomsday fantasies have now been proved to be wrong. The fact is that the impact of the 1974 oil price increase on the economies of the consumer countries ran its course within a year or two in most cases. In mid-1974 a World Bank estimate placed OPEC's investable surplus in 1980 at some $650 billion. Several other estimates at that time also pointed to a very large cumulative building of OPEC assets. However, the fact is that the aggregate surplus on the current account of major oil exporters fell from $68 billion in 1974 to as low as $7 billion in 1978. By 1978 the adjustment of world payments balances to the 1973-74 oil price increases had been largely completed. The surplus funds accruing to the oil exporting countries were placed in the financial markets of industrial countries, and from these markets they were available to the oil importing developing countries to finance their deficits. By 1978, the combined surpluses of West Germany, Japan and Switzerland were many times greater than the OPEC surplus. Also, by 1978, the industrial countries returned to their traditional surplus and the current account position of the oil exporting countries in real terms approached that of the period before 1974. The oil importing developing countries were also able to maintain the momentum of their economic growth. During 1974-79 the annual growth rate in these countries averaged 5.2 per cent. This rate was only slightly lower than the rate attained in the years up to 1973, but cannot be considered satisfactory, keeping in mind their development needs and the prevailing low standards of living. Sluggish demand for their products in the industrial world, trade barriers and lack of interest of the industrialized world in expanding direct transfer of resources are some of the principal factors contributing towards this situation. The oil importing developing countries have benefited substantially from large-scale grants and other development assistance from the OPEC countries. They have also received a greatly enlarged flow of remittances from their nationals employed in OPEC countries.

DEVELOPED COUNTRIES' RESPONSE DURING 1974-1978

In the five years from early 1974, OPEC countries exercised considerable restraint in their price policy, even at the cost of their own interests, in order to afford the consuming countries the time necessary for the required adjustments. From the beginning of 1974 until the end of 1977, the price of oil barely kept up with other prices, the real value of oil remaining steady or falling somewhat. In 1978 its real value fell sharply. The dollar price of oil remained nominally stable in the face of rising world inflation and the depreciating US dollar.

During this breathing spell, unfortunately, the oil consuming industrialized world failed to take adequate steps to promote the necessary level of conservation of oil and to develop alternative sources of energy. Instead, oil demand kept rising after the recession of 1974-75, and the governments of importing countries continued to press the oil exporters to raise their existing capacity to meet consumers' unrestrained demand. The outcome was predictable.

REASONS FOR AND IMPACT OF OPEC'S PRICE ACTIONS OF 1979-1980

By early 1979 only a small cut in output due to troubles in Iran sent the spot prices soaring to previously unimaginable levels. The oil price increases which have followed since then were, in fact, the direct result of market forces, a market largely created by the industrialized world.

Continuing troubles of the dollar were creating new problems. Confidence in currencies had weakened dramatically. More individuals, and some countries, were moving out of currencies into gold and commodities. Meanwhile, speculation on futures was becoming an important contribution to currency turbulence.

These points are nicely illustrated by the fact that the official sales price of Saudi Light crude rose by more than 90 per cent in US dollar terms between the end of May 1979 and the end of May 1980; by 70 per cent or more in all other major currencies; but in terms of gold the Saudi price actually declined by just over 1 per cent over this twelve month period. One ounce of gold bought about 19 barrels of Arab Light crude at each point in time.

This analysis makes it quite clear that escalating world oil prices were not directly the main reason for recession, demand deflation, inflation and trade deficits. On the other hand the oil prices had moved in response to the market situation created by the long standing adverse trends in the energy mix, the economic order and habitual dependence on unconstrained growth of oil supply at low prices. Unnecessary and competitive building of oil stocks in 1979 was largely responsible for the extent of oil price increases exceeding by a large margin those announced by OPEC in December 1978.

PARTICIPATION

Since its introduction to the petroleum industry, the word 'participation' has acquired new connotations and an unprecedented significance. This is so, in spite of the fact that the concept of joint ventures (namely, participation) is as old as the history of human economic activity. Normally only new concepts create a stir of curiosity and one might expect that the idea of joint venture being very old and commonly applied in the international oil industry, would not contain any element of excitement. But when this very concept was viewed as a possibility between national and transnational oil companies, it caused a great stir and even alarm. Evidently it was the participants that created the alarm, rather than the concept itself.

I reproduce a quotation from an essay written by Dr Edith Penrose in 1968(2) to sum up the apprehensions against participation that were felt at that time:

> "There are three broad grounds on which major international companies may fear partnership with governments. The first is political - an objection in principle to association with governments in this kind of business; or a fear that a government partner will interfere with operations on political grounds. The second relates to costs - a fear that payments to governments will exceed what the companies feel they can afford. The third relates to the effect of taking government partners on the industry generally - a fear that the companies' control over output would be so further weakened that prices would be severely affected and that the flexibility of the companies in managing their international operations would be seriously impaired."

The passage of time has already proved these apprehensions to be over-exaggerated.

Participation is one of the most significant elements in OPEC members' strategy to achieve effective control over their petroleum resources and maximize their economic growth. It is also necessary to prepare OPEC members to face the challenges of the future when the existing oil concessions will expire.

The concept of participation is based on three basic principles which are: the transference of full ownership to the national government; compensation on the basis of the net book value of the foreign companies' assets; and service fees for the foreign companies instead of profit sharing.

Saudi Arabia has never favoured a policy of nationalization, because we believe that it is an extreme measure which may discourage future foreign investments and invariably results in some disruptions in the smooth functioning and growth of the industry. Keeping in view the size of our international oil exports, it can be appreciated that the smooth and efficient functioning of our oil industry, free from any disruptions, however temporary, is not only a matter of paramount importance for our national economy but also for the economic well-being of the international community. It is significant to note that the imminent one hundred per cent take-over of Aramco in Saudi Arabia would represent a form of participation and not nationalization. Once ratified, this agreement would present a model of real co-operation between the national oil company and the transnational oil companies. It would facilitate the needed transfer of technology required by Saudi Arabia and open up to the companies vast new opportunities necessary for their growth and existence. Any fears or apprehensions on the part of the companies or oil importing industrialized countries that this measure may result in any disruptions or instability in our oil operations or may prove damaging to the long term interests of the companies are quite baseless. The compensation received by the companies has been more than adequate. The large profits earned by companies during 1979 resulted in considerable expansion in their 1980 exploration and production budgets compared with their originally planned levels. This alone is a clear sign of their continuing well-being.

INTEGRATION

Integration is one of those few words which has achieved considerable fame in the international petroleum industry. It has a kind of flair about it. To some it is a symbol of strength and security in the market and to others it is an aspired goal to have a certain worldwide national trade mark, something like a big international airline, a kind of national flag flying in many corners of the earth. These general illusions sometimes obscure objective considerations regarding integration.

For a correct appreciation, it is necessary to understand the basic factors covering the underlying market conditions and other considerations which determine whether integration is desirable or not desirable for a particular enterprise.

Integration can be full, from exploration and production to distribution, or it can be partial, consisting of two or more stages. The traditional divisions of the international petroleum industry into exploration, production, refining, transportation and distribution, combined with the transnational company's investment and management resources, gave these companies a qualitative strength to match their great size, thus making them commercially almost invincible.

National Oil Companies versus Private Oil Companies

A privately owned oil company operates according to private business incentives and objectives, whereas a national oil company owned by a government has for its frame of reference the government objectives and incentives that caused the national company to be created, and hence there is a major difference between the two.

The objectives of a private oil company are assumed to be long run survival and growth through generation of maximum profits. This constitutes its 'raison d'etre'. A microeconomic type of profit motive is assumed to be its incentive for existence and growth. A national oil company also looks forward to long term survival and growth. But it has a different survival and growth potential, as well as a different 'raison d'etre'. In essence it has a macroeconomic type profit motive. It operates, or should operate, on a national economic profitability basis. It is also directly subject to the macroeconomic as well as national and international political considerations of the government that caused its creation. A national oil company is one of the major tools in implementing the oil policy of its government. It also serves as one of the economic development institutions of the country.

It should be added that whenever national oil companies cross the borders of their countries they automatically shift to microeconomic decision criteria insofar as such considerations become consistent with their country's national or international oil policies.

It is also important to note that it is not always true that all companies in the oil business are long term oriented. Those having long term vested interests are the national oil companies and the integrated transnationals. The 'raison d'etre' of the first group is to protect the long term interests of the countries to which they belong, whereas the second group is not in a position to liquidate and go home like many smaller unintegrated companies.

The policies of OPEC members regarding integration should have regard to the basic objectives of their national oil companies which are quite different from those of the traditional transnationals. I personally feel that it would not be advisable to replace the existing integrated facilities of the transnationals, but rather to add to them on a strategic and selective basis to suit our definite requirements to meet well defined objectives.

APPRECIATION OF BASIC PROBLEMS OF OPEC MEMBERS

The growing general awareness of the expected worldwide shortages of oil and gas, and the global problems associated with such shortages, has encouraged numerous energy orientated studies, conferences, seminars and research centres in the advanced industrialized countries since the early 1970s. In these studies, debates and deliberations, there has been a general tendency to ignore completely, or give inadequate coverage to, the problems of oil and gas exporting developing countries. The problems highlighted have generally been those related to the advanced industrialized countries. The topics discussed have included security of supplies, stability of prices, inflation, recession, balance of payment deficits, development of alternative sources of energy, monetary problems connected with international liquidity and the stability of exchange rates. This misguided preoccupation with the industrialized countries' expected problems may be illustrated by reference to one example. The enormous growth projected in 1974 for the OPEC balance of payments surplus became a non-problem because of the rapid growth of OPEC imports and decline in real oil prices.

The solutions generally suggested to alleviate many of these problems are based on the highly unrealistic premise which considers OPEC as a sort of warehouse. The OPEC oil reserve often seems to be portrayed as being capable of meeting all the unconstrained and growing energy needs of the world, regardless of the current or future economic, social and political implications affecting the lives of over 320 million people living in OPEC member countries. It is important to realize that no lasting solutions can be found to the world oil and energy problems without an adequate understanding and appreciation of the real problems and policy options of all parties concerned.

The problems of OPEC members, like other developing countries, basically arise from their state of under-development. Prior to 1973 the exceptionally low price of oil combined with the unfavourable terms of the traditional concession system provided the oil exporting developing countries with inadequate revenues compared with their financial needs for development. This situation, coupled with unconstrained growth in demand, encouraged these countries to push for higher rates of oil production to offset the relative price deterioration they had been suffering for two decades. They had to do so in spite of their firm conviction that oil and gas were their dominant, precious, depletable resources, which needed to be conserved for the economic well-being of their future generations and for the establishment of viable diversified economies.

After 1973, these countries found themselves with surplus revenues, in certain cases far in excess of their current budgetary needs.

The most significant among variables which determine the real value of the oil and gas export earnings and exporters' accumulated surplus oil export revenues are the dollar exchange rate, imported inflation and the cost of technology and professional services. These factors are basically controlled by the policies of the advanced industrialized countries. The experience of the past decade has shown long periods of erosion in the real value of oil and gas export earnings and a chronic decline in the real value of the accumulated surplus revenues of the oil exporting countries.

One option for OPEC member countries to cope with this situation would have been to reduce their oil and gas production levels to suit their current revenue requirements. However, some of them have rejected this option so far because of the serious impact this action would make on the economies of the consuming countries. However, if lasting solutions are not found to alleviate the continuing adverse situation, the sacrifices made by some OPEC members may not continue in future.

Figure 3 illustrates one way of reviewing the relative oil and economic position of OPEC member countries which is based upon estimates for 1979. The experience of the last decade has shown that an oil producing country's position can move significantly in oil-economic space with changes along either axis even from year to year.

Another option to utilize the excess revenues is to plan for rapid economic development in OPEC members. However, there are limits to what can be accomplished in this respect.

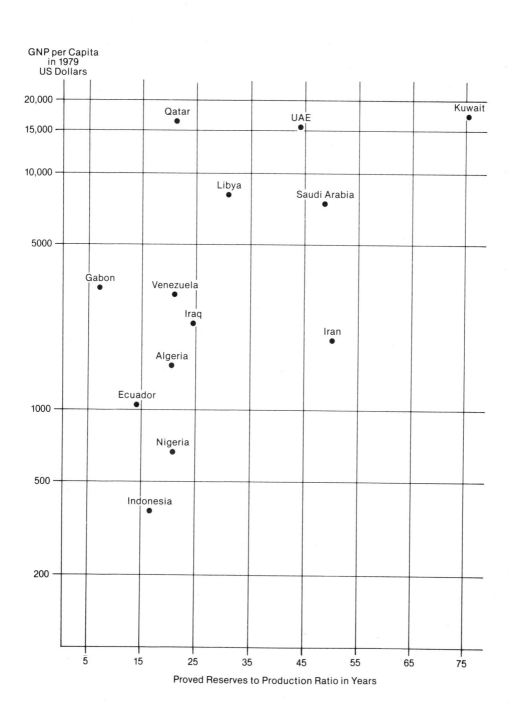

Figure 3

The Relationship Between GNP per Capita and Oil Reserves to Production Ratio among OPEC Member Countries

Most significant among these limitations are the need to modernize, or in some cases create, infrastructures, to develop the necessary technical and managerial skills, both in quantity and quality, and, last but not least, is the need to sustain the traditional social structure without undue strain.

It would be extremely unrealistic to adopt any economic development plans which do not provide adequate safeguards against serious problems of abnormal growth, such as deterioration of the social and religious traditions, rapid rural-urban migration which may result in serious social problems in the main urban centres, greater dependence on imports at higher costs, physical bottlenecks and higher rates of domestic inflation.

In order to transform their finite and fast depleting oil and gas resources into lasting assets, OPEC members require many things. They themselves must initiate sound domestic planning and follow prudent oil and gas conservation policies. There must be adequate price parities for their oil and gas exports and protection of the real value of their surplus accumulated oil and gas export revenues. OPEC member states need to plan for industrialization and diversification of their economies. This means that the energy base must be changed to a more balanced mix in which hydrocarbons are only one of two or three other major energy sources, such as nuclear, coal and solar, in order to take account of the need to remain viable after oil and gas reserves run down. They need access to the technology and markets of the industrialized countries.

PRESENT AND FUTURE OPTIONS IN OPEC'S PRICE AND PRODUCTION POLICIES

OPEC has moved a long way from the time when its Board of Governors was directed to establish a rational price structure in accordance with Resolution IV.32.(3) Since 1973 it has established oil prices, at times with a rational price structure and at times without one. One may wonder whether OPEC is trying to do the impossible, or whether its members' views are so diverse that rationality in oil prices means different things to different countries. The declared objective of OPEC, or at least some of its members, is a highly streamlined and unified price structure with well defined value differentials between various crudes and various locations to the point that the crude oil price system becomes predictable. What is so special about the international oil industry that makes this 'rationale' a necessity for OPEC and the rest of the world? Should one have such expectations of trade in other international commodities, whether those with established associations resembling OPEC or those other commodities that lack them, our world would be very different indeed. It would be a world economy that could be programmed by means of a mathematical model using a computer, a world where economic risk and uncertainty did not exist. Such theoretical exercises have been attempted and I am sure in the academic world such an approach is of great significance in developing skills that are very useful in aggregative international economic analysis. However, the validity of such models in solving real daily economic and international trading problems is questionable at least.

In the 1950s and the 1960s oil economists emphasized the fact that the international oil market is a natural oligopoly. That was and may still be a true description because of the relatively small number of sellers compared to the relatively large number of buyers. Oligopolists usually hesitate "to lower prices in the expectation that reductions will be matched" and they are usually also "reluctant to raise them for fear that rivals will not follow". Nevertheless, they incline over the long run to move toward relatively high prices and comfortable profit margins. These were Leeman's views in his book about oil prices published in 1962.(4) If one attempts to throw some light on the current and future international oil market structure, be it a natural or administered oligopoly or even one with a large competitive fringe, one needs to bring into consideration present and future supply and demand assumptions. Then one may explore the path of unifying oil prices and see whether it is a dead end street so to speak, or whether it leads somewhere desirable.

Implicit in the above oligopolistic description of the international oil market is a supply - demand equilibrium with a potential surplus of supply over demand. Furthermore, considerations of depletability of oil are not prominent in such analysis. However, considerations of prices and costs of alternative sources of energy such as coal and nuclear did not escape the consideration of oil oligopolists from the private international industry when they used to be the chief players in the international oil game.

It may be argued today that the so-called 'natural' oligopoly in the international oil market did not change, and that only the players have really changed. They are now basically governments, be they within or outside OPEC. So can one say, then, that these new players act in a truly oligopolistic sense, or do they have other motives that can apply only to governments? Of course, oil producing governments have motives and behaviour patterns, as well as modes of thinking, quite different from those that prevail in the corporate world of the international oil industry. However, should one accept the premise that the oligopolistic nature of the international oil market did not really change, then one may detect a certain inconsistency in the behaviour of the new players in the marketplace compared with the behaviour of the previous players. In the case of corporate players one can state that they acted and continue to act in the market within commercial rationality, watching their balance sheets and income statements very closely. It follows from this that some of them may lead price and production changes while others may follow, but all recognize the nature of the market forces and somehow ultimately accommodate such forces. Those players used to compete severely in the market-place, but never in a pattern that might be harmful to the basic oligopolistic nature of the market. That common bond, together with an administered degree of flexibility to accommodate market forces, was their insurance policy for survival with healthy balance sheets. They played both a short term as well as a long term game. In the short term they accommodated the changing market forces of supply, demand, market sharing, prices and production in a manner consistent with their profit motives and their long term established objective of survival and growth in an oligopolistic market.

The market behaviour of the new players since 1973 has tended to be formalized with highly publicized oil price and production administration. In the past oil companies unified prices as if by magic, without any meetings or conferences. Oil producing governments, and OPEC in particular, follow an opposite path.

So the big question is why were the oil majors so successful for decades in following a unified price pattern and the producing governments have been less fortunate? Let us see if we can find certain clues to this state of affairs.

The first clue might lie in the change in the oligopolistic nature of the international oil market. The producing governments control production only, whereas in the previous situation the oil majors controlled most of the production as well as most of the downstream facilities through their integrated worldwide systems. This meant at the time that a smaller percentage of the total oil produced was sold in what used to be called the third party market, and most of it through the affiliates of the oil majors. The third party prices merely followed, at least most of the time, those paid by the affiliates. I have used the word "followed" advisedly because in actual fact there was a myriad of prices to third parties with hidden price differences in the form of investment subsidies, longer credit periods, freight differentials and many other means. Furthermore, the so-called one oligopolistic price was not really a true arm's length price. It was a transfer price between affiliates and its true significance at the time was the use of those posted transfer prices for tax purposes. It did not mean much afterwards, since final consumer prices were the ultimate factors in the determination of the integrated profits of the oil majors. Such downstream prices of petroleum products may have been affected by the oligopolistic nature of the market, but certainly they were never truly unified. As long as the integrated profits of the up and downstream activities of the oil majors were comfortable it did not matter to them whether prices at the pump through their own networks and those to third parties were unified or not. They bought and sold in accordance with the market forces prevailing under the circumstances of each particular transaction, keeping in mind that within any one major oil company the posted unified crude prices were negotiated between only the seller and itself.

None of this applies to the new players. Most of the producing governments have no downstream facilities, except in their own countries. Furthermore, in most instances, prices are not really any longer relevant to the tax income of the producing governments since all oil income essentially belongs to these governments. Of course, in a few cases it is not so, but here I am referring to original producers and owners selling to third party buyers. The norm in third party oil markets has always been characterized by varied commercial patterns. Whether we have an essentially oligopolistic or a competitive international oil market, the forces in such a market are highly dynamic. This dynamic nature of the international oil market means that it is less amenable to streamlining and control in the absence of a very substantial control of downstream as well as upstream oil activities. Such comprehensive control is now highly unlikely. And even if it were so, for argument's sake, the dynamics of the refined petroleum products markets will determine many prices in many markets. Even the differentials of values in the market-place between various products will vary over time due to changes in the relative supply and demand position in the world markets for the various refined petroleum products.

So, any crude oil price unification with value differentials is bound to be short-lived unless it has appropriate dynamic components. To hope for a highly streamlined, predictable and unified oil price structure with well established crude oil value differentials that is characterized by dynamic market flexibility does not seem to be a viable proposition. Nevertheless, that does not mean that a decreed unified price structure such as OPEC is hoping to establish cannot be achieved. The essential requirement is that member governments are willing to agree and sustain their agreement in spite of adverse changes in the market-place.

So, let us assume that such a system is established by OPEC. We can go on to examine the costs and benefits of achieving it to OPEC and to others, and to assess its chances for survival.

We have seen in the previous analysis that the competitive nature of the international oil market, as well as the major players in this market, have changed. So today and in the foreseeable future we see an oligopolistic international oil market with a large competitive fringe at its downstream end, since the oil majors' control of that part of the market has also shrunk considerably. Today about 50 per cent of that world petroleum products market is not in the hands of these oil majors. As we have seen also, the oligopolistic structure of the upstream supply part of the international oil market has been modified by its separation from the rest of the market. Another feature of change has been the decreasing significance of the previous system of posted prices since its replacement by OPEC pricing resolutions. Also there is the structural change of corporate players being replaced by governments with new criteria for the international oil pricing game.

So, in order to achieve a unification of prices, OPEC members must first agree. This means politics first and market economics second. OPEC political economics, so to speak, may bring financial burdens to the oil consumers or to the oil producers unprecedented in market economics. The magnitude of such burdens is counted usually in billions of dollars. Expenditures are borne essentially by the various countries involved in producing or consuming oil entering international trade. The increasing marginal cost and price of oil is of great significance. First, because it affects the lives of hundreds of millions of people throughout the world. Secondly, because the methods of calculating such burdens and the complexity and accuracy of the results in a dynamic system are relatively obscure when we compare them with streamlined corporate accounts of the formerly dominant integrated transnational oil companies.

The case of some OPEC countries imposing premiums on top of official prices and the case of having different official prices are basically similar in nature. Both bring temporary gains and temporary losses to others. Those others who lose may very well be oil producers or consumers, while those who gain in this complex situation are also producers and consumers, and, at times, privately owned oil companies. However, owing to the difficulty of accounting for such gains and losses in the national accounts of various countries, one may find a diversity of views as to what actually happens. Be that as it may, one common denominator remains: namely OPEC price and production decisions carry with them nationally accounted for financial gains and losses to various producers and consumers. In most cases they carry with them financial gains to the privately owned oil companies which are disclosed in their annual financial statements. Such gains are a burden on the national economies of both producers and consumers. Oil politics at times can be very expensive in national economic terms. This only adds to the importance of creating sound energy policies which this book attempts to search for. It brings into focus the need to determine clearly the costs and benefits of dynamic energy policies rather than allowing obscurity to prevail. If a rational choice based upon financial parameters is applied in the selection of such policies both nationally and internationally, one may come up with oil policies quite different from those followed by OPEC and its member countries, or for that matter, by the IEA and its member countries. Admittedly the financial parameters are not the only criteria that govern energy policies at large or oil policies in particular. A complex fabric of political and market economics is at play.

What I am trying to explain here is that nationally accounted for costs and benefits may enter into the process of selecting between various pricing and production policies on the part of OPEC and its member countries. If the national costs and benefits of the proposed policies are known a more rationalized set of policies may be chosen, which does not necessarily mean the lowest or the highest cost ones. Should such an approach be followed, one may find that the chances for agreement among OPEC members will improve. On the other hand perhaps some of them may change their positions once the costs and benefits of such positions are quantified with respect to time and worldwide supply and demand prospects relative to price.

Maybe then, the road to more rationality and hence more harmony in the policies of OPEC and its member countries can be found. A harmonious OPEC does not by necessity mean unified prices. What it does mean is a higher capability to adapt to the international oil market dynamics at less cost and more benefit to its member countries. It also may lead to a less turbulent market-place, which can only be of mutual benefit to both the producers and to the consumers.

The need for moving towards a more balanced energy mix in both the industrialized and the developing countries for economic as well as for political reasons will become increasingly clear in this book. Many economically powerful countries command the required resources and technology as well as the political determination to accelerate the existing momentum toward reduced dependence on oil. This is so for most of the industrialized countries, many of the Communist countries and some of the developing countries. This accelerating momentum is expected to continue, in certain cases almost regardless of cost. The best that OPEC may hope for, if it follows more rationalized, harmonious and moderated price and production policies, is a slow-down in this process of change. Whether this objective is achievable through price unification or through alternative policies leading to a less turbulent international oil market, it remains a highly desirable one. Unreliable oil supplies and uncompetitive prices for oil entering the world market will lead to accelerated and perhaps uncontrollable erosion of exported oil's share of the world energy market. This risk is certainly not worth the high cost involved, perhaps even jeopardizing development plans of the national economies of OPEC member countries.

The real value of more rational and more harmonious OPEC policies lies in the fact that they will lead to more international co-operation in political relations and in economic advancement, and more sustainable changes in the energy sector. This, as we shall see later, is conducive to results mutually beneficial to the producers and the consumers.

The logical conclusion of this analysis seems obvious. If international economic relations move towards a co-operation scenario there will be net advantages for OPEC member countries as a whole, with fairly distributed national economic costs and benefits between OPEC member countries. Otherwise, the relatively short run of the next 20 years might witness the virtual death of the international oil market, as some exporters seek unrealistically high prices and come to experience economic lessons at first hand: namely, the laws of supply and demand and, in particular, the effect on those with the highest prices. Oil reserves and other parameters suggest this would be too short a time-scale for oil's demise and a far from optimal path for the evolution of the world economy, even though some of today's oil producers will no longer be exporting crude or products before the year 2000. In this context I should like to quote the remarks of His Excellency Shaikh Ahmed Zaki Yamani, Saudi Arabia's Minister of Petroleum and Mineral Resources, when responding to a question in early 1981 about the country producing more oil than its economy needs and selling at lower prices than other oil exporters.

> "As the price rises, consumption falls and capital is invested in searching for alternate sources of energy. Had you been with us on OPEC's Strategy Planning Committee, where we spent more than $2\frac{1}{2}$ years, assisted by some of the best qualified individuals and scientific institutions in the world, you would have been surprised to find out that raising prices excessively and without restrictions or limits will not be in the interest of certain OPEC members including Saudi Arabia and Iraq. You may have noticed that Iraq, which until 1977 or 1978 had pioneered the call for raising and amending prices, has begun to reconsider, inasmuch as the study, made with the benefit of computer data, indicates that we have reached a crucial point, and that going beyond it would jeopardize the interests of Iraq, Saudi Arabia and the UAE."

OPEC IN THE WORLD ENVIRONMENT

Up to 1973 OPEC oil production represented a rapidly growing share of the total energy consumption of non-Communist countries. The oil price increases of 1973-74 resulted in a sharp turn around in that trend. Since 1973, the OPEC share of the energy market has been declining year by year, except in 1976. The fall in the OPEC share was especially sharp in 1975 and 1980, and this will probably continue again in 1981. In these years the full effects were felt of the combination of economic recession in the major importing countries and the sharpness of the oil price increases. To some extent one can say that these years are below the normal post-1973 trend because of the natural rhythm of the economic cycle in industrialized countries and the fluctuations in real OPEC prices coinciding. The OPEC oil share of the energy market declined at an annual average rate of some 1 per cent after 1973. By the summer of 1981 the combination of a weak international economy, low seasonal demand, a strong dollar and some evidence of stock drawdown rather than the usual stock build up by the major oil companies demonstrated the specific weakness of demand for OPEC oil as the balancing contributor to world energy needs.

The medium term prospect for OPEC production and exports in 1985, say, is much more important than any temporary weakness of demand driving OPEC production down to something in the region of 20 million b/d.

If there is to be a realistic price regime envisaged by the Long Term Strategy Committee, it is also necessary that price must be responsive in order to maintain reasonable worldwide energy market equilibrium and to ensure that OPEC oil exports can be sold in acceptable volumes, having regard to the competitive challenge from other sources of energy. If OPEC prices are too high over a prolonged period, development of alternative energy sources will be artificially stimulated. In the very different world oil market of 1964, I expressed this same sentiment somewhat differently in my first book:

> "Nevertheless, the demand in the long run would expand, if crude oil prices should be kept consistently lower than those of other energy sources."(6)

Another factor which seems to have assumed considerable significance over the period since 1978 is the reliability of oil supplies from OPEC members. I have shown in Chapter 13 in Part II of this book that there was really no adverse effect on total OPEC oil supplies to the world market in 1979 arising from the Iranian revolution. Nevertheless, much of the media tended to dramatize the question of the security of OPEC oil supplies for this reason, but also because of subsequent events in 1979 and 1980. The fact is, of course, that if OPEC oil supplies to the world market are considered to be unreliable by politicians and decision-makers in the importing countries, then policy initiatives to develop alternatives to OPEC oil and to accelerate conservation in oil use in those countries will be given a further stimulus beyond those based upon economic influences alone.

The significance of the above paragraphs can best be seen in terms of medium term problems for other OPEC countries in exporting their desired volumes if the average OPEC price is too high in relation to the world market equilibrium, even with Saudi Arabian production at its official ceiling of 8.5 million b/d. My own researches lead me to conclude that an appropriate price for Arab Light crude at the beginning of 1981 was between $31 and $32 a barrel, compared with an actual government selling price of $32 a barrel. This estimate is based on criteria similar to those used by OPEC's Long Term Strategy Committee, though some of the index components are slightly different.

According to one estimate(7) the weighted average of OPEC crude oil official prices was a little less than $35 a barrel at that time for a gravity similar on average to Arab Light. This was some 10 per cent above the value which the proposed formula suggests as being appropriate. Since Saudi Arabian crude prices were actually quite close to those suggested by this formula, the implication is that the average prices of other OPEC members considered as a whole were more than 15 per cent above the proposed level.

The neutral scenario projections shown in Part III of this book and discussed in Chapter 14 of Part II are based upon the assumption of a relatively small, but steady, rise in real oil prices similar to that envisaged in the recommendations of OPEC's Long Term Strategy Committee. To the extent that actual rises are larger than recommended rises over the medium term, world economic growth is likely to be lower, and so too will be demand for OPEC oil. The principal numbers from these projections are shown below.

Projections		Neutral Scenario		
Million Barrels Daily	1979	1985	1990	2000
OPEC Oil Production	31.3	29.5	25.0	21.0
Worldwide Oil Production	65.7	67.0	61.7	48.0
Worldwide Oil Consumption	64.1	66.2	61.1	47.7

It is for this reason that the maintenance of significantly higher than equilibrium average OPEC prices threatens to jeopardize the return to what might be regarded as normal and desirable export volumes, at least for prolonged periods during the course of the next few years, though not at periods of peak demand. Assuming Saudi Arabian oil production is maintained at the preferred official ceiling of 8.5 million b/d, OPEC oil prices which are too high on average will make it difficult for other OPEC members to rebuild their export volumes to desired levels on a sustained basis. Also, the medium and longer term export volumes of all OPEC members are certainly likely to be less than their desired levels, even though the projections in Part III show these as declining over time anyway.

Another factor of importance underlying a healthy world economy and oil balance is the relative financial situation of oil importers and exporters. A review appears in the following table of the current external trade account of the balance of payments for selected years from 1972 to 1980 for significant groups of leading oil importing countries, both developing and industrialized.

Balance of Payments External Trade Account
In Billions of US Dollars

Oil Importing Countries	1972	1974	1978	1980
A group of five leading developing countries (Brazil, South Korea, Philippines, Thailand and Pakistan)	- 1.5	- 8.4	- 6.9	-14.3
A group of five leading industrialized countries (USA, Japan, West Germany, France and Italy)	+12.4	+ 3.9	+18.4	-44.3

Even though deficits on current accounts of the balance of payments can be offset by capital inflows and borrowing by oil importing countries, it would be unrealistic to think that large groups of the world's leading economies can go on indefinitely paying more and more for their oil imports in relation to their current exports.

Among the ten countries analysed above, Brazil was the only one which had a more favourable external trade balance in 1980 than in 1974. This was for a whole range of reasons, some economic and financial, such as the relatively low rate of increase in non-oil imports and the increase in unit value of coffee exports. Other reasons related to the energy sector. Brazil reduced the proportion of oil in its energy mix, started to increase domestic crude oil production and began the substitution of alcohol for gasoline as a motor vehicle fuel. If we examine the other four oil importing developing countries together, though, we find that their combined deficit on external trade rose from $3.7 billion in 1974 to $11.4 billion in 1980. This represents an annual average increase of over 20 per cent in nominal terms over the six years, or well over 10 per cent annually after allowing for inflation in terms of export unit prices of oil importing developing countries.

The problems of adjustment in the financial sector must be kept within manageable proportions, just like the inevitable adjustments in the real energy sector. Otherwise the policies of enlightened international creditors, whether countries or banks, are bound to come under increasing strain. If the credibility of the world economic and financial system were called into question because it was thought that intolerable strains were developing, then an unprecedented economic and financial collapse could ensue.

It seems to me that this is a prospect which OPEC members must avoid at all costs if they are to consolidate over the next two decades the progress made in the past. Just like any other countries on the face of God's earth, OPEC members have to live within the existing system whilst hopefully working to transform it for the better. A proper balance between political and economic pressures will ensure not merely the survival of OPEC and the world economic system upon which it must inevitably depend, but also an enhanced place in the world community of nations for OPEC members.

References

(1) Scott, Professor Bruce. OPEC, The American Scapegoat. Harvard Business Review, January-February 1981.

(2) Penrose, Dr. Edith. Government Partnership in the Major Concessions of the Middle East: the Nature of the Problem. Middle East Economic Survey, 30 August 1968.

(3) OPEC Resolution IV.32 of 1962 is quoted in full in Appendix I.1 of Part III.

(4) Leeman, Wayne A. The Price of Middle East Oil. Ithaca, NY: Cornell University Press, 1962.

(5) "Petroleum: A Look Into the Future" Unofficial translation of a lecture by H F Shaikh Ahmed Zaki Yamani, Minister of Petroleum and Mineral Resources, Kingdom of Saudi Arabia. Petroleum Intelligence Weekly, Special Supplement, 9 March 1981.

(6) Taher, Abdulhady Hassan. Income Determination in the International Petroleum Industry. Pergamon Press, 1964.

(7) Platt's Oilgram Price Report. 19 January 1981.

5
THE ORGANIZATION OF ARAB PETROLEUM EXPORTING COUNTRIES

FORMATION OF OAPEC

The Organization of Arab Petroleum Exporting Countries (OAPEC) was formed on the 9 January 1968. Though it is a younger organization than OPEC by several years, its antecedents can be traced to the Arab League which was formed in 1945. The Arab States recognized the problems and potentialities of oil as early as 1951 and formed a Committee of Arab oil experts which was charged with the task of submitting recommendations for the planning of an Arab oil policy and its co-ordination. Consequently an Arab Department for Oil was created inside the Arab League. The Committee of Experts throughout the years adopted many recommendations, the most important of which have been the following:

- The establishment of Arab national oil companies and support for those already existing. This includes the creation of an Arab oil tanker company; the study of the economic feasibility of the project for an Arab company for pipelines; and the possibility of the setting up of a fully integrated Arab oil company.

- The desirability of setting up joint Arab refineries.

- The need for the establishment of an Arab petrochemical and other oil-based industries.

- The need for the marketing of surplus oil products in Arab countries and the increase of oil trade and exchange among the Arab countries themselves.

- The unification of Arab oil terminology.

- The exchange of information and expertise in the oil sphere.

- The boycott of firms which smuggle Arab oil to Israel.

The Committee was also charged with the task of convening Arab Oil Congresses. The most important of them was the one convened in Cairo in April 1959 attended by about 420 delegates.(1)

OAPEC MEMBERSHIP

On 9 January 1968, Saudi Arabia, Kuwait and Libya concluded in Beirut (Lebanon) an agreement establishing the Organization of Arab Petroleum Exporting Countries (OAPEC). Algeria, Abu Dhabi, Dubai, Qatar and Bahrain joined the Organization in May 1970. Membership at this stage was limited to Arab countries where oil constituted the main source of national income. On 9 December 1971, the first paragraph of the Seventh Article of the OAPEC Founding Agreement was amended to read that membership is open to any Arab country for whom petroleum constitutes an important source of national income. This change permitted Syria and Egypt to join OAPEC in 1972 and 1973 respectively. In May 1972, Dubai and Abu Dhabi withdrew from the Organization as individual members and combined their memberships as constituents of the United Arab Emirates (UAE). Iraq joined in 1972. Thus, by 1973 the Organization's membership had increased from three to ten and included the quasi-totality of oil producers in the Arab World. These were: Algeria, Bahrain, Egypt, Iraq, Kuwait, Libya, Qatar, Saudi Arabia, Syria and the United Arab Emirates. However, on 17 April 1979, Egypt's membership was suspended, following the signing of the Egyptian-Israeli peace treaty.(2)

PURPOSE AND OBJECTIVES

Article Two of the Agreement of OAPEC defined its basic objectives. These are:

- Co-ordination of the petroleum policies of its members.

- Harmonization of the legal systems of the member countries to the extent necessary to enable the Organization to carry out its activities.

- Assistance to members in the exchange of information and expertise and providing training and employment opportunities for their nationals.

- Promotion of co-operation among members in working out solutions to problems facing them in the petroleum industry.

- Utilizing members' resources and common potentialities to establish joint projects in various stages of the petroleum industry.(3)

It has been remarked that OAPEC is distinguished by a thrust toward development, as compared, for example, with OPEC which is deeply involved in oil pricing and which acts as a protector of its more diverse membership's shared interests. Thus, OAPEC is able to concentrate its activities in serving the regional, political and economic problems of the Arab countries for which petroleum constitutes an important source of national income. It is capable of counteraction against any irrationally conceived policies affecting those countries which fall outside the objectives and terms of reference of OPEC.(4) OAPEC is both an outcome of the complex political situation in the Arab World, and at the same time an instrument for achieving cohesion among countries having a fundamentally similar political outlook but which are distinguished in having relatively diverse human and oil resources and means at their disposal for achieving their national objectives.

OAPEC members have on average oil reserves to production ratios, populations and GNP per capita which significantly differentiate them from non-Arab members of OPEC in the formulation of their economic policies as well as political decisions. A comparison of averages for OAPEC members, non-Arab members of OPEC and OECD members is shown below.

	Proved Oil Reserves to Oil Production Ratio*	GNP per Capita in US Dollars**	Population in Millions**
OAPEC Members	43.3 : 1	3660	53
Non-Arab Members of OPEC	29.8 : 1	840	281
OECD Countries	12.3 : 1	8610	769

* Oil reserves at 1 January 1980 and production in 1979 (Petroleum Statistical Review, Saudi Arabia; Oil and Gas Journal; BP Statistical Review)
** Population and GNP estimates for 1979 (World Bank Atlas)

From this analysis it can be seen that in spite of the unfavourable reaction of the Press in industrialized countries to the events of 1973-1974 involving large oil price increases and transfer of wealth to the Arab members of OPEC in particular, nevertheless even OAPEC members remained relatively poor compared with OECD countries. Also their oil reserves to production ratios were rather modest, having regard to their ambitious economic development plans and to their lack of both abundant alternative energy resources other than oil and gas, and of developed technology. Furthermore, the OAPEC countries represent no real industrial threat to the OECD countries. Their combined GNP is considerably less than that of either Italy or Canada although rather more than that of Spain. The OAPEC countries' collective GNP in 1979 amounted to well under 3 per cent of the OECD total, and that of other non-Arab members of OPEC to less than 4 per cent.

Following a period of what is sometimes seen as being marked by dramatic change, it can be a useful exercise for those expecting cataclysmic consequences to see how the relative magnitudes of the world economic landscape remain largely undisturbed.

OAPEC was established in 1968 in response to economic needs of the Arab States and reflecting basic developments in the petroleum industry. Its objectives include, "the co-operation of the members in various forms of economic activity in the petroleum industry and the realization of the closest ties among them in this field"(5). OAPEC is thus primarily concerned with the long term aspects of co-operation and interdependence among its members.

OAPEC's relationship to OPEC is clearly defined in its Founding Agreement. Article Three states that the "OAPEC Agreement shall not be deemed to affect the OPEC Agreement and especially in so far as the rights and obligations of OPEC members in respect of the organizations are concerned" and that "OAPEC members are bound by the ratified resolutions of OPEC even if an OAPEC member is not a member of OPEC"(3).

ORGANIZATION

The Council of Ministers: consists of the petroleum ministers of the member countries and is the highest authority of the Organization, responsible for drawing up its general policy, directing its activity and laying down the rules governing it. Its main responsibilities are:

- Deciding on applications to join membership.

- Taking resolutions, making recommendations and giving directives with regard to the general policy of the Organization or vis-a-vis a particular situation or issue.

- Approving draft agreements reached by the Organization.

- Approving the Organization's annual budget.

- Appointing the Secretary General and Assistant Secretaries.

The Council convenes in regular sessions twice a year or in extraordinary sessions at the request of one of the members or the Secretary General.

The Executive Bureau: consists of the Under-Secretaries to the ministries of petroleum of the member states. Its main functions are to draw up agendas for the Council; review the annual budget of the Organization; approve the staff regulations of the Secretariat and introduce appropriate amendments thereto; and to handle such tasks as may be assigned by the Council. Its work is carried out after consultation with the Secretary General and acts as intermediary between the Council and the Secretariat of the Organization. Its meetings usually precede those of the Council. The Chairmanship of both is by rotation according to the Arabic alphabetical order of the member countries, each for a period of one year.

The Secretariat: the Secretariat assumes the planning and executive aspects of the Organization's activities. It administers the programme and policies laid down by the Council and the Executive Bureau. It is headed by a Secretary General aided by Assistant Secretaries who cannot exceed three. The Secretary General can be appointed for a period of three years extendable for a further period or periods. He supervises the work of the Organization.(3)

The Secretariat is composed of the following departments: Economic, Hydrocarbon Industries, Energy, Legal, Information and Public Relations, Library, Training, Administrative and Financial. Besides the regular staff of these departments, a number of qualified advisers and technical experts is chosen by the Secretary General to assist with the work and development of the Organization.

Early in 1979, the General Secretariat established an Exploration and Production Unit. Functions of the new unit were defined as follows:

- Follow-up of world developments in oil exploration, drilling and enhanced recovery methods.

- Follow-up of oil and gas exploration, drilling and production operations applied in the Arab countries and comparing them with those applied in other oil producing countries.

- Follow-up of studies on exploration (geological, geophysical and geo-chemical) and reservoirs carried out or planned by the member countries. This is to identify common bases in those studies for the ultimate purpose of implementing them jointly to save time, effort and money.

- Follow-up of methods of collecting, analysing and documenting exploration and reservoir information in the member countries for purposes of standardization or co-ordination to facilitate the communication of such information among the member countries.

- Formulation of standards to measure national efforts in the fields of surveying, drilling and well servicing in order to follow up the development of these efforts and identify methods and means of support.

- Formulation of standards to measure the efficiency of exploration and drilling operations and discussing methods to improve them in co-ordination with the member countries.

- Organizing training courses in the fields of oil exploration, drilling, production and reservoir operations in co-ordination with the member countries.

- Organizing seminars on exploration, drilling, production and reservoirs for identifying national problems in these fields and studying them jointly.

- Assisting member countries in exchanging information on exploration and production and following up the implementation of such exchange.

- Working, in co-ordination with the Arab Petroleum Services Company (APSC), towards the establishment of a specialized petroleum laboratory to conduct research and studies in the fields of reservoir engineering, exploration and production.

- Working towards setting basic rules and regulations regarding the conservation of hydrocarbon resources in co-ordination with the member countries.

The Judicial Board: Established on 9 May 1979, the Judicial Board is meant as an arbitration council between member countries, or between any member and a petroleum company operating in the territory of the said member. The judgments of the Board are final, binding on the parties to the dispute and enforceable in the territories of the members.(2)

The Energy Department has two related functions: co-ordination and information. The latter involves accumulating and publishing data relating to world energy supply trends as well as energy consumption and supply in member countries.

Training has been emphasized in the Agreement of OAPEC as one of the major fields of activities. A unit for training and manpower was set up in March 1976.

SPECIALIZED JOINT COMPANIES

OAPEC has established four major specialized joint companies. Other projects for joint ventures are under study. However, profitability of some specialized joint companies is a matter of concern for the Organization. At a meeting of the General Secretariat and the directors of OAPEC sponsored companies held on 20 November 1978, it was agreed that the Secretariat would lay down guidelines for an extensive study on the concept of profitability for the companies as well as their requirements for support and protection. A working group was set up for this purpose.(2)

The Arab Maritime Petroleum Transport Company (AMPTC)

AMPTC was established on 6 January 1973, with headquarters in Kuwait. The company's authorized capital was fixed at $500 million. The subscribed capital is $100 million.

AMPTC's founding agreement provided that it undertakes all activities of maritime transport of hydrocarbons. It can acquire tankers for crude, liquefied gas, refined products and petrochemicals.

The Arab Shipbuilding and Repair Yard Company (ASRY)

ASRY was founded on 30 November 1974, with headquarters in Bahrain, for the purpose of developing the ship industry in the Arab countries and training Arab nationals in shipbuilding, repair and maintenance. The company's capital was initially set at $100 million, but in May 1976 it was raised to $300 million.

The company's dry dock was officially inaugurated on 15 December 1977.

Arab Petroleum Investments Corporation (APICORP)

The founding agreement of APICORP was signed on 11 July 1974. The corporation has an authorized capital of 3.6 billion Saudi Riyals (US $1 billion) and a subscribed capital of SR 1.2 billion. Its headquarters are in Dammam, Saudi Arabia.

APICORP was established to contribute to financing petroleum projects and industries and fields of activity which are related, ancillary or complementary to such projects and industries, with priority being given to joint Arab ventures. The Corporation also advises member countries on the investment of their capital funds in such projects as may ensure economic and financial growth.

The Arab Petroleum Services Company (APSC)

The founding agreement of APSC was ratified on 22 November 1975. The headquarters of the company are in Tripoli, Libya. Its authorized capital was fixed at 100 million Libyan Dinars and the subscribed capital at LD 15 million. The company's objectives are to provide services through the establishment of affiliated companies specializing in various petroleum services.

New Projects

The following projects are at various phases of preparation:

- Mediterranean Drydock Project.
- Lube Oil Production Project.
- Project for the Production of Catalysts.
- Carbon Black.
- Synthetic Rubber Industry.
- Detergents.
- Arab Venture for Engineering and Design in the Hydrocarbon Industry.

Oil and Gas Proved Reserves of OAPEC

	Crude Oil		Natural Gas	
At 1 January 1980	Million Barrels	Percentage of OAPEC	Billion cu. ft.	Percentage of OAPEC
Algeria	8,440	2.5	132,000	32.7
Bahrain	240	0.1	9,000	2.2
Iraq	31,000	9.2	27,500	6.8
Kuwait	68,530	20.4	33,500	8.3
Libya	23,500	7.0	24,000	5.9
Qatar	3,760	1.1	60,000	14.9
Saudi Arabia	168,390*	50.2	95,730	23.7
Syria	2,000	0.6	1,500	0.4
United Arab Emirates	29,411	8.8	20,500	5.1
OAPEC total reserves	335,271	100.0	403,730	100.0
OAPEC reserves as % of world reserves	51.8		15.7	

Sources: Oil and Gas Journal, December 31 1979, except * Petroleum Statistical Bulletin 1979, The Ministry of Petroleum and Mineral Resources, The Kingdom of Saudi Arabia.

Saudi Arabia appears to enjoy a dominant position on the world oil stage since OAPEC members account for over half the world's proved oil reserves and Saudi Arabia accounts for over half of OAPEC members' proved oil reserves. However, at the recent rates of oil production, the ratio of reserves to production in Kuwait is in fact very significantly higher than in Saudi Arabia.

Another point brought out in the above table is that for certain OAPEC members natural gas reserves are relatively much more important than for other OAPEC members. In terms of quantity natural gas accounts for 73 per cent of Algeria's total oil and gas reserves. For Qatar the proportion is also 73 per cent and for Bahrain it is 87 per cent. In contrast, natural gas represents only 9 per cent of the combined oil and gas reserves in Saudi Arabia, 8 per cent in Kuwait and 13 per cent in Iraq.

References

(1) "Organization of Arab Petroleum Exporting Countries" An address by Mr. George Tomeh (Consultant for International Relations to OAPEC) to a Meeting of CAABU at the House of Commons, Westminster, 22 February 1977.

(2) AOG Research Department. OAPEC Objectives, Organization and Joint Companies. Studies and Documents.

(3) OAPEC Department of Information. *Basic Facts About The Organization of Arab Petroleum Exporting Countries.* Kuwait, 1976.

(4) Penrose, Dr. Edith. Government Partnership in the Major Concessions of the Middle East: the Nature of the Problem. *Middle East Economic Survey*, 30 August 1968.

(5) The Secretary-General of OAPEC, Dr Ali Ahmed Attiga, writing in this publication. *The Voice of the Arab World* (London), 11 August 1976.

6
HISTORICAL REVIEW AND RATIONALE OF THE IEA'S CREATION, POLICIES AND ACTION

INTRODUCTION

Just as the formation of OPEC can be traced to the unilateral action of the major oil companies in reducing nominal crude oil prices, so the emergence of the International Energy Agency (IEA) can be traced to initiatives taken by OPEC and OAPEC respectively in October 1973 in raising oil prices and introducing a partial and limited embargo of oil supplies in the wake of the Arab-Israeli war.

Some of the fundamental questions and main principles of the IEA were discussed by the US Secretary of State in a speech to the Pilgrims of Great Britain in December 1973. At that time he proposed an Energy Action Group involving the nations of Europe, North America and Japan to initiate action in the following specific areas: to conserve energy through more rational utilization of existing supplies; to encourage the discovery and development of new sources of energy; to give producers an incentive to increase supply; and to co-ordinate an international programme of research to develop new technologies that use energy more efficiently and provide alternatives to petroleum.

The Washington Energy Conference followed in February 1974. On that occasion the Secretary of State summarized the US views on the major issues under the following headings.(1) First, the energy situation posed severe economic and political problems for all nations. Second, the challenge could be met successfully only through concerted international action. Third, the developing countries must quickly be drawn into consultation and collaboration. Fourth, co-operation not confrontation must mark the relationship with producing countries. Fifth, the United States recognized its own national responsibility to contribute significantly to a collective solution and was prepared to join with other nations in a truly massive effort toward the major goal of the assurance of abundant energy at reasonable costs to meet the entire world's requirements for economic growth and human needs. It was suggested that the conference consider seven areas for co-operative exploration: conservation; alternative energy sources; research and development; emergency sharing; international financial co-operation; the less developed countries; consumer-producer relations. In regard to the last of these, the US Secretary of State said that the ultimate goal must be to create a co-operative framework within which producers and consumers would be able to accommodate their differences and reconcile their needs and aspirations. Only in this way could the evolution and growth of the world economy and the stability of international relations be assured. He noted that future generations may not enjoy a permanent source of petroleum. The consumer-producer conference which he foresaw would need to consider what constituted a just oil price and how to assure the long term investments of the oil producers, based on their oil export earnings.

The full communique of the Washington Energy Conference is set down in Appendix II.1.

FORMATION OF IEA IN 1974

In November 1974 the OECD Council decided to establish the International Energy Agency as an autonomous body within the framework of the Organisation for Economic Co-operation and Development. The Governing Board of the IEA was charged to decide upon and carry out an International Energy Programme for co-operation in the field of energy, the aims of which were:

- Development of a common level of emergency self-sufficiency in oil supplies.

- Establishment of common demand restraint measures in an emergency.

- Establishment and implementation of measures for the allocation of available oil in time of emergency.

- Development of a system of information on the international oil market and a framework for consultation with international oil companies.

- Development and implementation of a long term co-operation programme to reduce dependence on imported oil including: conservation of energy, development of alternative sources of energy, energy research and development, and supply of natural and enriched uranium.

- Promotion of co-operative relations with oil producing countries and with other oil consuming countries, particularly those of the developing world.

A list of some of the more important articles of the IEA agreed between the original participating countries is shown in Appendix II.2.(2)

IEA'S PRINCIPLES OF ENERGY POLICY, 1977

Following the lack of success with the conclusion of the Conference on International Economic Co-operation and with US oil imports rising to a new all time record, in October 1977 IEA Ministers, in recognition of the need to promote a strengthening of member countries' energy policies, adopted twelve Principles of Energy Policy. These Principles provide a framework for national policies and programmes to expand indigenous energy supplies, lower energy demand and develop new energy technologies. Ministers also established an oil import Group Objective of 26 million barrels a day (excluding IEA bunker requirements) for 1985, and directed the Agency's Standing Group on Long Term Co-operation to review annually the contribution of member countries to the Group Objective, assess the continuing validity of the Group Objective, and consider the need to establish objectives for later years.

The procedure adopted to meet this directive has involved reviewing the performance, programmes and policy of each IEA member country every year. The criteria against which countries are reviewed include not only the Principles and the Group Objective for 1985, but also other relevant factors such as the Agreement on Principles for IEA Action on Coal and expectations regarding the evolution of the world oil market over the medium to long term.

The Principles for Energy Policy were adopted by the Governing Board in October 1977. These are listed below.(3)

- Further development by each Participating Country of national energy programmes and/or policies which include the objective, formulated as specifically as possible, of reducing in absolute terms or limiting future oil imports through conservation of energy, expansion of indigenous energy sources and oil substitution.

- Constant and careful attention to important environmental, safety, regional and security concerns to which the production, transportation and use of energy give rise, and improvement of the speed and consistency of public procedures for resolving conflicts which may exist between these concerns and energy requirements.

- Allowing domestic energy prices to reach a level which encourages energy conservation and development of alternative sources of energy.

- Strong reinforcement of energy conservation, on a high priority basis with increased resources, for the purpose of limiting growth in energy demand relative to economic growth, eliminating inefficient energy use, especially of rapidly depleting fuels, and encouraging substitution for fuels in shortest supply, by implementing vigorous conservation measures in various sectors along lines which include the following elements:

    - pricing policies (including fiscal measures) which give incentives to conservation;

    - minimum energy efficiency standards;

    - encouragement and increase of investment in energy saving equipment and techniques;

    - progressive replacement of oil in electricity generation, district heating, industries and other sectors by discouraging the construction of new exclusively oil-fired power stations;

    - encouraging the conversion of existing oil-fired capacity to more plentiful fuels in electricity, industrial and other sectors;

    - encouraging the necessary structural adjustments in the refinery sector in order to avoid an excess of heavy fuel oil;

    - directing efforts to the reduction of the use of heavy fuel oil as a primary energy source in those sectors where efficiency is low.

- Application of a strong steam coal utilization strategy and active promotion of an expanded and reliable international trade in steam coal, composed of the following elements:

    - rapid phasing-in of steam coal as a major fuel for electrical power generation and in industrial sectors;

    - further development of steam coal policies within producing, exporting and consuming IEA countries to support increased utilization by enhancing market stability through reliable and increased export and import flows under reasonable commercial terms;

- development of policies to remedy anticipated infrastructure bottlenecks.

- Concentration of the use of natural gas on premium users' requirements, and development of the infrastructure necessary to expand the availability of natural gas.

- Steady expansion of nuclear generating capacity is a main and indispensable element in attaining the group objectives, consistent with safety, environmental and security standards satisfactory to the countries concerned and with the need to prevent the proliferation of nuclear weapons. In order to provide for this expansion, it will be necessary through co-operation to assure reliable availability of:

  - adequate supplies of nuclear fuel (uranium and enrichment capacity) at equitable prices;

  - adequate facilities and techniques for development of nuclear electricity generation, for dealing with spent fuel, for waste management, and for overall handling of the back end of the nuclear fuel cycle.

- Stronger emphasis on energy research, development and demonstration, including collaborative programmes, more intensive national efforts and greater co-ordination of national efforts, in order to make energy use more efficient and to meet future energy requirements. Each Participating Country should contribute to energy technology development, with emphasis on: (a) technologies which can have relatively near-term impact, (b) policies which facilitate the transition of new energy technologies from the research and development phase to the point of utilization, (c) technologies for broadly applicable renewable energy sources, and (d) investigation of whether there are technological possibilities for significant contributions from other renewable resources, through:

  - providing the fullest possible financial support for energy research, development and demonstration;

  - increasing participation in international collaborative projects to extend the effectiveness of funds available;

  - encouraging investment in energy/technology development by appropriate incentives;

  - ensuring that R & D policies remain consistent with and supportive of the objective of ongoing energy policy.

- Establishment of a favourable investment climate which encourages the flow of public and private capital to develop energy resources by appropriate pricing policies, by minimizing uncertainties about the general directions of energy and other policies such as mentioned in the second principle above, and by providing government incentives where necessary, in order to:

  - give priority to exploration activities including those in offshore and frontier areas;

  - encourage rates of exploration and development of available capacities which are consistent with the optimum economic development of resources.

- Providing in energy policy planning for alternative means, other than increased oil consumption, for meeting any development of supply shortfall or failure to attain conservation objectives, taking into account the appropriate requirements of economic development and social progress.

- Appropriate co-operation in the field of energy, including evaluation of the world energy situation, energy research and development and technical and financial requirements, with developed or developing countries or international organizations.

THE IEA IN 1980

By mid-1980 the International Energy Agency's membership comprised 21 of the 24 member countries of the OECD, as follows:

Australia, Austria, Belgium, Canada, Denmark, Germany, Greece, Ireland, Italy, Japan, Luxembourg, Netherlands, New Zealand, Norway, Portugal, Spain, Sweden, Switzerland, Turkey, United Kingdom, United States.

The only OECD member countries which are not members of the IEA are France, Finland and Iceland.

The most recent statement of the basic aims of the IEA has been set down as follows:

- Co-operation among IEA Participating Countries to reduce excessive dependence on oil through energy conservation, development of alternative energy sources and energy research and development.

- An information system on the international oil market as well as consultation with oil companies.

- Co-operation with oil producing and other oil consuming countries with a view to developing a stable international energy trade as well as the rational management and use of world energy resources in the interest of all countries.

- A plan to prepare Participating Countries against the risk of a major disruption of oil supplies and to share available oil in the event of an emergency.

Following some eighteen months of sharply rising spot market prices for crude oil and products, the Governing Board of the International Energy Agency met in May 1980 and issued a communique which appears as Appendix II.3 in this book.(4)

REFLECTIONS ON THE NATURE OF THE IEA AND ITS PERFORMANCE

We have seen how the International Energy Agency was formed in 1974 as an autonomous body within the OECD as a direct response to the challenges mounted by OPEC and OAPEC to the economic and political power of the advanced industrialized countries. The very fact of the IEA's foundation can be interpreted as a recognition of the gravity of the challenge to the survival and future growth of the economies of the industrialized countries. These became increasingly heavily dependent on imported oil during the period of high prosperity culminating in 1973 and some continued to do so for several more years.

The OECD itself grew out of the Organization for European Economic Co-operation (OEEC) which was formed after World War II to revive the devastated economies of Western Europe and to act as an economic buffer (to complement the NATO military buffer) to possible Soviet expansion. The economic objective was achieved beyond all expectations during the 1950s with the West German economy in particular growing rapidly, based largely on a new and efficient infrastructure. From late in that decade onwards the momentum of economic growth was maintained in Western Europe with the creation of the Common Market among the original six members of the European Economic Community. The OEEC became the OECD in December 1960, some three months after the formation of OPEC. Other members joined the OECD subsequently, most notably Japan in 1963. Though influential voices, such as the West German Finance Minister, could be heard as early as 1961 urging the need to stimulate economic development in the less developed countries, the OECD remained curiously narrow-minded with respect to its own potential development as an institution, as far as an external observer could ascertain. The remarkable economic recovery of Europe in the 1950s had been consolidated by the unprecedented prosperity of the industrialized countries, featuring exceptional growth of the Japanese economy, remarkably high per capita growth in countries such as France and Italy, the growth of intra-EEC trade and the absence of a recession for nearly ten years in the USA.

The fact that this growth was fuelled very significantly by rapidly increasing but exhaustible oil supplies at very low prices seemed to have escaped the notice of all but a very few observers in the oil industry. It was perhaps not until the Club of Rome's 'Limits to Growth' was published in 1972 that it became widely recognized that discontinuities might interrupt high exponential rates of growth, whether of population, food, energy production and consumption, or other resources.

While the developing countries benefited from the rapid expansion of the industrialized countries' economies through the remarkable growth in the volume of world trade, the developing countries' (including OPEC members) share of the total actually decreased from 20.6 per cent in 1963 to 19.2 per cent in 1973. Over the same ten year period their share of world population rose from 46.7 per cent in 1963 to 49.7 per cent in 1973. Also over this ten year period the share of world trade represented by manufactured goods rose from 52 per cent to 61 per cent. Since 1973 the developing countries have accounted for rising and important proportions of manufactured exports from the industrialized countries, thus providing an important means of continued economic growth in these countries. Even OPEC's share of world trade fell from the high 14.5 per cent reached in 1974 to 10.9 per cent in 1978.

These statistics suggest that the remarkable growth in world trade and economic activity generally up to 1973 was concentrated to a remarkable and disproportionate degree in the industrialized countries. OPEC's initiatives in late 1973 were seen as a challenge to the existing economic order among developing countries generally. The IEA was perceived as a bulwark against change by some, and as an instrument of confrontation by others, in a sector of critical importance and great potential vulnerability for the industrialized countries. The fact that the attempt to broaden the dialogue in the Conference on International Economic Co-operation did not achieve effective progress can be represented as an attempt on the part of the industrialized countries to contain the spread of bargaining strength demonstrated by the developing countries in the oil sector to other important parts of the international economy.

In some respects the stance of some well-known public figures closely associated with the formation of the IEA seems to have become distinctly less co-operative over the recent years. For example, in May 1975, the US Secretary of State Dr Henry Kissinger is recorded as having told an IEA Ministerial Meeting:

> "We can examine co-operative efforts to accelerate the development programs in producer countries. New industries can be established,

combining the technology of the industrialized world with the energy and capital of the producers. Fertilizer is a promising example."

By June 1980 Dr Kissinger's thinking seemed to have undergone a rather fundamental change under a new set of pressures. The "us" and "them" mentality is crystallized in the following sentence from his address to the Conference on 'The Energy Emergency : Oil and Money 1980':

"We in the West agree that economic development automatically produces political stability but the fact is that economic development substantially produces in developing countries political instability because it is bound to undermine the traditional patterns of authority in which so many of these countries have operated for centuries."

The same Dr Kissinger was talking about giving producers an incentive to increase supply in December 1973, but by mid-1980 he was saying "that as the price of oil increases and resources multiply the incentive to reduce production will also be increased . . ." The reluctance of the industrialized countries to come to terms with their dependence on OPEC oil and make the necessary accommodations in economic and political terms seems to be leading them more in the direction of confrontation than of co-operation.

In its first real test in 1979 the IEA was shown to be a body without effective power to bring discipline to bear on the demand end of the oil market. The reality of increases in worldwide supply running well ahead of negligible growth in demand (discussed in Chapter 13 of Part II) should have proved quite manageable, in spite of the structural changes involving an increasing role in the world oil market for national oil companies of the producing countries, and a diminishing role for the transnational oil companies. In spite of the close link between the latter and the IEA, there seems to have been no obvious recognition on the part of the IEA or member governments of the responsibility of themselves and the companies for the damaging effects of competitive bidding for spot market supplies. Moreover, this competitive bidding was not even indulged in to ensure that unconstrained demand could be met, but simply in order to build up stocks. To a remarkable extent, the tight market seemed to be a result not of the Iranian revolution but of the fact that stock changes throughout 1979 and the first three quarters of 1980 contributed to a tighter oil market than corresponding changes in 1978.

One theme running through the articles of the IEA is the need for a co-operative relationship with the producing countries. It has never been very clear how these articles were to be institutionalized. The reference to the encouragement of a stable international oil trade and promotion of secure oil supplies on reasonable and equitable terms seemed to imply that the arbiter should be the IEA. However, a just price can only be determined in bargaining between willing buyers and willing sellers.

It may be useful to remind ourselves that certain notions are held in common. For example, in opening the Washington Energy Conference in 1974 Dr Kissinger observed that future generations may not enjoy a permanent source of petroleum. It was precisely because many oil exporting countries had begun to realize that their existing oil reserves could run out in less than one generation at existing prices and existing rates of increases of production that "excessively" high prices were introduced. Buyers always had the option of not accepting those prices. OPEC members did not merely countenance with equanimity but actually welcomed the prospect of "shrunken export markets for the producers in the future", since this would relieve them of the pressure to increase production to which they are still subject. It is remarkable that the "excessively" high prices of 1974 have risen further in real terms seven years later, and yet the average rate of decline in oil's share of the world energy balance has been much less sharp than the average rate of increase experienced previously. The explanation owes something to the long lead times required to develop alternative sources of energy, but more important still is the fact that the costs of developing alternative forms of energy in large volumes still appear high by comparison with oil priced at $30 a barrel in money of mid-1980. In this connection one might observe that in 1974 the industrialized countries had the considerable build up of production in the North Sea and on the North Slope of Alaska to look forward to, whereas the probable rise in production from newly exploitable fields looks likely to be much less over the next few years. Moreover, in 1974 the IEA seems to have had little perception of the likely deterioration in the incremental capital output ratio in producing more energy as the world's hydrocarbon resources were progressively exploited. At this point in time, it remains uncertain whether the full costs of the extraction and use of coal and the disposal and hazards of nuclear waste have been fully taken into account in the environmentally-conscious advanced economies, quite apart from the larger unknowns of developing more exotic forms of energy on a very large scale.

It is for these reasons that the recommendations of OPEC's Long Term Strategy Committee deserve serious consideration by the world community generally, and by the IEA in particular, as a reasoned basis for strategic planning. This would be compatible with paragraph 18 of the IEA's Communique of May 22, 1980, wherein Ministers of member countries noted that a smooth medium term transition away from an oil-based economy is a prerequisite for a prospering world economy in which all nations can pursue economic growth and development (see Appendix II.3).

Another aspect of the interaction between oil and the world's financial system which has done much to alienate informed opinion in the so-called capital-surplus oil exporting countries was the promise held out that stable oil earnings wisely invested and increasing by the principle of compound interest could provide a long term source of income. The problems of inflation, dollar depreciation and political risk associated with the holding of foreign government bonds have persuaded some transnational oil companies as well as some producing countries to recognize that oil kept in the ground is likely to be a more rewarding policy option. The lessons of experience in this department reflect less on the oil exporters' relationship with the IEA, but more fundamentally on the unsatisfactory incentives given in return for what amounts to altruism compared with a more self-interested policy.

Other observations which seem significant relate to what one saw as shortcomings in IEA or member countries' policies. The deregulation of gas prices urged by the IEA has still some way to go before being fully implemented in the largest consuming country. This action alone would bring about a significantly healthier balance in the world supply and demand for this premium energy source. Among the twelve Principles adopted by the IEA in 1977 was the concentration of the use of natural gas on premium users' requirements. Yet there has been a marked reluctance, especially in the USA, to recognize that the criterion for natural gas pricing might be number 2 fuel oil (heating gas oil in Europe) rather than number 6 fuel oil (residual fuel oil in Europe). Such a recognition would certainly expand the availability of natural gas in an energy-short world, especially from those countries with relatively large natural gas reserves in relation to oil and whose main export outlets are likely to be in the form of liquefied natural gas (LNG) rather than pipeline supplies. It is interesting that Japan has been the first major importer of LNG willing to pay an FOB crude oil equivalent price for supplies. Another shortcoming of most IEA members' policies has been their reluctance to increase excise duties and taxes in line with crude oil prices, or indeed to introduce such taxes where they have not been traditional. This particular failure has been the subject of critical review by both the IMF and the OECD, as well as by expert commentators in the countries concerned. The absence of real increases in motor gasoline prices for consumers over much of the period since 1973 has been an important explanation for the relative buoyancy of demand for this product.

The question of the building up of oil stocks seems to have become something of a fetish since the conception of the IEA. Yet only in June 1980, it seems, did the Agency recognize that stock building had to be controlled as a contra-cyclical activity, after the unhappy lessons of 1979 had made this evident for all to see. There is still no clear evidence that the concept of stock building in soft markets and stock drawdown in tight markets is being put into practical effect.

By April 1981 some transnational oil companies were declaring their intentions of destocking in the fast softening international oil market of that period in a manner rather reminiscent of their actions in early 1978. On the question of stock building and stock draw policy, there appears to be a still unresolved dilemma between the security of supply considerations which are of paramount importance for the IEA, and short term financial criteria and expectations of changes in the oil supply and demand balance which have a major influence on the stock policies of the transnational companies and traders. If the latter prevail in terms of what actually happens, then the inevitable fluctuations in demand seem likely to be exacerbated. The experience of the 1978 to 1981 period has provided strong evidence of remarkable swings in market sentiment moving from surplus to shortage to surplus to shortage and back to surplus again. These swings have been associated with developments on the demand as well as the supply side of the international oil market equation.

The Communique issued by the Governing Board of the IEA on 22 May 1980 carried the following passage:

> "Ministers expressed their concern about the level of oil prices which confronts the world economy with declining economic activity, having serious negative results for all countries. In particular, the price increases since the end of 1979 have occurred despite falling oil demand and appear to have been made without taking into account their adverse impact on the world economy."

This analysis and interpretation conveniently ignores the fact that it was largely competitive buying of oil supplies for stock building by IEA member countries and companies based in them that caused spot market prices to rise so far. OPEC prices lagged far behind the prices which IEA members were willing to pay when faced by what they expected to be a shortage, rather than an actual shortage. A shortage was averted, thanks largely to a big increase in OPEC production on the part of most members in 1979. In fact, had Saudi Arabia pursued a policy of revenue maximization in 1979, there is every reason to suppose it could have earned at least $40 billion more that year than it actually earned.

The OECD, the IEA's parent, seems never to have recognized the extent to which it was regarded by developing countries as a privileged club of the wealthy, except for members such as Turkey who seemed to qualify largely on the criterion of geography and strategic considerations. The opportunity was lost of turning the OECD into a more outward looking institution, working more closely with the World Bank, the IMF, the UN and the GATT to develop the world economy on a more balanced basis. The sharp differences in outlook on oil prices, supply and demand are directly related to the desire of the industrialized countries to avoid a fundamental change in the world economic balance with the developing countries which these latter are anxious to secure through improved terms of trade, more aid and better access to markets and industrial technology, etc. For this reason, prospects for a resumed dialogue under the auspices of the UN continue to appear very unpromising. Yet some common notions and ideas about the future do exist between the IEA and OPEC, as I have indicated. A review of them and their possible implications may prove a worthwhile first step in defining the probable impact of changes in the energy sector on structural shifts in the world economy over the next twenty years, and the likely development of relationships between the industrialized countries, the centrally planned economies, OPEC member countries and other developing countries. Among the existing international institutions, the World Bank with its proposed energy affiliate may be an appropriate neutral forum in which some constructive thinking could be concentrated, with some inputs from all the contending parties.

References

(1) The US Department of State. <u>US International Energy Policy October 1973 - November 1975</u>, Selected Documents No.3. Released December 1975.

(2) Decision of the OECD Council Establishing an International Energy Agency of the Organisation (of which Finland, France and Greece abstained), 15 November 1974.

(3) <u>Principles for Energy Policy.</u> Adopted by the Governing Board of the IEA, October 1977. Annex 1.

(4) Communique of the International Energy Agency. Meeting of Governing Board at Ministerial Level, 22 May, 1980.

7
IMPACT OF STRUCTURAL CHANGES ON THE INTERNATIONAL ENERGY INDUSTRIES

The recent changes in the international petroleum industry discussed in the earlier chapters signify a crucial turning point in its long history. The transfer of effective control over petroleum resources from the traditionally dominant transnationals to the newly emerging national petroleum enterprises of the producing countries is an irreversible evolutionary phenomenon. This will inevitably lead to significant changes in the traditional policy-making criteria in the international petroleum industry.

THE EMERGENCE OF NEW DEVELOPMENT CRITERIA

Traditionally the criteria for exploitation and development of petroleum resources of the developing countries served mainly the interests of the energy consuming countries. They resulted in the encouragement of wasteful uses of depletable and finite energy resources and unacceptable trends in the growth of the energy mix.

The most significant influence of national petroleum enterprises in the international oil and gas industry is emerging from their evolving policies. These are related to conservation of petroleum resources, elimination of wasteful practices, and diversification of the economic base in oil exporting developing countries.

CONSERVATION

The economies of most oil exporting developing countries are solely or heavily dependent on petroleum resources. These are depleting rapidly. In several cases their reserves to production ratios stand at little more than the desirable minimum. Accordingly, their governments have a great interest in slowing down the depletion of these resources, without, of course, reducing oil production in normal circumstances or harming the oil importing countries' economies. This desire to slow down oil and gas depletion can be attributed to two important factors. Firstly, these countries are concerned about the availability of substantial quantities of oil and gas for use in large-scale local industrialization in the coming decades, either as energy or as a feedstock for petrochemical industries. Naturally, these countries would like to ensure a reserve of oil and gas within their territories, especially when the expected future increase in the price of alternative energy sources is taken into account. The second factor is the extremely adverse effects that rapid depletion would have on oil exporting economies. Government policies of oil exporting states relating to oil and gas production levels, therefore, would be determined by two vital considerations. First, the needs of the world economy and second, the economic and strategic needs of the exporting countries. It will be imperative to strike the right balance to achieve optimum conservation of oil and gas resources. In this regard, international co-operation in the efficient use of oil and gas for essential and non-substitutable uses only would also be imperative.

ELIMINATION OF WASTEFUL PRACTICES

Until recently most natural gas and natural gas liquids produced together with oil in oil exporting developing countries were flared or re-injected. The transnationals had not paid enough attention to the development of this valuable asset consistent with their own medium and longer term interests. With the emergence of national petroleum enterprises, great efforts are being made to eliminate the wasteful flaring of this very valuable source of energy and feedstock for chemical industries. There are now numerous projects under planning or implementation to utilize the gas in local industries, or for export as LPG, natural gasoline and LNG. Saudi Arabia, for example, hopes to be producing for export about 12 to 14 million tons/year of LPG and 5 to 6 million tons/year of natural gasoline by 1985, depending on her crude oil production level. The next development step will be to open up and market the extensive onshore and offshore resources of non-associated gas.

DIVERSIFICATION OF ECONOMIC BASE

Oil exporting state governments are giving top priority to development plans designed to diversify their economic base. Refineries, petrochemicals and fertilizer projects to serve domestic and international export markets are obviously most prominent in such development plans. These projects are expected to provide many added-value benefits to the economies of these countries.

With the emergence of export refineries, exports of crude oil from the producer countries will be proportionately replaced by exports of refined products.

ENERGY TRANSPORTATION

The size of potential future exports of crude oil, petroleum products, LPG, LNG, natural gasolines, fertilizers and petrochemicals from the oil exporting developing countries to world markets indicates vast potential opportunities for these countries' participation in the huge international marine transportation industry. Keeping in view their medium to long term strategic and commercial interests, these state agencies are expected to continue vigorous tanker acquisition programmes, in spite of the current difficulties of tanker operators due to the low rates prevailing on the spot market.

The emergence of large-scale export refineries and NGL projects will also change the traditional size of products and LPG carriers. In the past 15,000 to 40,000 dwt were considered economical sizes, as the big export refineries were close to the consuming areas. The future requirements for products and LPG carriers will be in the range of 70,000 to 150,000 dwt which will also require modifications and expansions in present receiving facilities in the consumer countries.

EXPLORATION FOR OIL AND GAS AND DEVELOPMENT OF ALTERNATIVE SOURCES OF ENERGY

As a result of the work of the Energy Commission of the Conference on International Economic Co-operation (CIEC) held in Paris during 1975 to 1977, it became abundantly clear that the energy supply and demand balance in the 1980s and beyond could not be achieved unless massive oil and gas exploration projects are undertaken in almost all parts of the world to discover new oil and gas reserves. Concerted efforts must be made to develop all other non-renewable alternative sources of energy too. The oil and gas exporting developing countries are deeply interested in the research and development of all sources of energy, as they will also need a different energy mix in their domestic markets in the not too distant future. Saudi Arabia, for example, has already signed an agreement with foreign partners for joint solar energy research, and is also taking steps to develop nuclear energy.

POSITION OF FOREIGN COMPANIES IN THE NEW SITUATION

The oil exporting countries are quite determined to implement the policies discussed above. They have the necessary investment capital, or at least the capacity to borrow such capital. What these countries lack are the experience, technical know-how and the integrated networks which are equally important for successful achievement of their goals.

Foreign oil companies, which can be considered to be established international traders in oil and gas, including the transnationals as well as many independents, have the necessary experience, technical know-how and the integrated networks which can serve the needs of the producing countries. The UN agencies, other governmental agencies and contractors have scope to participate in this field, but their contribution would be relatively small compared to that of foreign oil companies. This is understandable considering the companies' obvious advantages. They have greater flexibility in the sense that they have better scope for integrated, continuing project involvement and project management, from the original design concept, through the construction phase, to the technical transfer and the maintenance of operations subsequently, involving a skilled manpower team.

Keeping in view the policy interests of the oil and gas exporting countries detailed above, the desirable specific areas of co-operation become quite apparent. These can be broadly classified into the areas of international trade, project investment and technical services in all phases of the oil and gas industries and industries related to alternative sources of energy.

POSSIBLE MODES OF CO-OPERATION BETWEEN OIL EXPORTING COUNTRIES AND THE FOREIGN COMPANIES

The issues which had led to confrontation between the oil exporting countries and the foreign oil companies are already amicably settled. The future holds new vistas of vast potential opportunities. Co-operation based on fair play, an honest appraisal of mutual interests and a pragmatic view of the general interest of the international community would ensure world stability and economic prosperity for all. For many decades these countries had been co-operating with the international oil companies under the concession agreements. Later on, partnership and sharing agreements were developed. Nevertheless, the oil industry remained essentially a company rather than a government concern. Price, production and exploration decision-making remained in the hands of the companies. One of the main reasons for such a state of affairs was the industry's need for the marketing outlets of the companies, as well as the need for their technical and managerial know-how, and, in certain cases, their capital or access to capital markets.

In recent years, many of these factors have assumed different values. Out of all these factors, only know-how still represents a valuable service that the companies can offer to the producers, although there still remains some room for co-operation in marketing gas liquids and petroleum products, as well as in shipping and terminalling, and, to a limited extent, in downstream activities.

This is why new forms of co-operation between oil and gas producing governments and oil companies are needed. Essentially both need each other, the companies need the crude oil and the governments need the know-how. A degree of mutual dependence is characteristic of other sectors too.

To start with, the extreme case of the 'concession agreement' era is completely over, at least in major oil exporting countries in the developing world. That was the era during which the companies got most of the exceptional benefits and had virtually unlimited decision-making power in exploration, production and prices. The other extreme is full nationalization without any participation for the international oil companies. This other extreme may not be the predominant form today and for the forseeable future, until the countries concerned acquire full command over know-how and management. In between the two extremes there may develop various forms of co-operation between the two groups.

One very simple form is for the company to act merely as a service contractor for money or for the right to lift some oil. Availability of oil is the main thing which the companies seek, especially if the cost of exploration and operations is borne by the government. This form is probably good for the richer oil exporters where capital needed for oil and gas exploration and production is available from indigenous sources. For those who need capital or find it advisable to let the companies risk some of their own money, an incentive plus repayment of the invested capital from successfully accomplished programmes might be necessary. Also, partial access to newly discovered crude is often essential to attract the companies to share in the risks involved.

Funds from the international financial institutions represent a third alternative for those who do not have the capital and do not want companies' investments, but need to explore for hydrocarbons in their own lands. The companies may act as contractors with limited rights to the discoveries in such cases. In essence it can be said that oil companies are very much concerned about crude availability in the future. They are willing to risk their own money in different degrees if the reward is crude supply, given a reasonable prospect of recovery of investment with a consideration for the degree of risk involved.

Hence, with good faith and goodwill, co-operation in exploration and production can take various new forms depending on the government concerned and its resources and policies. Companies should be ready to perform the usual exploration and production functions, provided they are given the right incentives, including access to future crude supply.

Figure 4

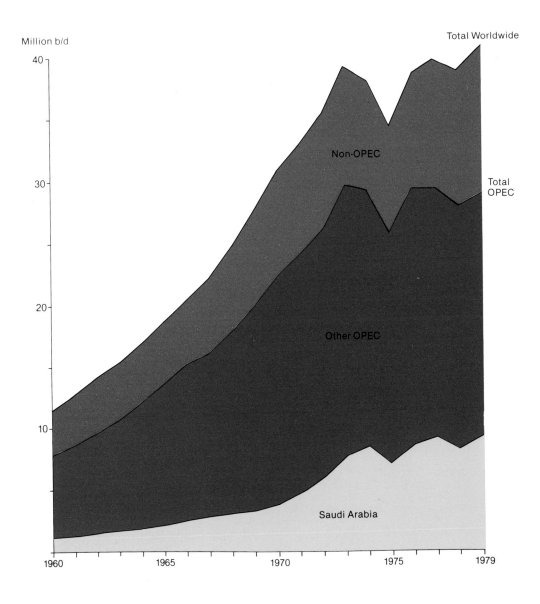

8
PETROMIN, SAUDI ARABIAN OIL POLICIES AND INDUSTRIALIZATION THROUGH JOINT VENTURES

PETROMIN'S INTERNATIONAL SIGNIFICANCE

The General Petroleum and Mineral Organization (Petromin) was established in 1962 as a public corporation wholly owned by the Saudi Arabian Government.(1)

The objective was to promote national participation in the development and exploitation of Saudi Arabia's petroleum and mineral resources and allied industries.

The significance of Petromin's future role in the world oil industry is basically due to Saudi Arabia's pre-eminent position in the world as the biggest oil exporter (9.3 million b/d in 1979) and to its having the largest crude oil reserves in the world (168.4 billion barrels at end-1979).(2) Saudi Arabia's proven remaining natural gas resources amount to 95.73 trillion cubic feet. Figure 4 shows the evolution of Saudi Arabia's oil exports in the context of world oil trade over the twenty years 1960 to 1979.

Petromin's role in the international oil industry is defined by Saudi Arabia's oil and energy policies. The prime considerations in the formation of these policies are the basic principles determining OPEC members' policies, and our national objectives.

NATIONAL OBJECTIVES

Saudi Arabia's national objectives for energy development are based on rational considerations of national development aspirations and the world's long term energy requirements.

These objectives can be listed as:

- Development of an integrated national oil and gas enterprise with special emphasis on: development of national marketing capability to serve international oil and gas markets; increased processing of petroleum in Saudi Arabia; and increased national participation in the international energy transportation industry.

- Augmentation of current oil and gas reserves and production capacities.

- Conservation of depletable oil and gas resources.
- Development of alternative sources of energy.
- Industrialization through joint ventures.
- International co-operation with special emphasis on assistance to and promotion of developing countries' interests in the field of energy.

These objectives are the main motivating factors which determine Petromin's future role in the world oil and energy industries.

POLICIES AND THEIR IMPLEMENTATION

Relationship with Aramco

The evolution of Saudi Arabia's relationship with Aramco (The Arabian American Oil Company) has always remained amicable and all issues have always been settled through negotiation.(3)

It was in 1950 that the Saudi Arabian Government and Aramco agreed on the so-called fifty-fifty principle. This meant a tax, including royalties of 50 per cent on oil production income. This arrangement stayed in practice until the mid-sixties, when royalties were accepted as an expense which changed the sharing of income a few cents per barrel in favour of the Government.

The Saudi Arabian Government's concern with prices, however, continued in the following years. The prices FOB Saudi Arabia or Sidon in Lebanon were determined by the then Aramco shareholders and the prices in essence were transfer prices between affiliates. They were not arm's length market prices. Furthermore, the major oil companies, as if by collusion, always established relatively uniform prices in various countries and reduced those prices from time to time without even consulting with the producing governments. It was imperial power at its peak. The real price of oil by 1970 was about 50 per cent of its price in 1950. Also, production shifts from one country to another used to take place without due consideration to the producing countries' economic development requirements.

Although this state of affairs was inherently unsustainable, the transnational oil companies failed to foresee the inevitable. I think it was the arrogance of their traditional power which blocked their vision. Saudi Arabia was one of the founding members of OPEC when it was established in 1960. By 1972, the first signs of the change in the balance of power between the transnational oil companies and OPEC were evident when the short-lived participation agreements were signed in December that year. Saudi Arabia at that time acquired 25 per cent of Aramco without a corresponding 25 per cent share in the lifting of oil. Prices for the rest of the 75 per cent were heavily discounted: another show of stubbornness on the part of the transnationals. In October 1973, OPEC took over the price and production decisions from the oil companies. By June 1974, Saudi ownership rose to 60 per cent, and by December of the same year the tax rate reached 85 per cent.

These changes may look drastic if viewed in isolation, but seen in the historical perspective discussed in the previous chapters they seem quite reasonable.

The Aramco shareholders were still left with substantial access to Saudi oil, and with very healthy profits. However, since then their attitudes and policies toward Saudi Arabia have been modified or in some cases totally changed.

Negotiations for one hundred per cent take-over of Aramco by the Saudi Arabian Government started in December, 1974 and were completed two years later. An agreement has already been tabled with the Saudi Arabian Government for approval. That agreement, although not signed yet, has actually been implemented, and its financial benefits to the Government have been obtained since January 1976.

Aramco, as an organization, has behaved mostly in a manner reasonably consistent with the country's domestic interests and traditions. It has always attempted, and still does attempt, to follow first class oil field practices.

Hence, our policy is to retain that organization and create the right atmosphere for it to survive and function efficiently. In essence this means a continuing relationship of one form or another with Aramco's previous owners and is based on mutual interest.

We feel that as long as these previous owners continue to make a substantial and positive contribution to the efficient functioning and growth of the Saudi Arabian oil industry, there is room for them to stay, mainly as service contractors rather than as equity holders. This means continued exploration for oil and gas, as well as a business relationship of a buying and selling nature. It implies further continued co-operation in the management of Saudi Aramco.

It can be seen that Saudi Arabia, in its relationship with Aramco and its owners, followed a rational and pragmatic approach. It followed a path of self-interest with due consideration to an amicable and stable form of relationship with the Aramco owners as long as they acted in favour of our country's interests and were not against OPEC policies and resolutions. In the main stream of events, Saudi Arabia, as an OPEC founder and influential member, adopts OPEC policies, whether they are in relation to the oil companies or to the oil consumers.

Oil Price Policies

Our oil price policies are basically derived from OPEC policies. Hence, our future price policies are expected to follow OPEC long term strategies, as and when such strategies become official OPEC policy. Such strategies are based on prices to be maintained in real value terms with due consideration to economic growth rates and other energy options.

So far as crude oil prices are concerned, Saudi Arabian policies are those of OPEC. However, our price policies within OPEC itself are coloured by moderation and a degree of objectivity in our efforts to achieve unanimous approval of OPEC's long term price strategy.

Domestic Oil and Gas Policies

Oil production level Our net production of oil for export is a matter of great international significance because of the anticipated notional gap between worldwide supply and demand for oil. Our policies in this respect have always been based on a balanced view of our national and international interests.

Oil and gas exploration Saudi Arabia intends to continue to spend large sums of money on various programmes of prospecting and drilling exploration wells wherever chances of success warrant such spending inside our country. We have even instituted lately a special natural gas exploration programme.

In the meantime, various new oil fields will continue to be developed within a framework of first class oil field practices with the objective of maximizing the oil to be recovered from these fields.

In other words, we anticipate making huge investments in simply maintaining current sustainable oil production capacity, on top of enormous investments to build additional sustainable capacity. Those investments will be consistent with prudent oil depletion policies to maintain all our oil fields in the best conditions for yielding the highest quantities of recoverable oil using the most up to date technology.

Significant Projects

For the last ten years Petromin has been thoroughly studying and implementing various projects, both upstream and downstream in our oil and gas industry. Petromin's five year plans covered all of our downstream investments, whereas Saudi Aramco's plans and programmes took care of most of Saudi Arabia's present and future developments in exploration for and production of oil and gas.

A total of thirteen major projects based on oil and gas have been approved for construction during the third five year plan (1980-85). Refineries to produce the main oil products, lube base stock refineries, lube blending plants, as well as ethylene, methanol and ammonia and urea plants, are expected to be on stream between 1983 and 1985.

Among Saudi Arabia's energy projects, the following, which are now at various stages of planning, construction and completion, are destined to have a very significant impact on the future international energy industry.

International petroleum marketing. After the 100 per cent take-over of Aramco, Saudi Arabia's national participation in international crude oil and products marketing will rise at a rapid pace. With the emergence of our export refineries the future availability of crude for export will gradually decline. This will be replaced by exports of products. Petromin had already launched its direct crude oil marketing operations in 1973. The volume of our direct crude sales had reached around 2 million US barrels per day by early 1981. In the case of petroleum products and NGLs, a total of over 800,000 barrels per day of various petroleum products will be made available to the world markets by 1985, as well as over 12 million tons annually of gas liquids, both from the east and west coasts of Saudi Arabia. Our price policy is already established and is based on world market prices. That is, if the market is soft, prices will tend to be lower than otherwise. If the market is strong, prices will move up.

Refinery projects. The following table shows the rated capacities of Saudi Arabia's present and future refineries:

MAINLY DOMESTIC	Thousand B/CD	MAINLY EXPORT	Thousand B/CD
Existing		New	
Aramco, Ras Tanura	450	*Petromin/Petrola at Rabigh (1981/83)	325
Jeddah O.R.C.	90	*Petromin/Mobil at Yanbu (1984)	250
Riyadh O.R.C.	120	*Petromin/Shell at Jubail (1984)	250
	660		
Luberef	4		825
New			
Petromin Yanbu (1982) Phase I	170	Petromin/Texaco Chevron Lube Oil Refinery At Jubail (largest single-train Lube facility in the world)	12
Petromin Yanbu (1985) Phase II	250		
	420	Petromin/Ashland Lube Oil Refinery at Yanbu	5
	1084		
		*Possible additional units at each of the above after 1985	750
			1592

In implementing our refining policy, we consciously chose the joint venture approach to our export-oriented refineries. Our decision recognized our limitations and the need to co-operate with the best to acquire the best. The joint venture partner needs a safe and secure supply source for his fuel and feedstock requirements. He welcomes attractive financing conditions, and an extremely modern, effective and extensive infrastructure support for his project. For us, as producers, we welcome a new relationship with these companies based on their potential contribution to our endeavours. In this case, the assurance of oil and gas supply to the joint venture projects brought about the establishment of major projects jointly with foreign companies that contributed their marketing and management expertise. Hence, by early 1974, joint venture negotiations were started to establish various projects in export refining, petrochemicals, steel, aluminium and fertilizers, which in fact offered us the necessary transfer of technology and technical management as well as access to these companies' worldwide marketing networks.

The refineries we are building will be the most technically advanced. They will take into account the changing market requirements and thus will be strongly white-product oriented, producing a high proportion of gasolines and middle distillates.

The joint venture partners will be entitled to 50 per cent of refined products available for export to be distributed through their own channels. The remaining 50 per cent will be marketed by Petromin. In all cases, the prices of products will be geared to international market prices.

The Saudi master gas system for the gathering, treatment and transmission of associated gas, previously mainly flared, is destined to stand as a milestone in the development of the international gas industry. The combined output will make Saudi Arabia the world's largest exporter of NGLs by the mid-1980s.

Since the early sixties, Petromin has been studying the best ways and means by which the flared associated gas can be exploited. In the past, because of depressed prices for oil, no scheme for such exploitation was deemed feasible, resulting in the total waste of billions of tons of gas. When the world oil price system was restructured in 1973, Petromin took the lead in conceiving the largest gas gathering and treatment system in the world to exploit 3.6 billion cubic feet per day of associated gas. By April 1980 we had started to operate the first plant, and natural gas liquids were loaded for export to the international markets.

The significance of such developments is not only in their net additions to the availability of hydrocarbon derivatives in the world market, but also in substituting natural gas for oil in the Saudi Arabian refining and petrochemical industry as well as water desalination and power generation plants.

Underground storage of residue gas and NGL products will complement the Saudi Gas Programme. This will also be a significant step towards conservation and development of national energy reserves.

The 1250 km trans-Arabian Peninsula crude and NGL Parallel Pipelines, from the Eastern Province to Yanbu on the Red Sea coast, will play a significant role in transforming the Red Sea coast into an important industrial centre and international export terminal for crude, products, NGLs and petrochemicals. This would take large quantities of oil and LPG about 5000 miles closer to Western Europe than other alternative sources from the Arabian Gulf. This would also mean added economic and other advantages for Saudi Arabia and the Western world.

Energy transportation industry. The tanker capacity needed to transport Saudi Arabia's production at maximum capacity rate of 12 million barrels a day is estimated as equivalent to one third of the total international tanker capacity, 3000 dwt and above. This reflects the size of Saudi Arabia's energy transportation requirements which are truly enormous.

Saudi Arabia has all the resources to participate actively in this enormous energy transportation market in the coming years. The role of Arab Maritime Petroleum Transport Company (AMPTC) in this field will also be very significant.

Long term projects. Among long term energy projects are the alternative energy programmes. The inherent finiteness of oil and gas resources makes it imperative that alternative sources of energy are developed to meet the long term national energy needs. However, we can use, for the time being, only oil and natural gas, as well as solar energy in a small way. Natural gas will supply over 30 per cent of our energy needs by 1985. Still, our oil consumption will be substantial by the year 2000. We estimate that by 1990 we ourselves will consume over one million barrels per day of oil and about half that much natural gas and LPG. I hope that other energy options like nuclear and solar will also make substantial contributions to our energy mix in the late 1990s to free some oil from having to meet incremental Saudi energy needs.

Whatever quantity of oil is substituted by alternative sources of energy in Saudi Arabia means improved availability of oil to the world markets.

INDUSTRIALIZATION THROUGH JOINT VENTURES

Historical Background

I have conducted or participated in negotiations with many different nationalities involved in doing business in Saudi Arabia. Aside from the serious part of negotiating joint venture agreements, there is always a great deal of pleasure and sometimes a great deal of amusement in dealing with people in general. This is even more true when one is exposed to dealing with people from other countries, since their philosophies and their negotiating techniques vary somehow in accordance with their cultural background as well as with their national histories.

In the early sixties we negotiated our first joint venture agreement to build the first petrochemical plant in Saudi Arabia to the point of having the agreements ready for signature. Before doing so, I felt I needed to consult with legal advisers who had better knowledge of the country where that particular company was based. My first great shock was to be told that in that particular country an agreement is only an expression of hope. In other words, if commitments cannot be met and the hopes cannot be realized, the parties have no more obligations to each other. By contrast, in our Sharia law, the contract is sacred and inviolable. The contract itself is the law that governs the relationship between the parties. Be that as it may, that agreement as such was not signed at that time; a modified one was signed in 1980. It took almost sixteen years; so one must have patience.

Another feature which can affect negotiations is the emergence of different attitudes to the principle and practice of national sovereignty over natural resources. This is a more common question in a way, and it applies to many situations. Depending on which country the other party comes from, one finds that a colonial outlook still prevails to the extent that these natural resources are looked at as if they are owned by that foreign country or that foreign company. This is a far more involved issue because it relates to control over such natural resources, be it resource utilization, production, prices or taxation. Although companies interested in the utilization of natural resources have no power to question such sovereignty in their own countries, they at times raise questions and make comments that reflect obsolete views or lack of appreciation for a government's sovereign rights over its natural resources. Nevertheless, usually in our system joint venture negotiations go on and contracts are ultimately agreed and signed. Afterwards, they become the law, respected by both parties in a country that has trading traditions that go back into ancient times.

This ancient tradition of amicable and fair trading practices with almost all parts of the world was inherited by the contemporary Saudi Arab and has flourished tremendously with the increasing stability and prosperity of Saudi Arabia in the last twenty years.

Philosophy

This brings me to the philosophy underlying our joint venture agreements in the light of world economic conditions as they evolved over the last thirty or thirty-five years after the end of World War II.

During the first ten years of this period of prosperity in the West, Saudi Arabia, in spite of its vast oil and gas resources, was not experiencing a high economic growth rate consistent with the rate of growth of a developing economy. Its oil was sold cheaply; its vast associated gas was flared; its mineral resources remained underdeveloped; its manufacturing sector barely existed; its agricultural sector was traditional and its trade and other services struggled for survival.

This state of affairs was totally unsatisfactory. The Saudi Arabian Government under the leadership of HM the late King Faisal started doing something about it with the help of the World Bank and some of the United Nations agencies. The Government started a process of reorganization and modernization, and established new agencies to care for certain sectors of the economy. Among these agencies was the establishment of the General Petroleum and Mineral Organization in 1962 (Petromin). Also in 1975 under HM King Khalid's Government, the Saudi Basic Industries Corporation (Sabic) was established. Many other agencies and plans have also been established, but I choose these two because of their relevance to the question of joint ventures.

For many years since its establishment, Petromin was trying to motivate foreign investors to join forces with it to establish various oil and gas related joint ventures in Saudi Arabia.

One basic principle in such ventures was a minimum of 50 per cent ownership by Petromin and the Saudi public. But until the early seventies Petromin had only moderate success in attracting foreign partners to establish joint oil and mineral service companies for surveying and drilling.

Its efforts to establish joint export refineries and petrochemical plants started in the early sixties. Those projects were found not feasible economically for one basic reason. We were competing with ourselves by selling our oil so cheaply to the point that a petrochemical plant was more economical if it was built in Japan, for example, based on cheap imported oil, than a similar plant based on free gas in Saudi Arabia.

It was a very disappointing state of affairs. At a time when we did not have the necessary expertise, technology and investment, the free abundant natural gas was not enough to establish any serious joint ventures in gas utilization. Nevertheless, with the experience we gained over the years up to 1973 in studying and evaluating oil and gas industrialization projects, we decided to go it alone, at least in natural gas utilization. So we conceived of a Master Gas Gathering and Processing system that makes various fractions of associated gas available in at least two locations designated as industrial cities; in Jubail on the Arabian Gulf and Yanbu on the Red Sea, where the Government, through the Royal Commission, is supplying the necessary infrastructure for the major industries to be established there. That Master Gas system and the infrastructure constitute the backbone of the Saudi Arabian industrialization programme, including joint venture industries in these industrial cities as well as in a few other locations in the Eastern Province of Saudi Arabia.

Supply Assurance and Resource Economy

The assurance of gas supply to the joint venture projects broke the barrier against the establishment of such major projects jointly with multinational companies that have the necessary know-how, marketing and management expertise. Hence, by early 1974 joint venture negotiations were commenced to establish various projects in petrochemicals, export refining, steel, aluminium and fertilizers.

To make the ethane and methane gas required for these major projects available, we developed a gas supply agreement that virtually guarantees supply of the required gas for the life of the project and established a formula to calculate the cost and price of the said gas to the joint venture. It is a unique formula that makes such gas available at cost during the early life of the joint project with prices subsequently reflecting the profitability of the joint venture. It has a built-in incentive to get the project started and yields profits at up to 25 per cent return on equity, at which point the price of gas will be increased up to an equal Btu (British thermal units) value with Arabian Light crude oil from incremental income on top of the said percentage.

So, the joint venture project is not burdened with the investment needed for infrastructure or for gas supply, which is a guaranteed quantity at predictable prices for the life of the project. This was the first attraction. To understand the underlying reasons for the gas supply pricing formula one needs to go back to the time when most of the gas associated with the production of crude oil was flared. Presumably, since it had been flared, it must have been worth nothing, which is difficult to accept. Nevertheless, the rules of the game at that time did not take into consideration the intrinsic value of such gas, nor did it correct the cost of producing oil by adding the cost of conserving rather than flaring the said gas. Of course, by today's standards in an energy hungry world, flaring of any associated gas that can be conserved is not only an economic waste but also it represents a reduction in the energy availability to a world that badly needs it.

Be that as it may, in the early seventies we came up with the concept that associated gas has an intrinsic value at the well-head and it should be dealt with as such whenever such value can be determined.

The determination of the intrinsic value of associated gas was a brand new issue in the Saudi Arabian oil industry where most of such gas was dealt with simply as a flared by-product. Even when certain fractions of such gas were utilized by extracting the gas liquid, such as C_3, C_4 and C_5, no value was assigned to the feed gas that constitutes the bulk of the ultimate product which was exported to world markets and commanded remunerative prices.

So, instead of being arbitrary and decreeing the intrinsic value of gas, we went back to the accounting and economics text books and produced a fair and practical bookkeeping as well as market oriented intrinsic value determination formula. Necessarily we had to use certain business criteria such as the rate of return on equity. Hence, our empirical solution was based on the premise that if the joint venture return on equity is less than 25 per cent in any one year, then the intrinsic value of the gas utilized by the joint venture is no more than the cost of gathering, treatment and transportation (GT&T). This is considered as the minimum intrinsic value derived from the books kept for the Master Gas system. The maximum value on a Btu basis is the price of an equivalent number of Btus in Arabian Light crude oil. Such price will be obtained only from profits exceeding 25 per cent return on equity. This means that owing to the absence of the intrinsic value of such gas in the income calculations of the joint venture, the joint venture makes abnormal profits that rightly belong to the original owner of such gas in the form of a price or royalty that does not exceed the market price for oil. Alternatively, should the joint venture be unable to make more than 25 per cent return on equity, it is postulated that in the market-place such methane and ethane gas, which are very costly to liquefy and export in liquid form, are not worth more than the cost of gathering and treatment. In the case of butane, propane and natural gasoline (which are gas products by themselves, unlike methane and ethane which are used either as fuel or chemical feedstock), we find that the same principle also applies.

The so-called incentive crude oil supply contracts are one-time sweeteners in these joint ventures. Because of the expected shortfalls in the world oil supply, the assurance of availability of a certain quantity of oil for fifteen years, although at world market prices, must be one great advantage to those partners. As I said, it is a one-time deal that has no adverse effect on us. Prices are equal to those paid by others, and oil is to be supplied from Saudi production at whatever level is authorized by the Government. No additional burdens are foreseen.

Finance

The second major issue in negotiating a joint venture agreement is financing. Most of these projects are in the billion dollar plus range. Each one of them by itself represents a significant financing issue to both sides. Hence, the joint venture agreements deal with this issue in more than one way. First and foremost in structuring the financing of such major projects are the extremely thorough technical, economic and investment confirmation studies. Such extensive economic, financial, marketing and engineering preparations are the basis for the final decision as to whether or not to go ahead with such major investment. It is the point of no return so to speak. Each party has the absolute right to withdraw from the project within a given time after the completion of such studies. Legally speaking, no reasons for the withdrawal are required, in order to give each party full freedom to use any decision criteria found appropriate.

The financing structure used in the studies is based on the premise that to be a serious foreign investor a minimum equity of 30 per cent of total investment is to be provided for by the two sides on a 50/50 basis. Another 60 per cent is to be provided as a government loan through the Public Investment Fund at interest rates that vary between a minimum of 3 per cent and a maximum 6 per cent per annum depending on the profitability of the project. It is apparent that there is a measure of subsidy involved in this scheme of financing. Many countries, and in particular some European countries, give outright grants to attract such industries that require the mobilization of a great deal of money and know-how. We did not find it economically advisable to do so. We took a different approach. The project should pay its own way as far as the joint venture partners reasonably expect it to do so. This is a minimum requirement within the given framework of availability and cost of infrastructure, financing and gas. We reckon first that we own 50 per cent of the project, the profits and subsidies of which are retained in the national economy anyway. Secondly, insofar as the foreign partner is concerned, a fair return on his equity investment, entrepreneurial risks and the transfer of technology is a calculated payoff from the joint venture, should the project realize what the studies show it can realize. If it does not do so, despite all of the project and investment evaluation that has been done, one might pass the buck, so to speak, and blame it on unforeseen circumstances. Alternatively, one might be candid and admit that risk is an inherent part of almost all investments. Winds blow in all directions and no investment guarantees are available. An old Arabic proverb says that trade is both a profit and a loss. Thus we are still guided by our age old Arabic traditions.

Now that does not mean taking anything other than calculated risks. We utilize sensitivity analysis, using the computer. So, if things go wrong, again we can blame the computer.

Readers may have noticed that 10 per cent of the total investment is still not yet covered. We left this for the banks; this represents another initiative to ensure objective analysis, the evaluation of risk and the follow-up commitment of funds.

There is more to the financing package than I have explained so far. There are important details that are beyond the scope of this book. They are dull, as almost all details are, but unfortunately business success depends a great deal on these dull details. So I omit them here only to avoid enlarging this section of the book unduly.

Now once we have the money, we move to the management phase of these joint ventures; management during construction and start-up, manpower planning and operating management, including marketing in particular.

Marketing

The marketing provisions in the Saudi Arabian joint venture agreements vary to some extent in accordance with the type of the end product or the home of origin of the second party concerned. Basically, however, the second party, namely the foreign joint venture partner, is responsible for the marketing of at least 50 per cent of the total production. At times, buy-back provisions for part or all of the other 50 per cent at competitive international market prices are part of such joint venture agreements. The more desirable pattern, of course, is to leave the marketing responsibility for the entire production with the joint company to be established in accordance with the joint venture agreements. Why the variations, one might ask? Our own national objectives are to establish ourselves as marketers of our share of the products available, either through the joint company itself, or if that is not possible, to market such share directly. Among the advantages of leaving the international marketing function to the joint company is the introduction of Saudi Arabian oil, petrochemical and mineral products as Saudi Arabian products per se, and not the products of the XYZ transnational company. The introduction of the joint venture company itself into the world markets gives it twice as much production under the Saudi Arabian label compared with the alternative case of marketing our own share directly through our parent organizations. Nevertheless, there are situations where the foreign partner cannot be a party to joint marketing because of anti-trust laws. This is true for all American companies. In other cases the said partner may require all or part of his share of the production and some or all of our share of the production for feeding into its integrated operations elsewhere. Another advantage of marketing through the joint venture company some or all of the production to third party buyers is the competitiveness and arm's length character of the price. Hence, the marketing options employed so far require a minimum share to be marketed in a truly competitive international market to provide arm's length prices for application to whatever share of the production is fed directly to the foreign partner's integrated operations outside Saudi Arabia.

So, a measure of an internationally competitive market price is always provided for in these joint venture agreements. Otherwise, the transfer prices that were characteristic of most of the multinational companies will be the only prices available to the partners for tax and profit calculations. By their very nature, transfer prices were always meant to be in favour of the buying partner. Hence, a competitive measure is always needed to correct such situations.

Prices are set by the Saudi Arabian Government from time to time as market conditions change, and in particular for refined petroleum products and liquefied petroleum gas. These government established prices are derived from prevailing international supply and demand market pressures. There is even a contractual provision for withdrawal from the sale-purchase contract to ensure that prices remain market-sensitive.

So much for marketing and prices. Let us now move to the questions related to management and manpower development.

Management and Manpower

The 'raison d'etre' for having joint venture industrialization and mining projects is the transfer of know-how in technology and management to Saudi Arabia to provide a base for further development. This is a most critical phase of negotiating joint ventures. The easiest provision for the foreign partners is to leave such matters entirely to their own organizations. This way they feel that they can design, engineer, build, commission and operate these complex industrial or mining facilities at least cost and in the shortest time possible. I am not so sure that this is always true, since these companies utilize their own subsidiaries for most of the design, engineering and project management. Whenever a company directly utilizes the services of its own subsidiary, the necessary competitive measures become obscure. That was never a satisfactory state of affairs for us. Nevertheless, we agreed to utilize the technical services of the said subsidiaries, providing as best we can for alternative measures of competitiveness and participation of our own technical personnel all the way through.

The construction of the plants is usually initiated through international competitive bids so that at least the major portion of the investment is done on a comparative and competitive basis most of the time.

Finally, we come to the question of people: the men and women needed to commission and operate these complex facilities in an efficient manner and on a profitable basis. That is the game we are just about ready to play. We do not have enough skilled players, however. In some cases we do not have sufficient skilled manpower to run certain facilities from either side. Almost always on our side very few skilled people are available from within Saudi Arabia to immediately fulfil the provisions laid down in the joint venture agreements requiring full Saudi participation in all phases of operations and management right into the executive boardroom. However, we are working hard at it by expanding our recruiting and training effort for Saudi technocrats and labour so that more and more Saudis over the next ten years will join these major projects and gradually assume the responsibility for running them successfully.

God willing, we are determined to succeed.

References

(1) Petromin Handbook 1382 H. - 1397 H., 1962 A.D. - 1977 A.D. General Petroleum and Mineral Organization, 1977.

(2) Petroleum Statistical Bulletin, 1979. Ministry of Petroleum and Mineral Resources, Kingdom of Saudi Arabia.

(3) Aramco And Its World, Arabia and the Middle East. Edited by Nawwab, Ismail I., Speers, Peter C., Hoye, Paul F., Aramco, Dhahran, Saudi Arabia, 1980.

9
NORTH-SOUTH: AN INTERNATIONAL ENERGY DIALOGUE

BACKGROUND

Fewer than 800 million citizens of the twenty-four industrialized developed OECD countries known as the North consume most of the world's resources, produce most of its manufactured goods and enjoy the highest standard of living, with a GNP per capita of well over $8,000 in 1979, on average.

In contrast more than two billion people in over one hundred developing countries known as the South* produce most of the world's raw materials, the bulk of which are consumed by the North. Even among OPEC member countries, only some 3 per cent of their populations, or about 10 million people, had a per capita GNP exceeding the average in industrialized countries in 1979.(1)

Historically, the vast natural resources of the South, including petroleum resources, remained for long under the domination and control of the North. This left the developing countries so deprived of the true benefits from the exploitation of these resources that most of their populations have been denied what are regarded as the barest necessities of human life in the industrialized countries.

The emergence of national control over indigenous oil resources in the developing countries is a comparatively recent phenomenon. With it has grown an awareness of the urgent need to conserve and regulate the exploitation of these precious resources to best serve their national development objectives and plans.

In order to undertake suitable development projects to raise the standard of living of their populations, the developing countries have identified their priority requirements which are considered imperative to achieve their aspirations. Some of the principal requirements are as follows:

---

*Developing countries including OPEC members, excluding centrally planned economies.

- Fair prices for their raw materials and preservation of the value of their export earnings.

- Conservation of natural resources and diversification of their economies.

- Transfer of technology to improve exploration and exploitation of natural resources for setting up indigenous industries and infrastructures.

- Access to the markets in the industrialized countries for their finished products.

- International monetary reforms.

Traditionally, technology, access to markets and the world monetary system have remained under the predominant control of the North.

In other words, the developing countries are seeking a reduction in the vast economic disparities existing between the North and the South on the bases of justice and rationality. It was in line with such aspirations that the 'Group of 77', which represents over 100 developing countries in the UN General Assembly, approved a resolution which demanded a 'New International Economic Order'.

In April-May 1974, the Sixth Special Session of the UN General Assembly devoted to the problems of raw materials and development called for the establishment of a 'New International Economic Order'. The first Summit Conference of Sovereigns and Heads of State of OPEC Member Countries in Algiers in March 1975 resulted in further progress in this direction.

Until 1974, far from accepting such demands, the developed countries were not even prepared to discuss them seriously. However, with the emergence of OPEC as a dominant force in international affairs after its historic actions in 1973-74, the situation changed.

The motivating force for convening the Conference on International Economic Co-operation was the mounting concern in the North about the future price and availability of oil from the developing countries. In March 1975 the Algiers OPEC Summit Conference also agreed in principle to the holding of an International Conference bringing together the developed and developing countries.

CONFERENCE ON INTERNATIONAL ECONOMIC CO-OPERATION (CIEC)

The Agreement on procedures to hold the Conference on International Economic Co-operation in Paris was finally arrived at in October 1975. The Conference was convened in December 1975.

CIEC provided the first opportunity to address international energy problems of increasing importance to all countries, together with the long outstanding problems related to raw materials, development and financial affairs.

Representation

The 'Group of 19' nominated by the 'Group of 77' to represent all developing countries comprised Algeria, Argentina, Brazil, Cameroon, Egypt, India, Indonesia, Iran, Iraq, Jamaica, Mexico, Nigeria, Pakistan, Peru, Saudi Arabia, Venezuela, Yugoslavia, Zaire and Zambia.

The 'Group of 8' representing the group of all developed countries comprised Australia, Canada, the EEC, Japan, Spain, Sweden, Switzerland and the USA.

The Four Commissions

The Conference decided to establish four commissions on energy, raw materials, development and financial affairs.

On the insistence of the 'Group of 19', and after considerable opposition from the 'Group of 8', the concept of 'linkage' between the work of the four commissions was accepted. This meant that the commissions would function in parallel, and that progress in any one of the commissions would have to be matched by similar progress in the others. The developing countries considered this necessary to ensure that their oil leverage in the Energy Commission could be used for the full benefit of all developing countries.

Participants chosen to serve on the Energy Commission included - Algeria, Brazil, Canada, Egypt, the EEC, India, Iran, Iraq, Jamaica, Japan, Saudi Arabia, Switzerland, the United States, Venezuela and Zaire. Representatives from Saudi Arabia and the United States of America were elected as the co-chairmen of the Energy Commission.

Inter-Governmental Organizations

A number of inter-governmental functional organizations, including the United Nations Secretariat, OPEC, OAPEC, IEA, UNCTAD, OECD, FAO, GATT, UNDP, UNIDO, IMF, IBRD and SELA participated in CIEC's four commissions as permanent and ad hoc observers.

CIEC Objectives

The objective agreed for the CIEC was to initiate an intensified North-South dialogue on all matters related to energy, raw materials, development and financial affairs, leading to concrete proposals for an equitable and comprehensive programme for international economic co-operation, including agreements, decisions, commitments and recommendations. This, it was agreed, should constitute a significant advance in international economic co-operation, and make a substantial contribution to the economic development of the developing countries.

Working Sessions

During the working sessions of the four commissions, in the period February 1976 to May 1977, the Energy Commission examined a number of proposals and studies related to the international energy situation submitted by the 'Group of 19', the 'Group of 8', some individual members or by the observer organizations (see list of proposals and studies in Appendix III). Based on these proposals and studies, an assessment of the world energy situation was made by the Energy Commission leading to conclusions and recommendations. However, members of the 'Group of 19' and the 'Group of 8' could not agree on uniform assessment and interpretations of the energy situation on certain issues and arrived at different conclusions and recommendations. (See Appendix III for the final 'Report of the Conference on International Economic Co-operation', and the Conclusions and Recommendations of the Energy Commission.)

Points of Agreement and Disagreement in the Energy Commission

After intensive studies and debates the participants of CIEC were able to agree on the following issues and measures regarding energy:

- Conclusions and recommendations on availability and supply in a commercial sense, except for purchasing power constraint.

- Recognition of the depletable nature of oil and gas. Transition from oil based energy mix to more permanent and renewable sources of energy.

- Conservation and increased efficiency of energy utilization.

- Need to develop all forms of energy.

- General conclusions and recommendations for national action and international co-operation in the energy field.

The participants, however, were not able to agree on the following issues and measures regarding energy:

- Price of energy and purchasing power of energy export earnings.

- Accumulated revenues from oil exports.

- Financial assistance to bridge external payments problems of oil importing countries, whether industrialized or oil importing developing countries.

- Recommendations on resources within the Law of the Sea Conference.

- Continuing consultations on energy.

Results of Other Commissions' Work

The only tangible results achieved in other commissions could be summarized as follows:

The 'Group of 8'

- Undertook to contribute, subject to the necessary legislative approval, $1 billion for a special action programme to help to meet the urgent needs of individual low-income countries "facing general problems of transfer of resources".

- Agreed in principle to underwrite a common fund for the purpose of financing buffer stock for certain raw materials exported by LDCs, with the details left for further negotiation in UNCTAD.

- Industrial donor countries pledged to increase their flow of official development assistance (ODA) "effectively and substantially" in real terms. The governments that had not yet accepted the UN target of 0.7 per cent of GNP as their annual aid objective, committed themselves to work towards that goal.

This outcome was certainly most disappointing for the developing countries. It failed to provide any improvement in solving the major issues that the North-South dialogue was set up to tackle.

Overall Assessment

The overall progress achieved by the CIEC could be judged from the following three reactions, quoted as expressed by the developing and the developed countries from the Final Report of the CIEC dated June 1977.

> "The participants from developing countries in CIEC, while recognising that progress has been made in CIEC to meet certain proposals of developing countries, noted with regret that most of the proposals for structural changes in the international economic system and certain of the proposals for urgent actions on pressing problems have not been agreed upon. Therefore, the 'Group of 19' feels that the conclusions of CIEC fall short of the objectives envisaged for a comprehensive and equitable programme of action designed to establish the New International Economic Order."

> "The participants from developed countries in CIEC welcomed the spirit of co-operation in which on the whole the Conference took place and expressed their determination to maintain that spirit as the dialogue between developing and developed countries continues in other places. They regretted that it had not proved possible to reach agreement on some important areas of the dialogue such as certain aspects of energy co-operation."

> "The participants in the Conference think that it has contributed to a broader understanding of the international economic situation and that its intensive discussions have been useful to all participants. They agreed that CIEC was only one phase in the ongoing dialogue between developed and developing countries which should continue to be pursued actively in the UN system and other existing, appropriate bodies."(2)

Since the conclusion of the CIEC the North-South dialogue on energy matters has continued in various informal seminars and conferences organized by the UN, OPEC and other independent organizations.

Being intimately involved with the CIEC and being a regular participant in numerous energy seminars and conferences since its conclusion, I am quite convinced that the lengthy ongoing dialogue has not been a fruitless exercise after all, as some circles tend to believe. The dialogue has certainly contributed towards a very clear cut identification of the adverse international trends and their future implications. The new awareness has been of great help for all countries in the formation of their energy policies. The new energy policy moves by OPEC members, the USA, Western Europe and Japan, and the decisions at the Venice and Tokyo Summit Conferences, are all expressions of this awareness. The report of the Independent Commission on International Development Issues entitled, 'North - South: A Programme For Survival' (also known as the Brandt Report) is another useful addition to the growing awareness of the problems and the urgent need to find solutions.(3) The only danger which persists is that, in their eagerness to solve their problems, countries may tend to lose the international perspective of the situation and start pursuing policies of confrontation rather than international co-operation. Such a course of action must be averted before it is too late.

References

(1) 1980 World Bank Atlas. World Bank, Washington DC.

(2) Report of the Conference on International Economic Co-operation. Ministerial Conference, Paris, 30 May - 2 June 1977.

(3) North-South: A Programme for Survival. The Report of the Independent Commission on International Development Issues under the Chairmanship of Willy Brandt. Pan Books, 1980.

10
STRUCTURAL CHANGES AND NEW STRATEGIES

STRUCTURAL CHANGES

From the analysis of the significant structural changes in the international oil industry it is clear that the traditional dominant role of transnational oil companies in the supply and distribution of oil and gas throughout the non-Communist world has declined. After being net sellers in 1970, most of them have become net buyers in 1980. In the meantime, the national oil companies of the producing countries have started to assume a major role in the supply of crude oil and natural gas liquids to consuming countries. This phenomenon is irreversible and in the coming decades the trend is expected to continue. Future energy policies in the industrialized countries must fully recognize this fact. Furthermore, political factors have assumed an increasingly important role in the marketing of oil and oil derivatives.

It is important to note that such changes are not the only ones that took place in the world economy in the last decade. Other structural changes in the world economy have also occurred during this period which have directly or indirectly contributed to or have been a consequence of the prevailing instability in the international monetary and financial systems. In turn these have affected the international energy markets, particularly oil.

Changes in oil prices and supply patterns have always been affected by dynamic and interrelated developments in the world economic order, such as potential trends in the world energy mix and the expected future changes in international oil trade. Their overall impact on the world economic order has become a question of great significance to the international community.

THE OIL AND MONEY RELATIONSHIP

The supply and demand for oil were kept in balance during 1979 and 1980, as always, through the price mechanism. Oil production in the non-Communist world is estimated to have risen by over 4 per cent in 1979, with OPEC production 3.3 per cent higher(1) and Arab oil producers increasing their output by over 14 per cent.(2) But non-Communist world oil consumption hardly changed from the year before level. In spite of this, because of anxieties associated with the Iranian revolution, remarkable stock rebuilding and structural shifts involving oil supplies, the official price of Arab Light marker crude rose by 89 per cent between the end of 1978 and the end of 1979. Not only was this a relatively modest price increase compared with those of many other oil exporting countries, including the UK, for example, but the official price of Saudi crude lagged far behind spot sales prices of crude oil and the value of the product demand barrel on the spot market. This can be seen in Fig. 5 opposite. This latter is estimated to have risen by 116 per cent in value between the end of 1978 and the end of 1979, FOB Rotterdam.(3)(4) Also, Saudi Arabia has exceeded its publicized oil production ceilings for nine of the ten quarters ending in June 1981. This provides a clear perspective of what moderation has cost Saudi Arabia in the recent past. Whilst some may argue that Saudi Arabia could not have realized spot market prices for all its 9.5 million barrels a day production, or more, it must not be forgotten that strict adherence to the 8.5 million barrels a day ceiling throughout 1979 and 1980 would have taken both official and spot market prices to considerably higher levels than those actually seen, on the basis of reasonable deduction from empirical evidence.

FUTURE ENERGY FORECAST

Assuming certain rates of growth in the availability of energy other than oil, and of certain energy/GNP coefficients to represent conservation savings, it is possible to derive estimates of the effect on GNP growth rates of alternative scenarios for oil supply.

This comparison of alternative scenarios is made to bring out the economic consequences in terms of maximum growth rates which might be achieved on the basis of different oil supply assumptions. It should also be interpreted as emphasizing the need to improve on the assumption made about conservation of energy use and substitution of oil by other forms of energy.

The growing supply shortages in relation to demand which could remain a feature of the current decade are resulting in a number of undesirable developments. With curtailed oil availability, we are already witnessing incidents where the transnationals are cutting back supplies to certain developing countries and at times are shifting supplies from one country to another. Many buyers now find that recourse to the spot market is necessary to ensure adequate oil supplies. We have also seen that during 1979 and 1980 the prime factor behind the increasing oil prices was the spot market. Only after April 1980 did the oil market start to show some signs of stability at a price of rather more than $30 per barrel, punctuated by another brief rise in October and November 1980 following the start of the conflict between Iran and Iraq.

The projected scenarios for oil supply and demand discussed in Part II indicate that the oil market is likely to tighten further in the future. One may wonder as to what will happen to prices in these situations.

NEW STRATEGIES

The situation needs new strategies. In a general sense this might be true for all markets and all goods and commodities, since markets are dynamic as are all factors involving economic behaviour.

First let us examine the meaning of the word 'strategy' in a market orientated situation. In the international oil market there are a number of players on the supply as well as on the demand side of the market. There are oil producing governments represented in most cases by their national oil companies. They have the main access to substantial quantities of crude oil. Many independent oil companies and many consuming countries have virtually no access to any quantities of crude oil that they can call their own. Each of these acts in the market accordingly. In between there are the major transnational oil companies, which have varying degrees of accessibility to a more or less assured flow of crude oil.

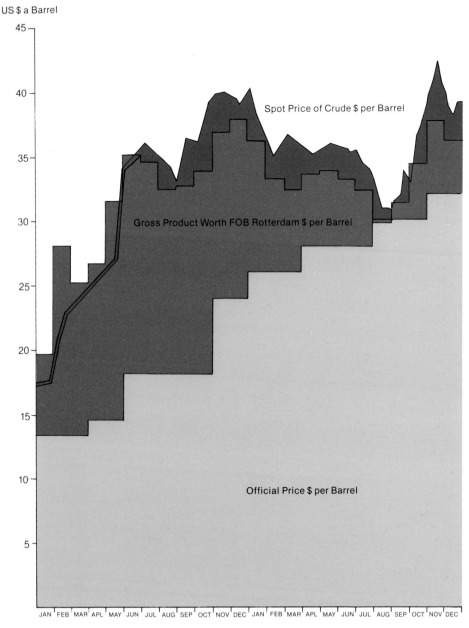

Figure 5

Official and Spot Market Prices of Arabian Light Crude

The very fact that, in a sense, this has already happened and is expected to continue in some fashion requires all three groups of players to re-examine their relative positions in the market and define their future buying and selling patterns accordingly. The existence of a possible conceptual gap between supply and demand of oil in the 1980s will make things easier in the market for those who have more reliable access to oil and oil derivatives. Those who do not have such a position will face varying degrees of difficulty in securing oil supplies and they may suffer economic and political consequences.

Hence, those who are in a surplus position need different market strategies compared with those who are in a deficit position. Others who are in more acute deficit constitute an important factor in the planning of market strategies for those who are in a surplus position. The oil producing countries, and in particular OPEC countries and their national oil companies, must take into consideration in their marketing programmes the fact that there are and there will be many countries, both developed and developing, which fall into this category. It is also clear that those countries likely to suffer most are those which do not command the resources to compete for the scarce energy supplies in the world markets, or those which do not have sufficient financial or technological resources to develop alternative energy resources within their own countries and so reduce their imported energy.

Supply, demand and prices usually define the market for many commodities. In the case of oil these are clearly not sufficient. Other considerations, discussed above, are equally significant in forming appropriate market strategies for OPEC national oil companies.

This discussion of strategies is inherently based on hypothetical assumptions because of the uncertainties involved. First, and most important, are the political considerations that may affect future price and supply decisions within OPEC itself and which may change from time to time. Nevertheless, it can be reasonably assumed that OPEC will follow a predictable path in its pricing and supply strategies. Secondly, there are many questions that relate to the financing of oil imports and the development of alternative energy sources for many developing countries that have not been answered yet. Nevertheless, these have a major impact on the planning of oil market strategies for OPEC national oil companies.

If it is assumed that the developed industrialized countries could achieve a greater than predicted change from oil and gas to alternative sources of energy as well as greater than predicted energy conservation rates, it would certainly make it possible to free more oil for export to the developing countries, while having the added benefit of alleviating the pressure on oil prices. It must also be assumed that enough funds can be generated for developing countries' energy development programmes and oil imports to keep their energy situations viable. These will be raised through the world financial institutions like the World Bank, the OPEC Fund and the various forms of government to government and private bank financing.

For OPEC national oil companies I think a high priority in their oil market strategy should be to see that the developing countries find access to a fair share of the international trade in oil. Otherwise, they will be put in a less favourable position in the market compared with others, and may wind up getting less than their minimum needs to achieve a reasonable rate of economic growth. This would be an undesirable state of affairs from the point of view of the governments of OPEC, as well as being undesirable from an international political standpoint in any case. The industrialized countries will continue to account for the bulk of world oil imports for many years to come.

If we continue to accept these assumptions, then OPEC national oil companies may find the magic solution to unify oil prices, and this may become their realistic actual market strategy. Otherwise, a variation in market strategies between various OPEC national oil companies will occur similar to what happened in the late seventies. There will be those who will use the spot market and there might be others who will not. Be that as it may, the feedback of the spot market to those who will not enter it will be a major factor in defining their market strategy. An important relationship exists between the price of a barrel of crude oil, whether it is a spot or unified OPEC price, and the value of the yield from that barrel in terms of prices of petroleum products. Since such products are usually sold on relatively short term contracts, and since some of these products, in most cases, are sold spot, like bunkers and aviation fuels, one finds that a spot market related strategy almost develops by default. One may still conceive of at least medium term contracts for petroleum products geared to the price of crude oil in some fashion. But, most probably, the strategy for marketing petroleum derivatives will still have a relationship with the spot market for such derivatives. These by themselves affect and are affected by the spot market for crude oil, although there may develop a degree of stability and predictability. Both markets react on each other to a certain degree, although in the past the basic influence has usually been derived from the spot market for oil derivatives.

The spot market prices for crude oil will become rather tempting should shortages in crude oil supplies develop in the eighties. It still seems more likely, though, that the larger volumes of crude oil will still move under term contracts at relatively stable prices, unless the supply demand gap becomes critical, either in terms of size or in terms of its relative effect on certain countries and certain companies.

The same cannot be said for products and gas liquids. A strategy already exists for these, at least in the Arabian Gulf, where most of these products move into the international market under what is known as government-established prices. These are set monthly in the light of world market conditions. A relatively small percentage of total supply is still sold as spot cargoes. There are those who think in terms of higher percentages to be sold in the spot market. But with the increasing volumes expected from producing countries' export refineries, and new gas systems, this seems most unlikely.

So, under the circumstances and assumptions described above, I suppose that the market strategies for petroleum products and gas liquids will be more uniform than the market strategy for crude oil.

So far we have examined the supply and demand relationships. Since notional gaps are not a feature of the real world, what in fact will happen will be lower economic growth and higher oil prices in some combination, subject to change through time at whatever levels of conservation and substitution by other energy forms are actually achieved.

The probability is that some countries will follow their financial needs as a criterion in determining their production policies. In others, consideration of international co-operation may influence decisions, and higher production than is necessary to meet their financial needs may be reached. Others may follow less predictable policies, especially at times of political change or instability. Finally, oil producers generally are likely to be reluctant to over-produce in relation to their financial needs, unless their accumulated surplus financial assets maintain their value in real terms, and yield a positive rate of interest after allowing for inflation.

If OPEC succeeds in implementing its long term strategy so that we can predict prices, and if consumers limit their demand for oil to the available supply, the solution to a highly complex situation could be greatly simplified. However, the probability of something like this on a sustained basis does not seem to be high. Hence, we have a potentially explosive situation that may affect the world economic system adversely, with serious consequences for both the developed and the developing countries.

All these scenarios sound quite pessimistic. We can certainly take some comfort from the fact that in spite of the drastic changes in the world economy during the last ten years, many industrialized and developing countries have been able to achieve reasonably healthy economic growth rates and manageable balances of payments. Although we have seen double digit inflation and wild monetary fluctuations, the system survives and the doomsday forecasters are still proved to be wrong. I firmly believe that the world economy has the intrinsic strength to adapt to changes in the long term. The system has self-governing mechanisms that help it to adjust to new pressures and survive somehow. The question of discomfort associated with drastic changes is a different matter and is certainly felt most by those who found the old system comfortable, in spite of its inherent inequalities and long term adverse implications.

Coping with the future energy situation will require an unprecendented degree of international co-operation to mobilize finance, labour, research and ingenuity in the future. In this connection it would be important to take note of the final conclusion reached by the Energy Commission of the Conference on International Economic Co-operation which says that "The world community requires an international energy co-operation and development program within the overall framework of an international economic cooperation program that would, recognizing relevant constraints, encourage and accelerate energy conservation and the development of additional energy supplies through, inter alia, facilitating and improving access to energy-related technology, expanding energy research and development and increasing investment flows into energy exploration and development. Without such a comprehensive program the world risks significant shortages of energy in the medium term and rapid depletion of oil and gas that will seriously jeopardize the economic progress of all countries. This comprehensive program would address financial aspects of energy development problems, energy conservation, exploration and development for non-renewable energy resources and technological research and development efforts related to both renewable and non-renewable energy sources. There is need to initiate measures promptly and simultaneously that will produce results in the short, medium and long term."

In the philosophy of business organization it is accepted that if optimum solutions are not available, all reasonably satisfactory solutions must be utilized so that the business organization may survive. By the same token, should an ideal optimum solution for the world's economic and energy problems be unattainable, less ambitious and more realistic schemes must be explored. Therefore, even though we may be far from achieving a new world economic order that supposedly would take care of all the world's economic and energy problems, we should not abandon efforts to correct the situation where immediate action is possible. The important thing to realize is that the world economic system has been moving in ways which make its continued survival and further progress precarious, though the system possesses features which make possible some self correction. Members of the system may need to make concerted efforts to enable the system to function with fewer problems in the future.

Energy is one area which has been well studied. Several positive and constructive solutions are on hand. All that is needed is the political will to start changing the energy mix, at least on a domestic level, towards more abundant energy resources. The process must start now, where it has not done so already.

A piecemeal national approach will not be the most satisfactory solution. A more internationally, or at least a regionally, oriented programme of development of energy alternatives to oil would definitely be more satisfactory. We all know that such actions require financial and technological resources allocated to them. This can be done in various forms, such as bilateral or multilateral aid, or more desirably, internationally co-ordinated efforts.

Should this happen, we may find ourselves closer to an improved world economic order rather than a new one. Efforts should also be made to modify monetary and trade measures and to improve the system even further so as to provide a better chance for survival. By using new or old techniques, or modified and improved ones, we should all try to achieve an increasing world real economic growth distributed fairly between all members of the world economic community.

THE FUTURE

In Saudi Arabia we believe firmly in the contribution of international co-operation and trade as the basis for development of the oil and energy sector, albeit in a changing environment. Although the role of national oil companies is developing rapidly in both the producing and consuming countries, there will be a continuing role for the transnationals. They will have a particular responsibility to develop oil reserves on whatever terms they are able to negotiate in different countries, but also to develop other sources of energy to reduce the burden on oil both in fuelling growth of the world economy and in maintaining their own corporate development.

The major oil importing countries have to show more responsibility than they did in 1979, if they wish to keep oil prices from rising faster in real terms than the rate recommended by OPEC's Long Term Strategy Committee. This rate is related to the expected trend rate of real economic growth in the industrialized countries, after allowing for inflation and currency fluctuations. It is based on the premise of the need for other energy forms to assume an increasing role in meeting growth of energy demand. The great merit of our economic system is that steady changes in relative prices to reflect scarcity of one energy form and abundance of others will enable a smooth transition to occur, provided that the system is allowed to operate effectively.

We entered the energy transition era, rather uncertainly, in 1974. The crucial point about the energy transition era is the switch from an oil industry based on historic costs to one based on the replacement cost of oil in terms of other energy forms. This has now been formalized in the proposal recommended by the OPEC Long Term Strategy Committee.(5)(6)

The exponential growth of demand in the years up to 1973 involved more or less a doubling in the volume of oil consumed every decade. In the context of the finite nature of a depletable natural resource concentrated largely in a few giant fields, the historic cost tradition was already giving rise by 1973 to forecasts of gaps developing between potential supply and demand. The long lead times and enormous investments needed to effect a transition to much greater reliance on alternative forms of energy and more efficient use, especially of oil, carried important implications for the role of price in maintaining equilibrium between supply and demand. The absence of the price increases introduced in 1973-74 could have led to far more economic dislocation than we have seen seven years later. The era of the energy transition started rather uncertainly in 1974, mainly because of widespread disbelief among leading politicians and eminent economists as to whether the new 'high' oil prices could be sustained. This was reinforced by widespread ignorance, even among well-informed members of the public in the leading oil importing industrialized countries, about the prospective shortage of oil in the medium term. It seems highly probable that the lower economic growth rates experienced since 1973, the further price increases of 1979 and 1980 and the publicized intention of OPEC to increase the real price of oil in the future have combined to dispel the illusions about abundant supplies of oil at low prices. They also hold out the promise of providing the basis for a reasonable equilibrium between oil supply and demand in the coming years, particularly in view of the significant improvements now evident in the energy/GNP ratio in most industrialized countries.

About twenty years ago a US academic wrote an historic article under the heading 'Marketing Myopia' in the Harvard Business Review.(7) He chastized the US railroads for not having recognized that they were in the transportation business, rather than just railroads. Thus they missed out not merely on new opportunities in road and air transport, but their own businesses declined. Subsequently, there was the sequel - the unhappy demise of Penn Central, the physical and financial decline of many railroad industries in many countries, and widespread subsidies to those in others.

Many of the major oil companies have taken these lessons to heart and invested in alternative energy forms. There were some false starts and caution induced by the further decline in real oil prices between 1974 and 1978. One may question whether the energy industry worldwide is ready to make the necessary investments at the necessary speed over the next twenty years to relieve the pressure on oil and then natural gas as balancing forms of energy. Remarkable investment programmes will be needed to sustain rising energy demand and to satisfy GNP growth rising at an average of say 5 per cent in the developing countries, and near 3 per cent in the industrialized countries, even allowing for generous and continuing improvements in the energy/GNP relationship.

The paper presented by the President of Royal Dutch Shell to the World Petroleum Congress in 1979 implied a doubling of the average real costs of exploiting new petroleum resources in each of the last two decades of the twentieth century.(8) If producers and consumers of oil and other forms of energy come to recognize the implications of such projections, the problems of the transition away from dependence on oil should be eased as technological changes arrive to reinforce the economic forces at work in the energy market-place.

However, the investment requirements look formidable by any standards. According to estimates put forward in a paper to the eleventh World Energy Conference, energy-related investments in the world excluding the Communist bloc may add up to about 16.5 trillion deutschemarks, or nearly 10 trillion dollars, over the period from 1980 to 2000. This exceeds the total value of world GNP in 1978.(9)

Co-operation rather than confrontation should be a happy consequence of the so-called second oil crisis. This is likely to happen only if the industrialized countries recognize more explicitly than they have done so far the fragile basis for OPEC aspirations. This is the heavy dependence of oil exporters on making maximum use of this unique, finite and non-replaceable natural resource for their own economic development.

The purpose of this book is to concentrate on the analysis of supply and demand for oil and other forms of energy, and to show how the changing balance of economic forces made OPEC so much more powerful in the 1970s than in the 1960s. It discusses how OPEC members were able increasingly to assert their sovereign rights over their precious natural resources in dealing with the world's most powerful companies, organized in a highly integrated, oligopolistic industry.

In spite of my preoccupation with economic influences in the oil market, this chapter on Structural Changes and New Strategies would be incomplete without brief reference to a major political determinant of Arab and, indeed, of Islamic oil policies. The continuing lack of decisive action on the return of Arab lands, and particularly Jerusalem, to the dispossessed Palestinians, will bring inevitable consequences, not only on the Zionists, but for those who are in a position to influence them and fail to do so. While Arab oil production rose dramatically to make up for the shortfalls in exports from elsewhere in 1979, in spite of lack of satisfactory progress on the Palestinian question, it is inevitable that the recent Israeli initiatives must bring retribution in terms of significantly lower Arab oil production and exports if such initiatives go unchecked. The political dimension of the medium term oil supply scenarios cannot be ignored. Oil importing countries will do so at their peril.

References

(1) BP Statistical Review of the World Oil Industry 1979. The British Petroleum Company Limited, 1980.

(2) Seventh Annual Statistical Report 1978-79. The Organization of Arab Petroleum Exporting Countries, Kuwait, 1980.

(3) Platt's European Marketscan, various issues.

(4) Platt's Oilgram Price Report, various issues.

(5) Middle East Economic Survey, 12 May 1980.

(6) Petroleum Intelligence Weekly, Special Supplement, 12 May 1980.

(7) Levitt, Theodore H., Marketing Myopia. Harvard Business Review, July - August 1960.

(8) Bruyne, D. de, President of the Royal Dutch Petroleum Company. Financing Problems in the Oil Industry. Opening Address to the 10th World Petroleum Congress, Bucharest, 9 September 1979.

(9) Diel, R., Radtke, G., Stossel, R. Investment Requirements in the Energy Sector and Their Financing. Dresdner Bank Federal Republic of Germany, 11th World Energy Conference, Munich, 8 - 12 September 1980.

11
TOWARDS AN INTERNATIONAL ENERGY DEVELOPMENT PROGRAMME

THE CURRENT PREDICAMENT

I have gone to some trouble in the previous chapters, and in Parts II and III, in trying to clarify the issues involved in a mass of data about the world's current energy situation before presenting the projected short, medium and long term scenarios which serve as guidelines for arriving at energy policies, plans and programmes. Such analysis, as we shall see, is absolutely necessary, whether we are considering international energy questions and policies, or those that relate to a certain country or group of countries. When we look at the short term outlook and compare it with the long term prospect, and when we study the global energy supply and demand balance carefully, we find that the world oil supplies are not moving in a direction commensurate with healthy economic growth rates on a global basis. Fundamentally this is because of the increasing limitations on future oil availability associated with declining reserves. One of the most important factors is the declining rate of finding new oil compared to the current or future demand for it. The other equally or probably more important factor is the budgetary needs of certain oil exporting countries in relationship to the rate of oil production within a framework of economic as well as political commonsense. If one examines the current and future production levels commensurate with a notional world supply and demand balance, one finds that certain countries like Saudi Arabia and Iraq would have higher oil production levels than is commensurate with their budgetary needs, especially since lower production levels will mean higher prices. Other countries, like Iran and Mexico, also have the potential to increase their production levels, though they have decided not to produce at their maximum capacity for national reasons.

It is evident that any oil producing country that generates income beyond its current financial needs is already suffering great financial losses. This is due to the fact that in most cases growth of earnings on such financial assets are lower than inflation rates in the countries in whose currencies such assets are held. Should one add the fact that oil prices are going to increase at least at rates not lower than inflation rates in the major consuming countries, then those who may be required to produce beyond their current financial needs must be motivated to do so. Incentives are not necessarily all economic or all political. It depends on which oil producing country or group of countries we are considering.

Evidently, there is a transition period which has started already. Countries likely to need increasing oil supplies will include those developing alternative sources of energy, as well as other countries which are unable to lessen their dependence on oil due to the lack of indigenous alternative sources of energy, or due to the time lag involved in this development. During this transition some countries will need oil which may have to come, if at all, from additional capacity in certain oil producing countries already producing beyond their current financial needs. The world may relax and depend on the wishful thinking that some oil exporting countries will always over-produce, regardless of what economic and political considerations may dictate. Even then, the potentially explosive notional gap between world oil supply and demand would only be postponed for a few more years. Such a gap may occur shortly after those countries reach their maximum sustainable capacity anyway. Current information indicates that this situation may occur before 1990. Alternatively, the world could take the other extreme and act in a purely rational manner: namely, developing an integrated global energy development programme commensurate with the world needs both in quantum and time.

Theoretically speaking, it is technically feasible to mine enough coal, to generate sufficient nuclear power, and to develop hydro and solar energy to fill in this notional gap that conventional oil and gas are unlikely to fill. A more realistic approach is probably a course of action somewhere in between that which is already being done today and this theoretical global energy development programme. However, before we examine such a possible course of action, let us see first what realistic considerations must be taken into account, and then proceed to examine at least one approach that seems to be more realistic than one that involves a global integrated approach.

COUNTRY BY COUNTRY ENERGY POLICIES AND PROGRAMMES

In the various scenarios about the short, medium and long term prospects in the world oil supply and demand balances, one historical fact seems both indisputable and significant. According to authoritative estimates, proven oil reserves at the end of 1979 were some 37 billion barrels lower than at the peak five years earlier. Figure 6 shows the rate of proving new oil reserves relative to the rate of oil depletion or production over four five-year periods through the 1960s and 1970s. The movement of total proved reserves between the end of each period is also shown.(1)(2) This must lead to a decline in oil production eventually, if the total proven reserves continue to fall. At the present rates of production the decline may take place in the late eighties or the early nineties, subject only to there being no dramatic large new finds, or to a remarkable and early rise in the recovery factor. Political developments in 1979 and 1980 suggest that even this hypothesis may prove too optimistic. It is obvious that unless developments of alternative energy sources are sufficiently large to substitute for such declining availability of oil, serious problems will arise for the world economy. Certain countries will be less adversely affected than others. It all depends on what kind of energy mix they will have at that time, their relative dependence on oil imports, and the speed at which they are reducing this dependence at the critical period.

In order to develop such alternative energy sources, many countries, especially developing countries, require assistance from developed countries in planning the most feasible energy mix suitable for their needs and indigenous energy resources. Furthermore, they require financial and technological resources to translate such plans into action programmes.

In the second part of this book, I present a kind of a general model for those countries which want to study their own energy situations as a first step towards defining their energy policies; and then to plan their individual programmes as to how to bridge the transition period; and ultimately to determine what specific means are required to reach a more balanced energy mix commensurate with their own energy resources in the light of the global energy supply and demand balance.

Many countries have been studying their energy problems very seriously during the last few years and even longer. In spite of this there are few if any which have been able to

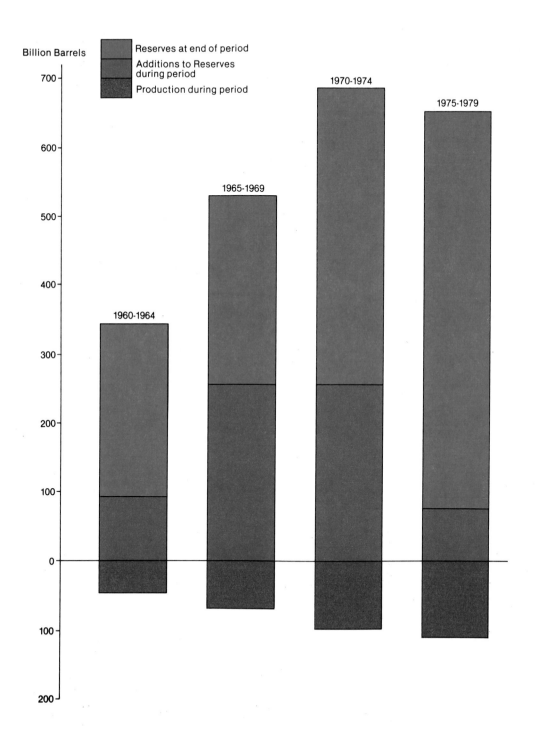

Energy Consumption Per Capita

Figure 7

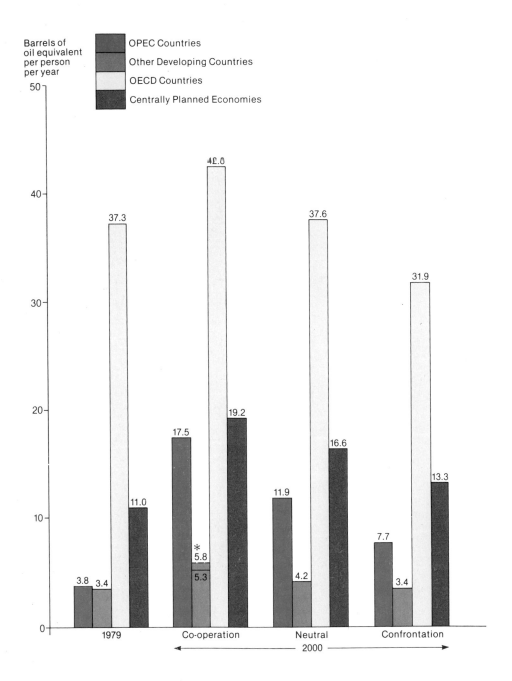

*See text, chapter 20 (Special Case Variant).

adopt energy policies and develop energy projects that may ensure sufficient changes in their energy mix consistent even with the most optimistic global energy supply and demand balance, as well as with their indigenous resources and exogenous capabilities and limitations.

Most if not all countries, at least within their actual institutions, seem to follow an immediate, or at best a short term, approach in handling their energy problems. What is needed is a longer perspective: energy projects that fulfil that country's needs, taking seriously into consideration the projected availability of various sources of energy from within and outside that particular country. That is to say, new energy projects of various types must be developed to minimize dependence on oil within a reasonable period of time. The problem, however, is that countries are highly interdependent and far from self-sufficient. Some of them, at least, need to co-operate with each other not only to develop and build their energy projects, but also to develop their energy policies and define their most appropriate energy mix. One may conceive of an international institution, like the World Bank, providing the necessary assistance in the development of energy policies for each individual country which asks for such assistance, and, with the help of specialized private or public enterprises, preparing the energy project proposals. This is one basic need that seems technically and politically feasible, and it could be met almost immediately, should the leading countries in the UN or the World Bank or UNCTAD decide to take prompt action. The cost of such energy policy studies and subsequent project proposals is only a small part of the ultimate cost of such energy projects. It is not unreasonable to assume that contributions for funding such costs are immediately available from international aid, or from OPEC or other funds. The question of the financing, engineering, construction and operation of such energy projects is a different matter that can be handled on a bilateral or a multilateral basis between various countries. What is needed, in the first place, is the creation of a practical and accessible institution where countries needing expert neutral assistance can find it, to help them in studying their own particular energy situation and developing their energy policies and projects in a manner consistent with the global energy situation.

In August 1980 the World Bank produced a new report, 'Energy in the Developing Countries' which set down the current as well as much larger desirable World Bank energy lending programmes.(3) Numerous studies have been carried out in recent years and reference to many of the more important of these is made in the bibliography. A review of the comparative tables in Part III illustrates the considerable diversity of outlook which exists about the global energy prospect. Figure 7 shows the extremely low level of energy consumption per capita in both OPEC and other developing countries compared with the average levels prevailing in the industrialized countries as recently as 1979 and the projected changes through to the year 2000 based on the alternative scenario assumptions discussed in Chapter 14.

INTERNATIONAL CO-OPERATION

So far I have identified one of the most important reasons that makes it necessary for almost all countries in the world to co-operate with each other, namely, to avoid major oil supply shortages from taking place in the short and medium term. In the past that has meant co-operation from one side only. Certain oil producers, for reasons of their own, have seen to it, so far, that such shortages are minimized, even to the point that a glut shows up in the world market from time to time. On top of that, Saudi Arabia and some of the Arabian Gulf States have exerted downward pressure on prices within OPEC, as well as through market forces.

The game of nations, so to speak, in the energy field includes many players. Each one, or rather each group of nations, may be willing to play and assume a certain role, provided the payoff is satisfactory. As usual, the interests of various players in the game conflict from time to time, and compromises of sorts are needed, if the game is to continue.

So far, a certain degree of international co-operation has taken place, by means of formal and informal contacts, between some members of OPEC and various industrialized and developing countries. This kind of co-operation is nothing more than a patching process, based on political expediency, to cater for the immediate problems of security of supply and prices. This probably will continue for some time and it may prove to be sufficient. Nevertheless, let us examine the medium and long term interests of various countries: oil producers and oil consumers, centrally planned or otherwise, developing or industrialized.

We have seen from the analysis of the energy situation in various groups of countries that, in the long run, all of them, including the current oil exporters, need an energy mix less dependent on oil and gas. Naturally, each country is a unique case that must seek, as we have said before, its own most appropriate energy mix. However, in the short and medium term, during what is generally called the transition period, the need for orderly international oil trade is most apparent, if the world community wants to avoid serious problems in international relations before the development of energy alternatives starts to make up for an actual decline in world oil production. Assuming that there exists sufficient appreciation of the energy problems in the short and medium term, then a common international position may be adopted to find ways and means to achieve the long term energy targets of as many countries as might be convinced of the benefits of international co-operation in this field. Energy projects resulting from such international, multilateral, or even, at worst, bilateral agreements between countries that can help each other in the energy field can be commenced in a very short time. As an example, an oil producing country, or even, say, the OPEC group of countries, may require the transformation of their domestic energy sector so that nuclear, solar and even coal sources might be used. This will be useful to those who will co-operate with each other, in more than one way. In the first place the oil exporting country or countries which introduce such other energy alternatives will acquire technology more suitable to the future when oil will be less available, or even totally depleted: in particular, technology that depends on renewable sources of energy. In this way, it ensures economic survival and long term benefits for all those countries recognizing the transitory nature of the age of oil.

Adequate protection of an oil producing country's surplus financial assets against inflation and other economic and political risks needs to be given to those countries which may have to produce beyond their current financial needs, if they are to have an incentive to do so. It should be self-evident that surplus oil producers (that is, countries exporting beyond their revenue needs) are more likely to make incremental oil supplies available to countries with viable alternative energy projects and advanced technology since their long run solvency and potential for reciprocal trading is likely to be greater than that of countries without such programmes. This pattern of international co-operation, if repeated and amplified, can help to minimize any future world energy problems.

Admittedly, there are other considerations of a purely political nature, whether domestic, regional or international, that have not been brought into focus. Salient among considerations of an international or regional nature are Jerusalem and the Russian offensive in Asia and Africa. Domestic, political and institutional considerations in many countries, whether in the oil producing or oil consuming countries, may stand in the way of those political decision-makers primarily concerned with rational and optimal energy policies.

Furthermore, the energy and financial problems of the developing countries carry both economic and political implications of an international as well as of a multilateral and bilateral nature. This is not sufficiently recognized on an international level. This is due to the complexity of the North-South economic and political relations and the preoccupation with domestic or even local issues in many countries. Many of the developing countries' oil-related problems have been resolved, at least on a short term and bilateral basis. Many oil producing countries have extended considerable financial assistance to many developing countries to alleviate their financial difficulties, whether related to oil imports or to general economic development. Furthermore, the World Bank, the OPEC

Fund, and several industrialized countries have also made their contributions. The world banking system has also made extensive loans to the developing countries.

During recent years official development assistance (ODA) from the member countries of the OECD has averaged some 0.35 per cent of the GNP of donor countries, or only half the target proportion. By contrast OPEC member countries contributed an average 1.8 per cent of their combined GNP to official development assistance over the five years 1975 to 1979 while for members of OAPEC the average proportion was almost 3.4 per cent and for Saudi Arabia alone it exceeded 4 per cent.(4)

What has proved adequate in the seventies is not necessarily adequate in the eighties, especially since the oil availability problems will be felt more and more in the future. Once again, self-interest, narrowly conceived for any one country, oil producing or industrialized, may lead to the comfortable conclusion that what is being done for the developing countries is sufficient for the time being. However, a broader view and especially one concerned for the future, indicates that more should be done. OPEC has adopted generalized resolutions to ensure oil supplies to the developing countries. That, in itself, is an important step taken by the OPEC Fund. Also there is considerable bilateral aid between many oil producing countries and oil importing developing countries. These developing countries also need the technical assistance to develop and transform their energy sectors. They also need the financial resources to implement their newly conceived alternative energy projects. What the World Bank and other institutions are advancing now will not be adequate to achieve satisfactory long run energy programmes. If we accept the premise that the health of the economies of the developing countries is important to both industrialized and oil producing countries on political and economic grounds, we need not hesitate in formulating similar energy co-operation programmes with them on whatever basis proves feasible, be it international, multilateral or bilateral.

In conclusion, I find it quite feasible to start the first part of an international energy development programme, namely, to provide technical expert assistance in studying the energy situation of the countries requiring such studies, by drawing a set of energy policies for them, and preparing energy project proposals for further implementation. The second part of the programme will entail the financing, engineering, construction and operation of these energy projects. It can be carried out on a case by case basis through the existing world institutions, or preferably through multilateral arrangement, if truly international ones are not possible.

To bridge the transition period problems of oil supply, price and finance is a very complex question. We have only touched upon it lightly in this chapter. Nevertheless, it is the cornerstone of any meaningful global energy co-operation scheme. Let us see if we can throw some light on these issues in the following chapter.

References

(1) Petroleum Statistical Bulletin, 1979 and earlier issues. Ministry of Petroleum and Mineral Resources, Kingdom of Saudi Arabia.

(2) Oil and Gas Journal 31 December 1979 and earlier issues.

(3) Energy in the Developing Countries. World Bank, August 1980.

(4) World Development Report 1980. World Bank, August 1980.

12
EPILOGUE

In the previous chapter I left out certain issues of a political and economic nature relating to international energy co-operation during the transition period.

Until now I have considered areas of probable, or at least possible, international agreement. I left out those issues that seemed more controversial. To resolve such issues a serious international dialogue and even negotiations are needed. However, effective co-operation in the critically important energy sector is likely to be achieved only in proportion to the degree of agreement on the broader economic and political issues. In several of these there is a latent conflict between the interests of the energy exporters and the energy importers.

I have tried throughout this book to follow an objective and analytical approach in order to identify problems and possible solutions free from subjective criteria. Unfortunately, our real world is far from being so objective. Therefore, in this final chapter, to make the analysis a bit more complete and realistic, it seems necessary to delve into the softer terrain of the political arena.

ECONOMIC DEVELOPMENT

It is well-known that one of the most important aspirations of the developing countries, including the OPEC countries, is economic development. In spite of this, authoritative and widely quoted voices in the industrialized countries can still be heard expressing the view that economic development which takes place in traditional societies, like some of the oil producing countries, might lead to social and political unrest in these societies. Attitudes of this type are detrimental to any international economic co-operation. They can hardly be brushed aside as superficial interpretations of what happened in one particular country. In fact, should attitudes of this type become policy in the industrialized countries to the point of seriously affecting the economic growth rates of the developing countries, such policies could very well lead to a more limited availability of oil in the international markets. If economic development were to be slowed or halted in the oil producing countries, this would lessen the need to raise income and consequently there would be a powerful incentive to reduce oil output.

What happened in Iran was not a revolt against higher per capita income for the Iranian people as a result of economic development. Provided that such development does not conflict with the basic traditions of the developing country, there is no need for alarm. A more equitable distribution of income arising from economic growth will be a stabilizing factor in these societies, as long as it does not conflict with their religious beliefs and their established traditions. On the contrary, any other course is a cause for alarm, whether it is a low rate of economic growth, a less equitable distribution of the benefits arising from economic growth, or economic growth inconsistent with established traditions and social behaviour. So let us leave each country to decide its own social and economic policies, and accept economic development as one of our major policy guidelines in international co-operation. This will have the obvious feedback to the industrialized economies in the form of increasing demand for their exports to the developing countries.

A further point which ought to be apparent to policy-makers in the industrialized countries is that, at this point in time, economic development and the improvement in living standards is a far more important aspiration of people and politicians in developing countries than it is in the industrialized countries where much higher standards prevail already. Preoccupation with concern for the environment is not a luxury which most developing countries will be able to afford for decades yet.

The 'Group of 77' developing countries (actually 120) took the initiative, once again, in the United Nations General Assembly in August/September 1980 of repeating their call for a massive transfer of resources from the industrialized countries on a predictable long term and assured basis. The debate is expected to be enlarged in 1981. Another initiative emphasizing the urgency of the need for early action to bring about an improvement in North-South relations was the Brandt Report: 'A Programme for Survival' by the Independent Commission on International Development Issues, to give it its full name.(1) This was an international study prepared by a group of statesmen, led by the former West German Chancellor. They were drawn from industrialized countries, OPEC countries and other developing countries. It is already something of an achievement that such an effective consensus message could be prepared by such a distinguished but disparate group of men. It will be much more of an achievement if some real progress is made in the directions they laid down. Writing under the provocative title 'Is Altruism in Retreat?' in Newsweek magazine, Varindra Tarzie Vittachi of the United Nations pointed out that several private institutions tried to act as honest brokers in the North-South dialogue. He went on to say:

> "At first their appeal was to philanthropy and goodwill. But since those ad misericordiam approaches have not worked, they are now arguing in terms of self-interest .... if poverty continues to increase at its current rate, the eventual explosion in the Third World will make recent events in Iran seem tame by comparison .... It seems to me that there is only one answer to the question of why we should care about another person's hunger, illness or homelessness: we are all human. That has always been the answer and always will be."(2)

MILITARY OPTIONS

As early as the 1950s, and perhaps even earlier, military options were considered as a means of establishing influence over the Middle East oil fields. The importance of petroleum in World War II is well-known. Then the Germans were forced to liquefy coal as a costly and highly complicated substitute. Later 'Operation Dropshot' was the title given to the American plan drawn up in 1949 against a possible Russian offensive. One of the most important elements of this plan was the weight given to the possible Russian offensive in the Middle East, as seen at the time, because of the importance given to the Middle East oil fields. Broader geopolitical reasons also weighed in the balance. More recent American contingency plans are not so secretive, as we hear from time to time about possible military intervention, American bases around the Arabian Gulf, and the American Rapid Deployment Force.

Sometimes one is rather puzzled hearing about such desperate and self-defeating plans, as if the minds of policy-makers are one track minds which fail to see other more constructive options to achieve the same targets. Perhaps this is a naive way of thinking. Nevertheless, one cannot help but wonder why so little attention was given for so long to implementing more effective energy policies in those countries where waste and slow development of alternative energy sources were characteristic, and where effective leadership could initiate worldwide energy co-operation schemes to fill in any transitional gaps in supply. This ordering of priorities would prove much less costly, more constructive, and would be far safer than military adventures.

THE SOVIET OFFENSIVE IN ASIA AND AFRICA

Maybe, if one extends the Asian and African offensive of the USSR towards the Middle East, one can think again in terms of military deterrence against the Russians as it was conceived in the 1950s. But to think of military action against some or all of the Arabian Gulf States, to force higher oil production and lower prices, seems to me a most ridiculous thought, unless a third world war is in the making. Should that kind of war get started around the Arabian Gulf, then one of the first facilities that would be destroyed would be the oil terminals, tankage, pipelines and other oil producing and exporting facilities. If the Russian objective in the Gulf is to secure oil for their own and their allies' use, then there is room for them in international economic co-operation. They command energy technologies useful for many countries around the world which they can trade in return for certain quantities of oil from various producers. These might be allocated to them or to their allies at times of critical shortages. At least, for the time being, they do not seem to need any more oil; hence one may conclude that they may only want to deny some of the oil from the Gulf to the West. This could very well be one of their objectives. But they could achieve this through trouble-making in countries controlling strategic marine locations, without recourse to outright war. Therefore, if one assumes that neither the Americans nor the Russians want to start a third world war, it seems much in the interest of the Free World to keep military options away from the Arabian Gulf, as long as the Russians are at a reasonable distance from it.

ISRAELI EXPANSIONISM

Accordingly, if the Russians are kept at bay from the Gulf, and knowing that almost all the Arab oil producing countries in the Middle East are not Russian allies, there is no need to fear that these exporters will deny their oil to the West, except when the West acts against their very survival. Supporting Israeli expansionist policies in Arab lands is an obvious example. Arab oil producing countries are not bordering Israel as such, but they cannot stand idly by whilst Jerusalem is annexed, and then most probably the West Bank. Jerusalem is holy to Muslims, just as Mecca and Medina are. For that reason, other Muslim oil producing countries might be involved in any future oil-related defence measures against Israeli expansionism. To put it mildly, it is in the best interest of the United States, Western Europe, Japan, the developing countries and the centrally planned economies to establish just and equitable peace in the Middle East as a major step in establishing security of supply for their oil imports.

Countries that are able to take the initiative in launching important international action to help solve significant global issues, such as the energy problem, must take into account at the outset all of these complex political, military and economic considerations. In other words, the world should not ask for moderate oil prices and high and increasing levels of oil production from the Arab countries, and possibly from the other Muslim oil producing countries, without offering substantial political moves to make it possible for the Palestinians to recover their occupied land and determine their own political future free of Israeli domination. Inevitably this means that in any settlement all other lands occupied in 1967 should be returned as well.

Epilogue

SOVEREIGNTY OVER NATURAL RESOURCES

The sovereignty of nations over their natural resources is a right long accepted and supported by the United Nations. Nevertheless, from time to time, we still hear cries about the good old days when the Western oil companies had the final word on oil production, marketing and prices.

Mr J B Kelly, the author of a book about the Arabs and their oil policies, has made such a cry, as if the Saudi Arabian take-over of the Arabian American Oil Company (Aramco) is a disaster for the West.(3) Saudi Arabian oil policy is notable for a great deal of moderation in prices, to the point of producing at higher rates to influence the market. Its oil and gas marketing policies are characterized by stability and excellent contractual performance.

Many other screams come from the Western press characterizing OPEC as some kind of a monster crying for Western blood by continuously raising oil prices. A more objective view would have seen clearly that if it were not for OPEC, and in particular certain OPEC members such as Saudi Arabia, oil prices might have been much higher than they are today. It was OPEC that issued a solemn declaration in 1975 to show its good faith towards the developing world. It was OPEC that initiated a long term price strategy that should bring stability and predictability in the international oil market.

It is a kind of a puzzle to read at times that some Western writers would rather see incremental income that accrues to the Saudi Arabian people stay with the major oil companies, which already realize staggering profits.

These extreme attitudes on the part of some Western writers, and sometimes on the part of Western policy-makers, are counter-productive. They tend to influence public opinion and policy-making in the oil producing countries adversely. They can create reactions that may lead to less moderate oil policies.

CONCLUDING COMMENT

In this chapter some comments were made about diverse political and economic issues related to international co-operation in energy. These comments were not intended to be exhaustive or complete. They are meant to provide some of the broader political perspective which has to be taken into account if there is to be a successful attempt to resolve the complex global energy problem. Much of Part II of this book may be regarded by some observers as painting an unduly pessimistic view about world energy prospects over the next twenty years. If confrontation were to prevail, the energy outlook would be bleak indeed. Even if co-operation characterizes this period, concern about the energy sector seems likely to dominate the world economy and international relations. An important function of such projections is to point to the need for a more fundamental reappraisal of policies, to plan for more realistic international political relationships and more enlightened decisions in the energy sector. Appropriate actions should lead to our having more control of the transition away from excessive dependence on oil, with political, economic, social and technological agents of change receiving a steady stream of consistent rather than conflicting signals.

So, let there be light.

References

(1)　North-South: A Programme for Survival. The Report of the Independent Commission on International Development Issues under the Chairmanship of Willy Brandt. Pan Books, 1980.

(2) Vittachi, Varindra Tarzie, Chief of the Information Division of the United Nations Fund for Population Activities. Is Altruism in Retreat?. <u>Newsweek</u>, 25 August 1980.

(3) Kelly, J.B. <u>Arabia The Gulf and The West. A Critical Review of The Arabs and Their Oil Policy.</u> Weidenfeld and Nicolson, 1980.

Part II
Energy Scenarios, Historical and Regional Analysis

13
GLOBAL STATISTICAL REVIEW OF PRIMARY ENERGY

HISTORICAL TRENDS

The World

In the six years ending in 1973 the global average annual growth rate of primary energy demand was 5.6 per cent. Since 1973, the average rate of increase up to 1979 was 2.6 per cent each year for the world as a whole, though over the four year period since the trough of the recession which affected industrialized countries in 1974 and 1975, the growth rate in primary energy demand has averaged nearly 3.6 per cent annually (see Table 4).

Another way to express this deceleration in the rate of growth of demand is to compare the incremental amounts of energy used up to and since 1973. In the six years from 1967 to 1973 the world used an extra 1.6 billion metric tons of oil equivalent, and in the six subsequent years the increase was slightly under 1.0 billion metric tons of oil equivalent. However, in the four years culminating in 1979, the extra primary energy used was only 100 million metric tons or 2 million barrels a day oil equivalent less than over the four years from 1969 to 1973.(1)

Worldwide, the share of oil in the total primary energy demand has started to go down from the peak of 47.1 per cent reached in 1973. However, the decline in the share since 1973 has been much less steep than the rise up to 1973. Over the four years to 1973 the oil share of primary energy demand rose by 4.3 percentage points, an average of nearly 1.1 percentage points each year. During the six years since 1973, the oil share of world primary energy demand declined only 1.9 percentage points or little more than 0.3 percentage points annually. Most significant of all within the most recent period, the oil share of world primary energy demand was actually higher in 1978 than in 1975. This reflected the erosion of the real oil price over this period, and it provides powerful testimony to the strong preference for oil among energy users. Worldwide oil demand in 1979 was 7.16 million barrels a day higher than in 1973, implying an average annual growth rate of 2.0 per cent, though in practice, it was very uneven from year to year, and in different parts of the world (see Table 4).

It is not the purpose of this chapter to examine the recent historical trends in the major regions and countries of the world, either of energy in general, or of oil in particular. Reference is made to the subsequent chapters of Part II where the reader will find the appropriate discussion. However, brief reference is made below to an analysis of the countries of the world outside the centrally planned economies as an aggregate (some

Non-Communist World

The non-Communist countries as a whole are still considerably more dependent on oil to meet their energy needs than the USSR, Eastern Europe and China as a group. In 1973 oil accounted for 54.0 per cent of primary energy demand in this large aggregate of countries. A similar pattern to that for the world as a whole can be found regarding the less steep decline in the oil share of primary energy demand since 1973 compared with the previous increase. Also, the oil proportion of primary energy demand was higher in 1978 than in 1975. Primary energy demand growth decelerated from an average annual rate of 5.4 per cent in the years ending in 1973 to only 1.7 per cent subsequently, though the rate from 1975 to 1979 recovered to a yearly average of 3.3 per cent (see Table 3).

Industrialized Countries

An analysis of the performance of the developed industrialized countries shows that up to 1973 their primary energy demand grew at an average yearly rate of 5.2 per cent, falling to only 1.0 per cent on average during the six subsequent years. After the recession of 1974-75 the growth in primary energy demand recovered to 2.8 per cent on average over the four years to 1979. In order to keep these growth rates in some sort of perspective, it is important to remember that the population of the industrialized countries as a whole has been growing at an average annual rate of well under 1 per cent during the last decade. Not only has this rate of population growth been some two percentage points less each year than for developing countries as a whole, but it also clearly implies annual increases in per capita energy consumption in the industrialized countries from existing high levels. In 1978 primary energy demand in industrialized countries averaged 4.3 metric tons of oil equivalent per person, compared with 0.4 metric ton in OPEC countries and 0.3 metric ton per person in all other developing countries.(2)

The industrialized countries' increased use of primary energy over the six years to 1979 amounted to only 226 million metric tons of oil equivalent compared with 944 million tons during the preceding six years. But because of the recession in 1974-75 the extra primary energy used over the four years ending in 1979 was 401 million metric tons of oil equivalent. This compares with an extra 562 million metric tons in the four years ending in 1973. This reduction was achieved at the cost of a lower average rate of growth of GNP, some 4 per cent over the most recent four year period. Some improvements were made in the efficiency with which energy was used. The industrialized countries reduced their use of primary energy per unit of GNP at a rate of 1.2 per cent on average each year from 1975 to 1979.(1)(3)

Just like the other aggregates discussed before, the industrialized countries were increasing the proportion of oil in their total energy consumption more quickly up to 1973 than they were reducing it subsequently: the rate of increase averaged 1.2 percentage points each year until 1973, with the subsequent decline averaging only 0.4 percentage point. This decline conceals an increase of 0.7 percentage point in the industrialized countries' oil share of primary energy consumption over the three years from 1975 to 1978, reflecting a natural economic response to falling real oil prices, the strong consumer preference for oil, and the absence of adequate development of alternative energy supplies. A further relevant consideration is the build-up of oil supplies from the North Slope of Alaska and the North Sea during the latter part of this period. Though total oil demand in 1979 was higher than in 1973 among industrialized countries as a whole, a considerable rise in the USA was partly offset by falls in Western Europe and Japan.(1)

Oil Reserves and Discoveries

Having discussed the recent changes in the growth of energy demand since 1973 and compared them with the higher rates prevailing previously, it is now relevant to examine briefly the supply of energy with special reference to oil.

One of the main reasons that a potential oil shortage was not widely or clearly perceived during the period when oil demand was doubling each decade can be attributed to the remarkable finding rate of new oil reserves. These were mainly in the form of giant fields in the Middle East. Though new oil reserves continue to be discovered and proved, the chances are low of repeating the extraordinary period of large oil discoveries achieved through much of the sixties and early seventies. The critical point when the rate of oil consumption exceeded the finding rate of new oil reserves has been passed. The likelihood is that new oil discoveries will exceed the volume of oil consumption only in exceptional years, or for brief periods in the future. Thus the reserves to production ratio seems likely to trend downwards in the future for the world as a whole from 27 years to 1 year which prevailed at the end of 1979. Five years earlier it was nearly 34:1. Two important features of oil reserves and discoveries are not universally understood outside the oil industry. These are that a high proportion of all the oil discovered in the world so far has been located in so-called giant fields, and that these giant fields tend to have been concentrated in relatively few, and geographically limited, areas of the world's surface. These considerations weigh heavily in the relatively pessimistic outlook regarding future oil discoveries among the best informed of the world's petroleum geologists and engineers.

There is an apparent paradox in high rates of growth of world oil consumption associated with relative lack of concern about future availability of oil supplies, followed by the more recent period of much lower growth in oil demand but widespread anxiety about the prospect of oil supplies running down. The relationship between the finding rate of new oil reserves and the rate of oil consumption can be summarized conveniently in the estimates of 'published proved' reserves at different points in time over the last twenty years, together with the average annual rates of change in these reserves for successive five year intervals.(4)(5)(6)(7)

World Proved Oil Reserves

	Billion Barrels		Average Annual Rate of Change
1979	651.1	1979 - 1974	- 1.1%
1974	688.2	1974 - 1969	+ 5.3%
1969	530.8	1969 - 1964	+ 9.1%
1964	342.9	1964 - 1959	+ 2.9%
1959	297.2		

Given that worldwide oil production amounted to 112.2 billion barrels over the last of these five year periods one can infer that the additions to proved reserves between the end of 1974 and the end of 1979 was only 75 billion barrels. This is equivalent to an annual average rate of additions to proved reserves of only 15 billion barrels per year and compares with an average but rising production rate of over 22.4 billion barrels a year through the period. This sort of relationship must be a cause of worldwide concern.

In order to see what could have happened to Middle East oil reserves and the reserves to production ratio if pre-1974 rates of growth in production had continued, it is useful to compare the actual situation at the end of 1979 with this hypothetical scenario. At the end of 1979 the ratio of published proved reserves to production stood at over 45 to 1. Had the 13.5 per cent annual rate of increase which characterized Middle East oil production from 1968 to 1973 continued, then Middle East proved oil reserves would have lasted only another 20 years at the hypothetical production rate for 1979 before being completely exhausted. In fact, of course, it is physically impossible to exhaust such reserves completely. Thus the continuation of the high historic rate of increase in oil production for only a few more years beyond 1973 could have led to a much more catastrophic energy transition during the last part of the twentieth century than the one confronting the world today. Furthermore, it could have led to such profound disruption throughout the economic system, that the shock waves would have been felt by mankind as a whole and not just among those engaged in oil-intensive and oil-dependent forms of economic activity.

Changes in Oil and Energy Demand

One of the most significant changes which has occurred since 1973 as a result of oil supply constraints is that, almost everywhere, the gasoline and middle distillates proportions of the demand barrel have increased considerably. The proportion of demand accounted for by residual fuel oil fell between 1973 and 1979 throughout the world (excluding the centrally planned economies, for which data are not available). In the industrialized countries demand for residual fuel oil fell by nearly 1.2 million barrels a day over this period. The reduction was concentrated almost entirely in Japan and Western Europe.

The table below shows changes in the amounts of various forms of primary energy used in 1979 compared with 1973, and the similar changes over the previous six years.(1) Comparative compound average annual growth rates of change together with the percentage distribution by type of primary energy in 1979 and 1973 are also given.

Changes in World Energy Demand

	World Increments* Million metric tons o.e.		World Growth Rates %		Shares in World Energy Balance %	
	1973-79	1967-73	1973-79	1967-73	1979	1973
Oil	326	1004	2.0	7.9	45.2	47.1
Natural Gas	162	316	2.3	6.2	18.0	18.3
Coal	303	226	2.8	2.1	28.6	28.2
Nuclear Power	106	38	21.1	28.4	2.2	0.8
Hydro Power	83	65	3.8	3.8	6.0	5.6
Total Primary Energy	979	1649	2.6	5.6	100.0	100.0

*Numbers do not necessarily add exactly to total because of rounding to nearest million metric tons o.e. for each energy form.

WHAT HAPPENED IN 1979 AND 1980

Oil Prices, Supply and Demand

In order to review recent historic trends it has been convenient to discuss some statistics for 1979 in the preceding section. This section discusses the rather extraordinary set of circumstances prevailing in the world oil market during 1979 and 1980.

The really remarkable feature of 1979, in an aggregate sense, was the combination of oil price increases, negligible growth in estimated oil consumption in the world outside the USSR, Eastern Europe and China, and the much more significant increase in production and total oil supply available to non-Communist countries as a whole. The average official price of Arab Light marker crude rose to $17.28 a barrel in 1979 - an increase of only 36 per cent from the 1978 price, but it rose 89 per cent between the end of 1978 and the end of 1979, followed by a further 33 per cent increase through 1980. The average official price of Arab Light crude was $28.67 a barrel in 1980. Even at the end of this period, the price of Arab Light crude remained well below the spot market prices paid for both crude oil and the weighted average product demand barrel based upon it.

The average spot market value of the demand barrel CIF North West Europe rose by some 117 per cent in 1979 over the average 1978 level, and some 126 per cent between the end of 1978 and the end of 1979. The average of oil exporters' official prices rose more than that of the marker crude in 1979, and considerably more if surcharges or market premiums could be calculated accurately. There was also a significant migration of oil supplies from official-priced contract volume to the spot market, but even so it can safely be assumed that on average crude oil prices in 1979 rose significantly less than the rise in the spot market value of the demand barrel in North West Europe. The US Department of Energy estimates that the average crude oil acquisition cost of imports for US refiners rose 48.7 per cent 1979 over 1978 and 93.5 per cent through the year 1979. Both of these increases are higher than the comparative increases in price of the marker crude, and indeed, of the average rate of increase for all OPEC official prices. These have been estimated at 44.4 per cent for 1979 over 1978 and a further 65.3 per cent in 1980 over 1979.

Superficially, it is difficult to reconcile these price increases with the volume estimates for supply and demand. These show OPEC oil production, inclusive of Iran, rising 3.3 per cent in 1979 over 1978 to nearly 1545 million metric tons (nearly 31.3 million b/d on average), and world production rising 4.6 per cent to 2510 million metric tons, excluding the USSR, Eastern Europe and China.(1) Net exports of crude and products from these Communist areas are believed to have fallen by some 100,000 b/d. This leads to a net increase in oil supplies available to the non-Communist countries of +4.4 per cent, some 2.2 million b/d more in 1979 than in 1978. Non-Communist world oil consumption is estimated to have risen only 265,000 b/d, little more than a tenth as much, again excluding the USSR, Eastern Europe and China. It should be noted that conventionally it was usual to allow some 3 per cent of supplies for unallocated demand and losses through evaporation, transportation etc, but this figure was almost certainly too high for 1979, given the low growth of consumption. This highly unfavourable combination of circumstances (from the point of view of oil importing countries) occurred in spite of Saudi Arabia's enlightened policy in respect of both prices and production levels, designed to alleviate the pressure from the oil sector bearing upon the level of world economic activity, inflation and external trade deficits of the oil importers.

The explanation has to be sought first in the extent of stock building which occurred in 1979. OECD official statistics show an increase of 34.6 million tons for crude oil, NGLs, feedstock and products in the year as a whole. This was the largest increase recorded in any year since 1974, except for 1977 (38.5 million tons). There was a drawdown of stocks in 1978 (15 million tons), virtually no change in 1975 and a small increase in 1976 (5.6

million tons). Throughout 1979 and the first three quarters of 1980 OECD quarterly oil stock movements were more conducive to a tight international oil market than throughout the corresponding quarters of 1978, as the table below shows. Furthermore, as soon as a significant stock drawdown started in the fourth quarter of 1980, spot market prices of both crude oil and products started to fall.(8)

OECD Oil* Stock Changes
Million Metric Tons

Quarter	1978	1979	1980
1	-46.9	-40.2	-5.6
2	+11.7	+24.6	+26.6
3	+18.4	+40.6	+20.3
4	+1.7	+9.4	-22.4

* Crude, NGLs and Products.

In spite of the disruption in supplies from Iran, with the benefit of hindsight and the estimates of supply and demand for 1979, it seems reasonable to conclude that had three conditions been fulfilled, the price of oil entering world trade might have increased no more than the 10 per cent average for 1979 and the 14.5 per cent through the year, as scheduled by OPEC in December 1978. These three conditions are discussed below.

1. No building of oil stocks in 1979. This would had to have included both developing and industrialized countries, as well as no increase in crude at sea or stored in tankers, or strategic stocks.

2. No run down of stocks in 1978. The major oil companies permitted themselves this luxury in early 1978, following the absence of any OPEC oil price increase at the end of 1977, in an effort to improve their cash flow positions. This can be seen as having been a short-sighted and misguided policy from a macroeconomic standpoint, if not from a corporate profit point of view.

3. Some re-allocation of oil supplies to alleviate pressure on those most affected by the disruption of Iranian supplies. These included Japan and West Germany among countries, and Shell, BP and some Japanese among the companies. Politicians generally, and the IEA and US Administration in particular, either failed to identify this problem, and its consequences, or lacked the political will to take an appropriate initiative. The companies and economically strong governments were quite prepared to pay spot market prices for growing supply volume regardless of the feedback effect through tightening of the market on official prices and the emergence of surcharges, market premiums and 'front-end loading' practices. This was because, for any individual company, inability to supply contract customers was potentially far more damaging than the acquisition of marginal supplies of crude or products on the spot market at very high prices.

US oil production (inclusive of NGLs) resumed its decline in 1979 with a -1.0 per cent fall following an increase of +4.5 per cent in 1978, attributable to output from the North Slope of Alaska reaching its peak. Worldwide production outside the USSR, Eastern Europe and China rose 2.33 million b/d in 1979, of which Saudi Arabia accounted for 51.3 per cent of the net increase or 53.4 per cent including its share of Divided Zone. The full list of country increments exceeding 100,000 b/d on a comparison of 1979 with 1978 is shown below. It should be noted that the average drop in Iranian production in 1979 is estimated at -2.11 million b/d, 220,000 b/d less than the aggregate net increase noted above.

Readers may care to note that the three leading Arab producers alone more than made up for the decline in Iran.(1)

Production Increments 1979/1978

	Thousand b/d
Saudi Arabia	+1195
Iraq	+850
UK	+505
Nigeria	+380
Kuwait	+340
Mexico	+290
USSR	+275
Canada	+255
Venezuela	+190
Divided Zone	+100

Saudi Arabia's share of world oil production rose to 14.5 per cent of the total in 1979 after 13.2 per cent in 1978 and 14.8 per cent in 1977. It was still behind the USSR at 18.1 per cent in 1979, and the USA, 15.0 per cent inclusive of NGLs. Just ten years earlier, US production had represented 24 per cent of the world total. OPEC production in 1979 was just over 1 million b/d more than in 1978, in spite of the disruptions in Iran.

It seems reasonable to infer that without the large increases in the UK, Mexico and Canada (which aggregated 1.05 million b/d) as well as those in Saudi Arabia and most other OPEC member countries, the price increases in the market during 1979 would have been much more than those actually seen. In this sense oil importing countries might be grateful that the events in Iran occurred during 1979 (if they had to occur at all), rather than in 1976, for example.

In 1979, the USA made a significant contribution to alleviating pressure on world oil supplies since product demand (net of processing gain) fell by 435,000 b/d, or -2.4 per cent, compared with a fall in production there of only 60,000 b/d, including NGLs. The US share of world oil consumption fell to 27.7 per cent in 1979 from 28.9 per cent in 1978. A further sharp fall occurred in the US share in 1980 as demand for products fell by another 1.5 million b/d, or about -8 per cent.(1)

Net Changes in Oil Consumption 1979/78

	Thousand b/d	Percentage
USA	-435	-2.4
OECD Europe	+220	+1.5
Japan	+75	+1.4
Canada	+60	+3.3
Australasia	+5	+0.6
Total	+75	
Total Non-Communist World	+265	
Developing Countries	+190	
Middle East	-120	-10.3
South Asia	-5	
The highest percentage growth was seen in South East Asia	+235	+10.9
Followed by Latin America	+215	+5.2

South East Asia is also notable in being very oil dependent: oil accounts for 63.7 per cent of primary energy, nearly as high a proportion as in Japan, 69.7 per cent.

Oil demand fluctuates sharply with the economic cycle in the industrialized countries, but in the rest of the world it increases more smoothly.

In spite of improvements in the fuel efficiency of automobiles already effected in the USA, the transportation sector there still accounts for over 50 per cent of oil product demand in that country. This is about one seventh of the total world oil consumption. It will be necessary to intensify the quest for continuing improvements in the efficiency of oil use in the US transportation sector in an oil-short world. Evidently an increasing real price for gasoline was having an effect in reducing demand for this product in 1980.

In both Western Europe and Japan, transport and chemicals together account for well under 50 per cent of total oil demand. Indirectly, this suggests considerable scope for substitution of oil by other forms of energy in the other sectors of consumption over the next ten to twenty years.

In 1979 Middle East exports continued to dominate inter-area world oil trade, accounting for 57.8 per cent of the total compared with 58.3 per cent in 1978, in spite of the decline in exports from Iran.(1)

Following a doubling of net oil exports by the centrally planned economies between 1968 and 1978, provisional estimates suggest that there was a decline of about 100,000 barrels a day in 1979 followed by a further fall in 1980. This will have had some marginal effect in reducing the growth of oil supplies to countries outside the Communist bloc, though these appear to have increased by some 4.4 per cent between 1978 and 1979. One of the main explanations for this change in the pattern of growth in net exports from the centrally planned economies seems to have been the sharp fall in the rate of growth of oil production in both the USSR and China in 1979.

Other Forms of Energy

A comparison is shown below of the percentage rates of change in demand for forms of primary energy other than oil in 1979 against the average annual rate of change for the preceding five years.(1)

	Natural Gas		Coal		Hydro Electricity		Nuclear Electricity	
	1979	1973 to 1978	1979	1973 to 1978	1979	1973 to 1978	1979	1973 to 1978
	%	%	%	%	%	%	%	%
Developing Countries	-1.1	+5.3	+7.7	+5.0	+1.4	+9.4	+7.7	+21.1
Total OECD	+1.8	-0.1	+7.4	+0.4	+2.2	+2.3	+3.1	+24.5
USA	-1.1	-2.5	+8.1	+1.2	+0.4	+1.1	-4.4	+28.2
OECD Europe	+6.4	+6.2	+6.0	-0.6	+4.0	+2.3	+14.4	+16.8
Japan	+30.0	+26.3	+8.5	-2.4	+14.4	+0.1	+17.6	+40.3
Non-Communist World	+1.4	+0.5	+7.4	+1.3	+2.0	+3.8	+3.2	+24.4
Communist Bloc	+13.9	+8.6	+3.2	+3.3	+9.2	+5.1	+11.6	+29.6
World	+5.3	+2.6	+5.2	+2.3	+3.0	+4.0	+4.0	+24.8

The most remarkable feature of this analysis has been the quite sharp acceleration in the use of coal in all main countries/groups of countries outside the Communist bloc during 1979. Indeed, there was a reversal in the declining use of coal in both Europe and Japan in 1979. This had been expected as a natural consequence of the oil price increases. The problem for the oil importing countries will be whether they can sustain an increase in coal use in recessionary economic conditions, and a rapid increase during periods of faster than normal economic growth. There is limited flexibility in existing fuel burning plant, and an absence of clear indications that really large sums are being invested in new coal using plant or conversions of existing equipment to substitute use of coal for oil in the electricity generating and industrial sectors.

As far as natural gas is concerned, the most noteworthy observation is that the decline in US production and consumption was evidently less in 1979 than it had been in most recent years. Doubtless this was due in part to the fact that US natural gas prices have been allowed to rise, although they remain far below the prices of gas imported from Canada and Mexico, and even further below the price of imported crude oil on a heat equivalent basis.

In any one year hydroelectric production is likely to be erratic because of atypical rainfall conditions in one or more regions. 1979 seems to have been noteworthy for the limited growth of hydroelectric production in developing countries, though this must also be seen in the context of some large schemes becoming operational in previous years which probably did not happen in 1979. This is another important cause of erratic annual growth of supply and demand for hydroelectricity.

One incident dominated the nuclear sector in 1979: the accident at Three Mile Island. This caused a major disruption in the growth of nuclear electricity supply and demand in the USA, -4.4 per cent down from the 1978 level, compared with average growth of over 28 per cent annually in the preceding five years. This involved a 3.3 million metric tons decline in oil equivalent on an input basis last year, compared with an average 10.7 million metric tons oil equivalent nuclear increment in each of the five preceding years, or a difference of 280,000 barrels a day oil equivalent. This was one of the most important reasons for US coal demand rising so rapidly in 1979. A further point about nuclear energy is the deceleration of growth everywhere compared with the high average compound rates of growth over the previous five years. This is almost inevitable as nuclear energy starts to become a more significant form of energy, though other factors which retarded potential growth included the non-operation of one completed nuclear station in Austria (because of the negative result of the referendum) and environmental opposition which retarded programmes in some other countries.

FUTURE PROSPECTS

The earlier sections of this chapter have discussed the sharp changes which have occurred in recent years compared with the trends which prevailed until 1973. Some explanation as to why changes seemed likely to become inevitable sooner or later has been given in the form of a combination of factors: the rapid increase in dependence on cheap oil to sustain high rates of economic growth and the prospect of dangerously rapid rates of depletion of proved oil reserves.

The changes which have occurred since 1973 look quite dramatic in terms of the way they have been analysed above. Nevertheless, the scenario numbers I discuss in the next two chapters represent what amounts to another qualitative change in the future of world energy supply equally if not more significant than the events of 1973-74. These represented a watershed in the history of the oil industry associated with initiatives by OPEC and OAPEC which were discussed in Part I of this book. The changes in prospect now assume their present significance because they seem to represent a relatively sustained reversal of the experience of the twenty-five year period which ended in 1973. It may be that the period from 1973 until 1979 will come to be regarded as a transitional phase between the old era and the new. Alternatively, it is possible that it will be interpreted as the preliminary period in a new energy epoch when many energy users and oil importing countries started to think in terms of whether or not future oil supplies and prices represented a serious economic problem.

Some readers will be tempted to interpret the events of 1979 and 1980 in terms of a discontinuity because of the way much of the media has chosen to interpret the impact on the oil industry of both the Iranian revolution and the conflict between Iran and Iraq. However, as we have shown previously, the net effect of the Iranian revolution in 1979 was a considerable net increase in OPEC and world oil supplies. Disciplined management of the international oil industry and abstention from abnormal building of stocks by the powerful industrialized countries could have avoided or at least reduced the pressure on supplies and the sharp rise in prices which occurred. Although the conflict between Iran and Iraq will have served to tighten world oil supplies, the initiative of other Gulf oil producers in increasing output and the end to stockpiling had almost as much offsetting effect as the combined reduction in exports of Iran and Iraq. Also, the coincidence of economic recession in most industrialized countries alleviated the tight oil market conditions. Prospects for the next two years seem likely to depend on the relationship between the resumption of oil exports from Iraq and Iran and the strength of the economic recovery in industrialized countries.

Figure 8 shows the economic growth rates which seem consistent with the energy supply and demand projections, together with the conservation savings and improvements in the efficiency of energy use which seem possible.

The changes which emerge from the projections discussed in Chapters 14 and 15 are summarized below in number terms in what I call the 'neutral' scenario. The term neutral is meant to convey the idea of no fundamental change in the political and economic relationship between the industrialized and developing countries, or, for that matter, in the international relationship between the centrally planned economies of the Communist bloc and other countries.

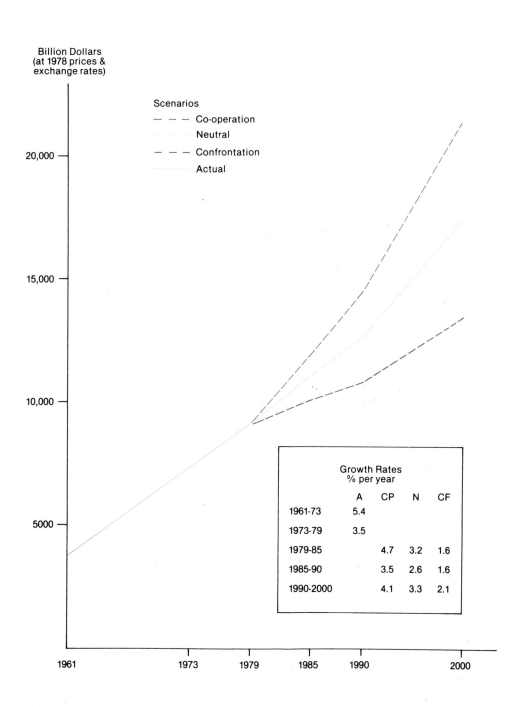

Figure 8

Since many observers and commentators on the energy scene tend to allow themselves to be dominated by aggregates, I have chosen to focus attention on change and rates of change in this selective introduction to the next two chapters.

The first interesting point to notice is something which may surprise some of those seasoned energy economists who normally confine their attention to non-Communist countries. The table below implies that the centrally planned economies of the Communist countries are expected to account for about 47 per cent of the growth in primary energy consumption between 1979 and the year 2000. However, in the six years from 1973 to 1979 these countries accounted for some 53 per cent of the apparent growth in energy consumption worldwide.(1)

Projected Additions and Rate of Change in Primary Energy Consumption 1979 to 2000

	Million b/d o.e.	Per Cent Average Annual Rate of Growth	
	'NEUTRAL' SCENARIO 1979 to 2000		ACTUAL 1973 to 1979
Centrally Planned Economies	37.6	3.0	4.8
OPEC Member Countries	14.9	8.3	N.A.
Other Developing Countries	16.3	3.5	N.A.
All Developing Countries	31.2	4.8	5.1
Industrialized Countries	11.5	0.7	1.0
Non-Communist World	42.7	1.7	1.7
Total World	80.3	2.2	2.6

Another estimate which some readers may regard with scepticism, in the light of disturbances affecting some Gulf member countries of OPEC, is the apparently high estimate of 8.3 per cent per annum average rate of growth of energy consumption in OPEC member countries. However, this is in fact considerably below the estimated growth of 11.1 per cent from 1973 to 1978 in commercial energy consumption recorded in United Nations statistics.(2) An important explanation for the rapid rate of growth foreseen is the scale of industrialization, which will tend to be energy intensive, but using formerly flared gas to the maximum possible extent.

A further point is that though the addition to energy consumption in OPEC member countries may appear remarkably high compared to the addition among industrialized countries, the fact is that per capita consumption in OPEC countries even in the year 2000 is likely to be little more than one third that of the OECD industrialized countries. This is partly attributable to the fact that per capita consumption in the industrialized countries was nearly ten times as great as in OPEC countries in the recent past, and secondly, population growth in the OPEC countries is expected to be very much faster than in the OECD countries over the next twenty years.(9) Another feature of this neutral scenario is that oil consumption in OPEC countries is projected to be much less in the year 2000 than in an alternative projection prepared in 1980 by the OPEC Secretariat.(10)

Average growth of only 3.5 per cent each year in energy consumption of non-OPEC developing countries is significantly lower than desirable, though it does make it feasible for real GNP per capita to grow by at least 1 to 2 per cent each year, on average, for this large group of countries. However, because demand may well grow faster than this in the more dynamic economies and in countries with an expanding energy producing sector, it is probable that many other developing countries will be hard pressed to maintain, let alone improve, the living standards of their people over the next twenty years. This problem constitutes an important rationale for resuming the North-South dialogue in a constructive way. The energy sector is likely to be of critical importance for many developing countries. Exploration for new energy resources (particularly oil and gas) is an urgent need for many of them. In this respect the World Bank energy affiliate is a promising initiative. The less fortunate and less well-endowed developing countries are likely to find themselves substituting non-commercial fuels (not included in these estimates or projections) for commercial forms of energy at the margin of consumption as the real price of the latter rises.

Though the predicament of the industrialized countries looks bleak, they will be able to maintain per capita energy consumption in gross terms in this neutral scenario. Furthermore, the scope for technological change and improvements in energy efficiency is considerable. Smaller and more efficient automobile engines are already making an impact in the USA. Wider bodied jets, better insulation standards in buildings, and the widespread adoption of equipment such as heat pumps are capable of bringing about conservation savings which make it possible for growth of energy consumption to match population growth and sustain a rise in living standards. Nevertheless, it is becoming widely recognized that industrialized countries will need to devote a higher proportion of their resources to energy production in order to improve living standards and take over from oil the role of meeting growth in demand.

Having said all this, the projections discussed in the following chapters which seem bound to attract the most attention are those relating to oil production, consumption and net exports from OPEC member countries. Indeed, these represent the central and certainly the most controversial feature of all the energy projections discussed below. At this point in time there is obviously great uncertainty attaching to the projections of oil consumption in OPEC countries twenty years ahead. In my view, though, there is even more uncertainty about the volume of oil production, since both political and economic factors are likely to have a major influence on the extent to which OPEC members are prepared to produce at what is likely to be near their maximum capacity of about 32 million b/d in the year 2000, rather than at a rate of perhaps only half that volume.

In the different scenario combinations envisaged in these projections, even with the development of harmonious international relations and a constructive resumption of the North-South dialogue, OPEC production of 32 million b/d in the year 2000 might be associated with total OPEC exports of no more than about 13 million b/d. This assumes high growth of domestic demand such as Japan experienced up to 1973, though with lower population growth than is likely in OPEC countries during the next twenty years. At the other extreme, I believe a continuous period of deterioration in the economic relationship between industrialized and developing countries could lead OPEC to produce no more than about 16 million b/d in the year 2000. Even with fairly modest growth of domestic demand for oil products in this scenario of less than 7 per cent annual average, OPEC members would probably have available for export less than 7 million b/d at the end of the twentieth century. Even with international relationships staying in their present state ('neutral' scenario) OPEC production is quite likely to be down to about 23.5 million b/d by the year 2000, leaving only some 10 million b/d available for export after meeting what I regard as the most probable level of demand which implies an average growth of rather less than 9 per cent annually.

One outcome of these projections in all scenarios is the relatively low level of dependence on OPEC oil among industrialized countries for their energy needs by the year 2000.

References

(1) BP Statistical Review of the World Oil Industry, 1979 and earlier issues. British Petroleum Company Limited.

(2) World Energy Supplies 1973-1978. United Nations, 1979. Statistical Papers Series J No.22.

(3) Main Economic Indicators. Organisation for Economic Co-operation and Development, 1980.

(4) Petroleum Statistical Bulletin, 1979 and earlier issues. Ministry of Petroleum and Mineral Resources, Kingdom of Saudi Arabia.

(5) American Petroleum Institute.

(6) Canadian Petroleum Institute.

(7) Oil and Gas Journal, 31 December 1979 and earlier end year statistical estimates.

(8) Quarterly Oil Statistics 1981/No.1 and earlier issues. Organisation of Economic Co-operation and Development.

(9) United Nations Population Statistics and Forecasts, 1950-2000. Selected Demographic Indicators, 1980.

(10) Domestic Energy Requirements in OPEC Member Countries. OPEC Papers Volume I No.1., August 1980.

14
ENERGY SCENARIOS FOR 1985 AND 1990

INTRODUCTION

The order of discussion of the energy supply and demand projections made in this chapter and the next will be first for the World as a whole, followed by the non-Communist part of it and then the industrialized countries as a group, concluding with some remarks about the Communist bloc. The reader is referred to the statistical estimates in Tables 35 to 45 inclusive in Part III which complement this chapter. The consideration of energy scenarios for the USA, Western Europe, Japan, OPEC countries, other developing countries, etc., will be found in the appropriate parts of Chapters 16 to 23 inclusive.

Within each of these geopolitical aggregates I discuss the projections for oil and energy demand on the basis of certain assumptions about supply and the degree of international political and economic co-operation which may or may not characterize the next twenty years. I have in mind particularly the debate and action regarding what has come to be called the 'North-South issue'.

SCENARIO DESCRIPTIONS

The basic hypothesis underlying the scenarios described below is that the world political and economic system may benefit from an improvement in international relations or suffer through a deterioration in such relationships. Alternatively, there may be no significant change from the existing situation.

This gives rise to three scenarios which I have called:

1. Co-operation.
2. Neutral.
3. Confrontation.

In practice, of course, international relations are likely to vary between all three of these scenarios over a period as long as twenty years. Furthermore, the relationship between the USA and the USSR may be different, at any one point in time, from that existing between Western Europe and Japan, or any of these and the Gulf States of OPEC. The way in which the relationship between the Arab states and Israel develops will be an extremely significant component of the scenarios, influencing the probable volume of oil available for export, and consequently, oil prices. Another specific feature of international relations which seems to be not sufficiently recognized in either the industrialized countries or the Communist bloc countries is the importance which OPEC countries attach to the improvement of economic conditions in other developing countries. I believe this will become an increasingly important feature of international relationships over the next two decades, whatever the outcome of any resumed North-South dialogue.

The substance of the scenarios I shall describe can be regarded as general stereotypes. I am concerned here to examine primarily their oil and energy aspects in some detail, the relationship to the world macroeconomic scene, and the possible alternative ways in which the North-South relationships could develop.

The features of the three scenarios are described below in qualitative terms.

Co-operation Scenario

- Improved international relations.

- Specific incentives to oil exporting countries to produce more oil than immediate budgetary needs require. Willingness to produce at high rather than low levels will depend on the satisfaction of reasonable political as well as economic criteria.

- Increased rates of exploration, production, research and development of all forms of energy worldwide, including conservation through the elimination of waste, flaring of associated gas, etc.

- Significant conservation of energy at the point of distribution and use, through technological improvements becoming increasingly attractive economically.

- Improvement in the flow of aid and technology to the oil importing developing countries.

- A more stable and predictable international financial climate likely to favour saving and investment relative to current consumption. This is likely to be particularly important for ensuring that there is an adequate real transfer of resources to the energy sector where long lead times tend to be characteristic. It is also important to protect surplus revenues generated from higher than necessary oil production and exports.

- Reduction in trade barriers, of both a tariff and non-tariff type, specifically to encourage oil importing developing countries to exploit their real comparative advantages and expand the export sector of their economies so as to achieve reasonable GNP growth and reduce the chronic current account deficits from which many developing countries suffer.

- Larger capital account flows, especially to developing countries with energy resources which appear to offer promise of economic development over the next twenty years. The World Bank and UN must be encouraged to develop their initiatives further, but there is scope for other institutions such as the OPEC Fund, the commercial banking system, multinational and bilateral projects, as well as those which can be financed by the leading transnational oil companies.

- Substantial progress by the industrialized countries in moving towards their targets for energy which they have pledged at summit conferences, notably at Venice in 1980.

- Implementation by OPEC of a long term strategy (after re-unification of the price structure) which makes oil production and price levels more predictable in the medium term. This is likely to include an element of flexibility to cover deviations from likely medium term trends, whether originating at the supply or demand end of the oil equation. Modest and predictable rises in real oil prices would be a stimulus to change in the energy mix.

- A more effective and sincere initiative, whilst recognizing differences and being prepared for compromises, to make real progress on the North-South issues, most of which are featured above.

The main specific results of this scenario which could be expected to materialize over a prolonged period would be as follows: higher finding rates of oil and gas reserves worldwide; higher volume of oil production and exports from 'discretionary' oil producing countries; faster development of coal, nuclear, hydro, solar, synthetics and other energy sources; a low energy/GNP growth coefficient by historical standards, related to rising real prices stimulating faster improvements in the efficiency of energy use; faster growth of worldwide GNP and world trade than in recent years.

Neutral Scenario

I envisage this scenario involving no significant policy changes either for better or for worse as compared with those which have generally prevailed during recent years. Nevertheless, the rate of growth in energy availability and consumption would tend to be lower than during the period since 1973. Thus the outlook for all the major economic groups of countries analysed in this book would tend to show a worsening of unfavourable developments seen during recent years which have been a focus for comment by the Brandt Commission and others. As far as oil and energy availability and use is concerned, this scenario would be more or less intermediate between the co-operation scenario discussed above and the confrontation scenario which will be described below. Increasing political strains might be expected to develop, both within and between countries.

Confrontation Scenario

As its name suggests, this scenario might be regarded as the opposite of the co-operation scenario. Its main features are listed below.

- Deterioration of international relations.

- Lack of any effective action to provide incentives to 'discretionary' oil producers to export larger volumes of oil than their domestic budgetary needs necessitate.

- Reduced rates of exploration, production, research and development of all forms of energy due to pervasive pessimism about international economic prospects, sharp cyclical swings and uncertainty and risks associated with long lead time and high cost energy projects.

- Lack of systematic policies to promote conservation and improved efficiency of oil and energy use; nevertheless, growth in demand is slow or even negative due to supply constraints.

- Some relative decline in the flow of aid and technology to the oil importing developing countries from the industrialized countries which become

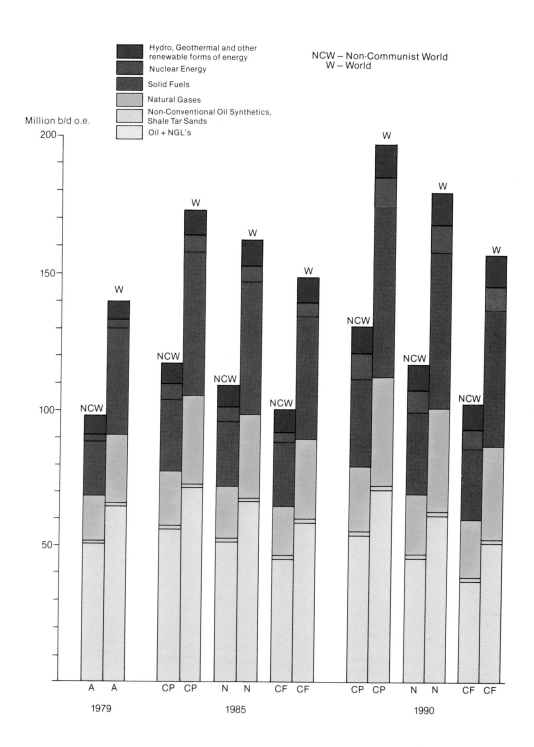

Energy Scenario Projections for 1985 and 1990

Figure 9

increasingly preoccupied with their domestic economic problems and friction with other industrialized countries.

- The development of an international protectionist mentality and increased trade barriers as more and more countries become preoccupied with the under-utilization of domestic resources, particularly of manpower in the industrialized democracies.

- An unstable and unpredictable international financial climate in terms of sharp fluctuations in exchange rates and security of funds invested abroad. The climate will militate against major international investment in capital projects as an increasing number of countries show signs of political instability and/or problems of credit-worthiness.

- A relative fall in international capital flows as a consequence of the uncertain and risky financial climate. This will exacerbate international balance of payments problems and limit the growth of world energy supplies severely, with feasible projects remaining undeveloped.

- The growth of alternative forms of energy slips increasingly from currently expected levels and this is both partly a cause and a consequence of low economic growth and relatively low levels of investment in the energy sector, high cost and technological problems.

- Lack of any long term strategy for oil prices or production among major exporters with diverse interests, time scales and objectives, leading to contradictions in policies, with the international oil market suffering from periodic crises. The outcome is likely to be sharp fluctuations in the volume of production and exports either side of a declining trend, and, similarly, fluctuations in real oil prices either side of a significantly more sharply rising trend than in the co-operation scenario. Uncertainty would be the fundamentally important characteristic of the international oil market with rather sharp swings between tight and slack conditions.

- A deterioration in North-South relations, to the economic and political detriment of both. There is an absence of political will to appreciate the problems and opportunities of the other side, the complementarity and dynamics of change in the international political and economic system.

The main results of this scenario which could be expected to materialize over a long period would be as follows: lower finding rates of oil and gas reserves worldwide; lower volumes of oil production and exports from discretionary oil producing countries; slower development of coal, nuclear, hydro, solar, synthetics and other energy sources; a very low rate of growth of GNP and declines in GNP per capita over part of the period in many countries, due partly to energy supply constraints.

Oil Prices

The experience of the last decade might be regarded as a sufficiently good reason for omitting any specific reference to forecasts of oil prices over the next decade or two. However, to ignore the issue in a book of this sort would be rather irresponsible, since the projections shown in Tables 35 to 55 of Part III are without meaning unless set in an oil price context. For this reason I set down my current thinking about the trend of real oil prices likely to be associated with each scenario over the period from mid-1981 to about 1990, based on the marker crude, Arab Light. This has regard to what happened during the past decade and the outlook for energy supply by type and region and demand by

Trend Rate of Change in Real Oil Prices 1981 - 1990

Scenario	% per year
Co-operation	up to 3
Neutral	3 to 6
Confrontation	more than 6

It does not seem reasonable to attach a high level of confidence to these figures, having regard to the many factors which can have a significant impact on them, either in a relative moment of time or progressively over ten years. To hazard a guess as to how oil prices might move through the 1990s seems unreasonable at this point in time bearing in mind the uncertainties relating to technological change, new oil discoveries, supply of other forms of energy and developments in the energy/GNP growth relationship. It seems sufficient to remark that in qualitative terms they will be similar to the pattern observable in the table above. That is to say that they will be softest or lowest in the co-operation scenario and hardest or highest in real terms in the confrontation scenario.

THE WORLD

Co-operation Scenario

Over the period from 1979 to 1985 the maximum average annual growth in the supply and demand for primary energy of all types seems likely to be about 3.5 per cent. This amounts to an increase of some 33 million b/d of oil equivalent over the six years. The realization of anything like this scenario will involve not only a rapid and significant shift toward the scenario stereotype described above, but also, and especially, a strong though steady cyclical upswing in the economies of the industrialized countries, peace worldwide, and the return to higher levels of oil production and exports in countries producing well below their capacities in early 1981.

The chances of all these things happening do not appear to be high from our present perspective. Perhaps even less likely is the potential rate of growth of the world economy such as might be associated with an average 3.5 per cent annual growth in primary energy consumption up to 1985. A review of the trend since 1973 and future prospects suggests that GNP growth could be as high as 4.7 per cent annually from 1979 to 1985.

Over the period from 1985 to 1990 some deceleration in the rate of growth of primary energy consumption appears inevitable in this optimistic scenario. Something slightly less than a 3 per cent annual rate of increase seems to be the best that can be expected realistically. Nevertheless this should be sufficient to sustain world economic growth of about 3.5 per cent on average each year, given a harmonious development of international relations and other features of this scenario described above.

In this scenario the oil share of the energy mix would fall from about 46 per cent in 1979 to 42 per cent in 1985, but then it would probably decline more sharply to about 35 per cent in 1990 as world oil availability starts to fall very gradually.

Neutral Scenario

This scenario assumes no significant change in international relationships. The prospect is that supply and demand for primary energy will grow by about 2.5 per cent annually over the six years from 1979 to 1985, virtually the same as the average for the preceding six years. This amounts to an increase of about 22 million b/d oil equivalent over the current period. The average rate of economic growth which might be achievable in these circumstances could amount to as much as 3.2 per cent annually.

Some slow-down in the rate of growth in primary energy supply and demand seems inevitable for the period after 1985. It might average 2 per cent annually or a little more in this scenario, with supply constraints being the most probable factor limiting growth. For this reason it seems possible that prices for oil and other forms of energy entering international trade may have to move up more sharply between 1985 and 1990 to keep supply and demand in equilibrium than during the next few years or after about 1990. Necessarily, this judgement is a little speculative depending as it does on the incidence of the international economic cycle and assumptions about a relatively 'surprise-free' political atmosphere prevailing more or less throughout the period to encourage the maintenance of fairly stable conditions in the international energy market. In these circumstances the world economy might be able to achieve an average growth of about 2.6 per cent annually from 1985 to 1990. If achieved, this would yield significant improvements in per capita living standards, though, of course, less than in the co-operation scenario. In this sense the energy supply constraint seems likely to represent a rather less dangerous threat to world political and social stability than it seemed to a few years ago on the neutral scenario assumptions. Another significant factor is that world population growth is now expected to be slower than previously. According to the UN population statistics published in 1980 the rate of world population growth which averaged almost 2 per cent annually from 1960 to 1970 was estimated to have fallen to 1.8 per cent in 1980. On the most likely medium variant projection it will fall to little more than 1.6 per cent over the last decade of the twentieth century and to 1.5 per cent by the year 2000.(1)

In this neutral scenario the oil share of the world energy balance would fall rather more than in the co-operation scenario. It is expected to fall from 46 per cent in 1979 to 40 per cent in 1985 and 34 per cent in 1990.

Confrontation Scenario

Whilst the neutral scenario may hold out the promise of a tolerable though uninspired future for the world as a whole and most countries in it, the confrontation scenario is likely to involve absolute declines in per capita living standards in many countries over prolonged periods. There is a real danger of a downward economic spiral developing, threatening political, economic and social institutions, and international relationships. In this sense negative departures from the quantitative scenario picture described below must be regarded as more likely than such departures from the projections which are a feature of the other two scenarios described above. This confrontation scenario is inherently more unstable, but not, unfortunately, any less likely to occur for that reason. This is the important qualitative reason for this scenario and the explanation for the widespread publicity given to the Brandt Commission's 'North-South : A Programme for Survival'. The real danger of this scenario is that international political relationships could go out of control, involving any or many of the following: nuclear war between the super-powers; very sharp cyclical fluctuations in economic activity in the industrialized countries; a trade conflict between Japan and some other industrialized countries; endemic conflict between Israel and the Arab states; increasing Soviet encroachment on the oil fields of the Gulf region; direct Western military intervention in this area; a systematic policy on the part of the industrialized countries to minimize their trade, aid, capital flows and technology transfer to the developing countries.

During the whole of the period to 1990 world supply and demand for primary energy is projected to grow at an average rate of about 1 per cent each year. This would be well below the expected rate of about 1.8 per cent population growth. Worldwide economic growth might just about match this, though some countries would achieve positive growth in per capita real incomes while others suffered prolonged absolute declines. This in itself is something of a recipe for international political tensions and a deteriorating prospect for world trade.

Another source of tension in this scenario is likely to be a sharper reduction in the volume of oil entering world trade. In spite of, or because of, higher unit prices per barrel of oil exported, those countries with surplus funds would tend to have less and less incentive to export oil as the security and income derived from those surplus funds seemed likely to be increasingly subject to political hazards and erosion in terms of future real purchasing power. The comparison is with oil reserves maintained in their natural reservoirs and subject to depletion through sovereign decisions of their national governments. Development of new oil reserves outside OPEC countries would also be adversely affected by the unfavourable economic and political climate. In consequence, the oil share of the world energy balance would probably fall to less than 40 per cent by 1985 and to something like 32 per cent by 1990.

THE NON-COMMUNIST WORLD

Co-operation Scenario

The prospect is that increased energy supplies might make it possible for energy demand to grow at an average rate of 3.2 per cent annually from 1979 to 1985, though subsequently the rate of increase would probably fall by about a percentage point to about 2.2 per cent on average over the five years to 1990. This would be a significantly higher rate of growth than the average of 1.7 per cent recorded from 1973 to 1979, but much less than the 5.4 per cent annual average over the preceding six years ending in 1973.

The oil share of the energy balance in this scenario might decline by about five percentage points over six years to 48 per cent in 1985, but then at an accelerating rate to 41 per cent over the next five years to 1990. These estimates presuppose a significant rise of nearly six million barrels a day in oil availability and demand from the 1979 level to about 57 million b/d in 1985. Over the following five years there might be a fall of nearly 3 million b/d owing to the likelihood that falls in production in the older producing areas (where reserves to production ratios are already low) would more than counterbalance increased volumes from newly developed fields.

Neutral Scenario

The outlook in this scenario is that energy demand growth will be limited to an average similar to that recorded over the six years from 1973 to 1979. It is likely to average slightly less than 2 per cent annually from 1979 to 1985, and then about 1.5 per cent each year from 1985 to 1990.

The oil share of the energy balance will fall slightly faster than in the co-operation scenario. Oil supply and demand is expected to be almost the same in 1985 as in 1979, or significantly higher than in 1980 and 1981. A fall of some six million b/d to about 45 million b/d is projected over the five years to 1990.

Natural gas supply and demand is projected to rise continuously with increased exports from the USSR expected. Solid fuels are likely to contribute at least half the 20 million b/d oil equivalent increase in the supply and demand for total primary energy projected for the eleven year period ending in 1990. This implies an average rate of growth of 4.2 per cent annually. Nuclear energy may contribute to the increase less than 6 million b/d fossil fuel input equivalent.

Energy Scenarios for 1985 and 1990

Confrontation Scenario

This scenario presents a depressing prospect. Primary energy supply and demand is likely to rise at an average rate of only about 0.5 per cent from 1979 to 1990, or much less than the average rate of growth of population. Almost inevitably, it seems to imply falling rather than rising standards of living as being typical in the great majority of countries over most of the period to 1990.

The oil share of the energy balance is projected to fall at an accelerating pace from 53 per cent in 1979 to 45 per cent in 1985 and 36 per cent in 1990, due to limited availability and more rapidly rising real prices. Sharp fluctuations in both are more likely than not, as the international oil market is characterized by disorder. Availability of crude and demand for products might fall to about 45 million b/d in 1985 and then to some 37 million b/d in 1990.

INDUSTRIALIZED COUNTRIES

Co-operation Scenario

The reader from one of the industrialized countries may be surprised by the low growth of energy demand projected over the 1985 to 1990 period and the rather rapid transition involving a sharper reduction in the oil share of the energy balance than in the world as a whole. In order to derive these estimates, essentially on a judgemental basis, a detailed analysis has been made of the trends since 1973, including cyclical fluctuations, the prospects for availability of oil and other energy forms, conservation of oil use both by sector and in terms of the energy/GNP growth relationship.

In consequence, a strong growth of the OECD economies is seen as being feasible between 1981 and 1985 associated with a potential average growth in energy demand of about 2.2 per cent over the six years from 1979. Average economic growth exceeding 4 per cent annually should be possible over this period, if there is a strong cyclical upswing from the recession in most industrialized countries in 1982. Nevertheless, such a rate of growth appears somewhat improbable, after GNP growth estimated at 1.3 per cent in 1980 and probably less in 1981. Other essential pre-conditions for this scenario to materialize are likely to be OPEC oil production returning to about 30 million b/d or more, real oil prices not rising until 1983, and then only moderately, together with realization of the co-operation scenario features discussed above. Assuming this happened, then a fairly sharp drop in the rate of growth of energy demand to about 0.7 per cent each year seems inevitable for the 1985 to 1990 period. This is because of supply constraints and lead time problems even in this optimistic scenario. Nevertheless, this seems likely to be compatible with average GNP growth of between 2.5 and 3.0 per cent, or about 2.0 per cent per capita.

Oil demand may be marginally higher in 1985 than in 1979, but would probably fall to about 34 million b/d in 1990 or as much as 7 million b/d less than five years previously. At this level oil is projected to account for only 37 per cent of primary energy consumption, compared with 52 per cent in 1979 and 46 per cent in 1985.

Neutral Scenario

This scenario incorporates growth of primary energy demand averaging only 1 per cent between 1979 and 1985, declining to only 0.3 per cent on average over the next five years to 1990. Put into the perspective of growth which averaged only 1.1 per cent from 1973 to 1979, and followed by the fall recorded in 1980, this does not appear to represent much of a shock to the system. The average rates of economic growth likely to be possible with these increases in primary energy are about 2.5 per cent during the six years to 1985, then about 1.5 per cent annually over the five years to 1990. These rates of increase represent significant rises in real incomes per capita, since population growth in the OECD is most likely to average only about 0.7 per cent annually until 1990.

While the oil share of energy demand is projected to fall some seven percentage points over the six years to 1985, it will probably decline by about eleven percentage points over the following five years to only 34 per cent in 1990. This implies a fall of more than 3 million b/d in oil consumption between 1979 and 1985, followed by a more dramatic decline of more than 8 million b/d over the next five years to less than 30 million b/d.

Natural gas supply and use should continue to rise slowly to 1990. In view of the projected decline in oil use, the growth in coal consumption alone is likely to be greater than the rise in energy use over the decade. Thus its share of the total should increase from about 19 per cent in 1979 to about 27 per cent in 1990. The nuclear share will also rise significantly, though it will still account for no more than some 9 or 10 per cent of total energy consumption in 1990.

Confrontation Scenario

This scenario involves an absolute decline in energy use in the industrialized countries through the 1980s at an average rate of about 0.5 per cent annually. Nevertheless it should be possible to achieve economic growth averaging about 0.5 per cent each year, though this implies a marginal average decline in per capita GNP over the eleven years from 1979 to 1990. It is probable that there would be strong year to year fluctuations around these projected trend rates of change. Assuming this scenario evolved over such a long period, it must become increasingly likely that political and social institutions in some industrialized countries undergo some fundamental changes as a result of the pressures on them.

This scenario would be accompanied by an even more dramatic reduction in the oil share of the energy balance to little more than 30 per cent by 1990. A very sharp decline in OPEC production and export volumes could mean OECD oil consumption falling from more than 40 million b/d in 1979 to less than 33 million b/d in 1985 and then to some 23 million b/d in 1990. Such a rapid rate of decline in oil availability and use would lead almost inevitably to severe economic dislocation, especially since it would not be a planned run-down but the result of a haphazard and unfavourable evolution of international relations. Increased friction between the industrialized countries would be likely to occur, not least because of the competition for scarce oil supplies. One feature of this scenario is that industrialized countries as a whole might be importing less than 10 million b/d of oil as early as 1990. The move towards independence from OPEC would have been bought at high cost.

COMMUNIST BLOC

Co-operation Scenario

Growth in energy consumption is projected to be significantly higher than in the non-Communist world, but the rate of growth is likely to continue to slow down through the 1980s as it did during the previous decade. It should be possible for energy consumption to grow at an average of rather more than 4 per cent annually to 1985, and then at slightly less over the next five years. This assumes that the Comecon countries are able to buy all the technology they require from the industrialized countries, and that they allocate adequate resources to the energy sector to realize the considerable increase of production -- 27 million b/d oil equivalent -- in conditions likely to prove increasingly difficult.

Oil consumption is projected to go on rising until 1990, though production may well plateau at about 16 million b/d during the second half of the 1980s. This implies that by 1990 oil supply and demand would be in balance, with no net exports of oil from the centrally planned economies as a whole. As an offset to this, net exports of natural gas are expected to be on a rising trend as a foreign currency earner. Similarly, an increased volume of coal exports should make a contribution to the foreign exchange requirements of the Communist countries, though at present the prospects are not looking too promising. The oil share of the energy consumption mix is likely to fall from 30 per cent in 1979 to 24 per cent in 1990.

Neutral Scenario

In this scenario energy demand is projected to grow at about 3.8 per cent on average to 1985, decelerating to about 3.2 per cent in the next five years. These are about twice the rates of increase expected to occur in the non-Communist world as a whole. Thus, in this scenario, the Communist countries are likely to account for about 35 per cent of world energy consumption in 1990 compared with less than 31 per cent in 1979. These relatively high rates of growth in energy consumption combined with population growth averaging little more than 1 per cent annually (even including China) should permit significant per capita improvements in living standards even if not too much progress is made in conservation of energy use and priority continues to be given to the productive sectors.

Oil use would continue to rise, but more slowly, up to 1990, though production might start to decline some years earlier without adequate resources being devoted to the sector, and with a lower level of access to the technology of the industrialized countries outside the Communist bloc. In consequence, there could be a swing in the net oil trade position. The centrally planned economies are projected to move from being small net exporters in 1985 to requiring net imports of some 1 million b/d of oil in 1990 in this scenario.

Pressure on the relatively low level of proved oil reserves should be relieved both by some new discoveries, but also through improvements in the recovery factor. However, both of these assumptions remain somewhat speculative. The chances are that both natural gas and solid fuels (mainly coal) should each contribute some 7 million b/d oil equivalent energy to the 20 million b/d growth in Communist bloc consumption projected for the 1979 to 1990 period in this scenario.

Confrontation Scenario

Significantly lower rates of growth in energy consumption are projected in this scenario: an average of 2.5 per cent until 1985 and then 2.0 per cent from 1985 to 1990. Though these rates of increase are still much higher than the average in the non-Communist world, they represent a dramatic reduction from the average of well over 4 per cent recorded over the years to 1979. It is reasonable to expect that the slow-down would be both a cause and a consequence of significantly lower economic growth, as well as a possible cause of friction between countries inside the Communist bloc and some of those outside it.

The Communist bloc would be a marginal oil importer by 1985 and require about 1.6 million b/d by 1990 in order to sustain an increase in internal consumption of no more than 0.5 million b/d during the second half of the 1980s. Total oil production might actually fall by about 1 million b/d between 1979 and 1985 and by another 1 million b/d between 1985 and 1990.

Reference

(1) <u>United Nations Population Statistics and Forecasts, 1950-2000.</u> Selected Demographic Indicators, 1980.

15
ENERGY SCENARIOS FOR THE YEAR 2000

INTRODUCTION

The reader who has arrived at this point without passing through Chapter 14 is advised to look at the introduction to that chapter and scenario descriptions in order to obtain an understanding of the assumptions we are using to survey the year 2000 time horizon. Statistical estimates which the reader may care to consult while reading this chapter can be found in Tables 35 to 45 inclusive in Part III.

The main point to be made here is the fundamental qualitative difference in terms of uncertainty which must be a feature of any projections for some two decades ahead, as compared with those for less than five and ten years ahead which I discussed in the previous chapter. Without suggesting that the current and next decades will in any sense parallel the last two, one need only remind oneself how relatively smooth and predictable was the decade of the 1960s compared with the turbulence and discontinuities of the 1970s in the oil sector, the energy scene in general and in the world economy as a whole. Yet, even in the 1960s, oil economists were consistently under-forecasting the rate of growth in oil demand. This was associated with falling real prices, unconstrained oil supply, the increasing share of oil in energy demand in these circumstances, and the consequential stimulus to economic growth among oil importing countries. Some commentators had identified the unsustainability of that earlier combination, given the finite character of oil reserves, the declining rate of finding new reserves, and the nature of exponential demand growth. Nevertheless, the extent and timing of the discontinuities of the 1970s were not clearly identified far in advance of their occurrence. A general weakness of most forecasts and projections made during the last ten years has been the insufficient ability to recognize short and medium term supply and demand elasticities for oil. Thus, the inevitable movement of price to maintain equilibrium seems not to have aroused the discussion it deserved. The long term elasticities are likely to be significantly greater, though reserve limitations represent a constraint on supply. Moving from these more theoretical aspects of oil and energy forecasting, it seems worth recognizing, too, that many medium and long term forecasts published in recent years seem politically naive in relation to what happened in 1973-74, and also over the 1979-80 period. Risk analysis came into vogue, but its implications seem not to have been fully assimilated.

It seems necessary to state explicitly that in all the scenarios presented here world oil production declines through the 1990s, since there is not sufficient evidence to take a more optimistic view at present. It should be understood that this does not preclude the possibility of a move towards confrontation and lower oil production during the 1980s, followed by a shift towards co-operation and high oil production in the 1990s. That is the essence of the scenario approach and its relevance to the review of prospects for the global oil and energy outlook in the uncertain world of today. In other words, actual experience over the next 20 years may well follow a composite mosaic of the scenarios discussed in this book, because that is the way the real world tends to progress over time. This means that something such as the profile of world oil production may follow a fairly erratic course over an extended period, as indeed it has done during the last ten years, yet remain explicable in politico-economic terms. This should serve to reinforce the central message of this book: namely, that a constructive international dialogue leading to harmonious international relations should facilitate an orderly rather than disorderly transition away from oil dependence. This is inevitable sooner or later, even if surprisingly large new oil discoveries should be made and exploited during the course of the next twenty years or so.

These considerations should make everyone engaged in looking at the misty 'energy 2000' horizon very chary of making strong claims for his own estimate or series of projections. However, there has been no shortage of volunteers, as can be seen in Part III of this book. I show there some comparative estimates from different sources published during recent years. I deliberately make no attempt at detailed comparative comment. It seems sufficient to notice that one of the most highly esteemed of these sources, namely World Alternative Energy Strategies of 1977, produced minimum estimates of OPEC oil production in the year 2000 which now seem optimistic.(1)

One of the encouraging features of the experience since 1973 in the industrialized countries has been the very significant reduction in the rate of growth in energy demand in relation to economic growth. In the USA and Japan it averaged about 0.3:1 over the six years to 1979, whilst in Europe the average was about 0.4:1. Even though these encouraging experiences include the once-and-for-all elimination of wastage in 1974-75, there is much scope for technological change facilitating big improvements in the efficiency of energy use through the 1990s. Therefore, growth averaging 2 per cent annually in energy consumption worldwide might be associated with significant per capita growth of GNP in a way which seemed improbable a few years ago. By contrast, the experience of recent years in terms of the incremental contribution from alternative forms of energy does not give one too much encouragement for the future. Perhaps particularly disappointing has been the political/ecological/technological slow-down in the nuclear sector. Even at mid-1981 oil prices there are few signs that more exotic and renewable energy forms are becoming competitive with petroleum and its derivatives. However, it does seem reasonable to assume that with increasing resources being devoted to the energy sector there will be some technological and economic breakthroughs during the next ten years which could start to have some effect in terms of commercial energy availability and prices during the 1990s. Some provision has been made for this in the projections, though it is not possible to be more than speculative about the prospects at this stage.

The order of discussion of this chapter is the same as that adopted in the previous one, namely, the World as a whole followed by the non-Communist part of it, and then the industrialized countries as a group, concluding with some remarks about the Communist bloc.

THE WORLD

Co-operation Scenario

The year 2000 horizon is approached in this scenario at the same rate of primary energy supply and demand growth during the 1990s as in the preceding five years, an annual average of about 2.8 per cent. The realization of this relatively high rate illustrates the

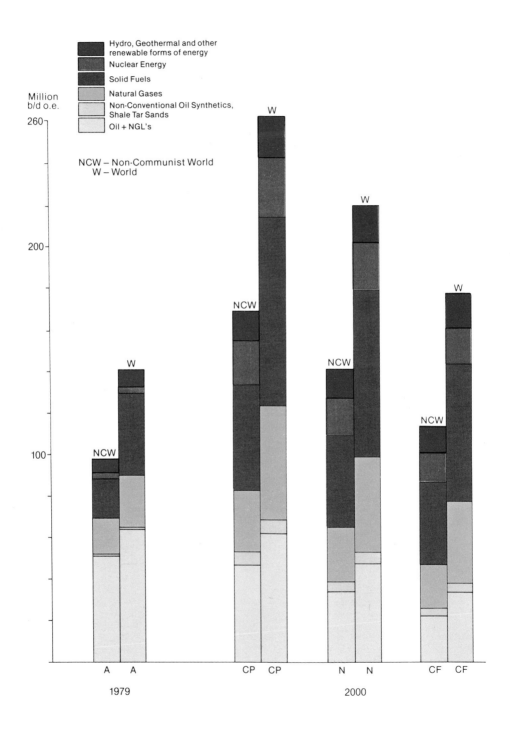

nature of exponential growth which implies an increment of some 63 million b/d oil equivalent over the ten years compared with only 25 million b/d equivalent during the preceding five years in the late 1980s.

Assuming a relatively smooth transition up to and through this period, the world economy might have the potential to grow at about 4 per cent on average through the 1990s associated with this growth in energy availability and use. This would make it possible for per capita GNP to grow at significant rates practically everywhere, as the latest most probable UN projections suggest that the world population is likely to grow at about 1.6 per cent each year through the 1990s.(2)

Even in this scenario the conventional oil and NGL share of the world energy mix is likely to fall rather dramatically by some twelve percentage points over the last ten years of the twentieth century to about 23 per cent in the year 2000. The fall in absolute volume of more than 8 million b/d is largely offset by a rise foreseen in non-conventional oil, synthetics, shale and tar sands, the supply of which is projected to rise to about 7 million b/d in the year 2000.

Solid fuels, primarily hard coal, are projected to make the largest incremental contribution to the rising supply and demand for energy during the 1990s: about 28 million b/d oil equivalent in the world as a whole. Because of the relative abundance of coal reserves and prospective costs of production, transportation (on a greatly increased scale) and use as an industrial boiler fuel, to generate electricity and as a base for synthetic oil and gas it should be possible to phase out large volumes of refined oil products based on conventional crudes in many parts of the world. With nuclear energy projected to increase by about 17 million b/d oil equivalent and natural gas by about 14.5 million b/d oil equivalent as well, residual fuel oil and distillate fuel oil will be produced in much reduced volumes in many parts of the world. Hydroelectricity, geothermal power, solar energy and some wind and tidal energy projects are expected to contribute another 6.5 million b/d oil equivalent to the increase in world energy supplies through the 1990s.

Solid fuels should account for over 34 per cent of the world primary energy mix in the year 2000 in this scenario, compared with 28 per cent in 1979.

Neutral Scenario

World primary energy supply and demand is projected to rise at 2.0 per cent annually through the 1990s in this scenario. This is similar to that foreseen in the preceding five years. Ten years of growth at this rate implies 40 million b/d oil equivalent extra in the last decade of the twentieth century.

In this scenario some growth in per capita GNP should be possible in most parts of the world during the 1990s, assuming a reasonably efficient world energy market and relatively efficient use of energy worldwide which a prolonged period of high costs and prices should induce.

The conventional oil share of the energy balance is projected to fall to only 22 per cent by the year 2000, implying a decline in volume by more than 13 million b/d during the 1990s. Non-conventional oil supplies might rise by about 4.5 million b/d during the same period, so that aggregate world oil supplies would decline by an average annual rate of less than 1 million b/d over the decade. Nevertheless, this rate of change would almost certainly give rise to problems of management of the energy transition, and these, perhaps, would be worldwide at times of strong economic growth during the ten year period. There would be some danger of an unseemly scramble for scarce oil supplies, and perhaps of sharp increases in oil prices if adequate precautions are not taken in good time. As oil supplies decline absolutely, there will be increasing pressure to back out oil products from the non-premium use sectors and to improve further the efficiency of oil product use where effective substitutes remain to be developed.

Solid fuels, especially coal, are projected to account for more than half the growth in world energy supply and demand during the 1990s, and nuclear energy for another one third. Natural gas, together with hydro, geothermal, solar and other renewable energy forms are also projected to make significant additional contributions. In this scenario, too, the former king, coal, returns to the premier position in the world energy mix, accounting for over 36 per cent of the total in the year 2000, if other solid fuels are included.

Confrontation Scenario

Growth of energy supply and demand in this scenario through the 1990s is limited to an average rate of 1.2 per cent each year. It is significantly less than world population growth. Countries which are relatively less well endowed with energy resources and technology may well encounter great difficulties in achieving growth of per capita GNP over the ten year period, unless they enjoy special strengths in other sectors. Nevertheless, a low growth of per capita GNP should be possible worldwide on average during the 1990s in the absence of energy wars or other catastrophes. The inherent strains in the world energy/economic system appear to be marginally less in this scenario during the next decade than in the current one. It should be remembered, though, that it will follow a period of chronic energy shortage and high prices in relation to potential economic growth, having regard to the availability of other factors of production. Thus it seems reasonable to suppose that very low growth of net energy supplies associated with a sharp reduction in oil availability might prove to be more destabilizing during the 1990s than the confrontation scenario mixture described in the previous chapter for the 1980s.

In this scenario conventional oil might account for no more than about 19 per cent of the world energy mix in the year 2000, with the volume at about 34 million b/d. Non-conventional oils might account for another 4 million b/d, but their development would be retarded in this scenario. This is little more than half the 1979 level of conventional oil. There would be little scope for oil use outside the premium market sectors in this scenario, except perhaps in a few OPEC countries.

Coal and other solid fuels would be the largest form of energy, accounting for about 37 per cent of the world total in the confrontation scenario, even though the volume of solid fuels projected as being available and used is some 14.5 million b/d oil equivalent less than in the neutral scenario and nearly 25 million b/d oil equivalent less than in the co-operation scenario. Natural gas is projected to be the second most important form of energy, representing about 22 per cent of the total. Nuclear, hydro and other renewable forms of energy are expected to be subject to delays in development so that they might make a combined contribution of under 20 per cent of the total in 2000.

THE NON-COMMUNIST WORLD

Co-operation Scenario

Energy supply and demand is projected to grow at 2.5 per cent over the ten years to 2000. This represents an increase of some 37 million b/d oil equivalent, though supply and demand for oil is likely to fall marginally even after allowing for an increase in non-conventional supplies. The conventional oil share of the energy balance is projected to fall from 41 per cent in 1990 to 28 per cent in 2000.

The share of solid fuels is projected to grow to some 50 million b/d oil equivalent in 2000, up from some 32 million b/d oil equivalent ten years earlier, and to make a 30 per cent contribution to the energy balance in the horizon year. Natural gas is expected to rise to nearly 30 million b/d oil equivalent, and nuclear energy to over 20 million b/d oil equivalent. Hydro, geothermal, solar and other renewable energy forms are projected to contribute over 15 million b/d oil equivalent, or 9 per cent of the total in the year 2000, compared with 7 per cent in 1979.

Neutral Scenario

In this scenario the outlook is for energy demand to grow at about 1.8 per cent annually from 1990 to 2000, an average rate similar both to that from 1973 to 1979 and that projected up to 1990.

The conventional oil share is foreseen as falling more rapidly than in the co-operation scenario to only 24 per cent in the year 2000, implying consumption of about 34 million b/d. Another 5 million b/d may become available from non-conventional sources by that time.

Solid fuels are projected to be the leading form of energy, accounting for 31 per cent of the total. Natural gas supply and demand should continue to grow, but only slowly, with a 3 million b/d oil equivalent increase over the ten years to 2000. Nuclear energy is expected to double and account for nearly 13 per cent of the energy balance in the horizon year.

Confrontation Scenario

Growth of energy supply and demand in this scenario might average about 1 per cent annually through the 1990s. This is rather more than that projected for the current decade. Combined with a falling rate of population growth, this means that limited per capita growth of real living standards may be possible in most countries. In practice, this could depend more on the absence of major political and economic shocks to the system, in spite of the strains likely to develop under these scenario conditions, rather than on any absolute shortage of energy. However, the probability of avoiding such shocks must be regarded as low in this scenario.

The speed of transition away from conventional oil is likely to impose special strains in most importing countries as its share of the energy balance declines to only 19 per cent in the year 2000 from 36 per cent ten years earlier. Even adding in the non-conventional oils, the total is projected at only 26 million b/d out of a total energy consumption of about 114 million b/d oil equivalent.

In this scenario solid fuels are expected to account for some 35 per cent of the energy balance and to grow by some 13 million b/d oil equivalent. This is more than the total rise in energy consumption over the decade.

INDUSTRIALIZED COUNTRIES

Co-operation Scenario

Energy supply and demand in the industrialized countries is projected to grow at an average rate of some 1.0 per cent each year through the 1990s. This implies an increase of nearly 10 million b/d oil equivalent over the ten years. Given the probable improvements in the efficiency of energy use in this scenario, the OECD economy might have the potential to grow at about 4 per cent on average each year. Combined with low growth of population projected at about 0.6 per cent annually, this could yield a high growth in per capita real incomes not much lower than the rate which was typical in the years up to 1973.

In order to achieve this, an orderly economic structural transformation will be necessary with high investment in the energy sector to ensure a big reduction in dependence on conventional oil. This is projected to fall from 37 per cent in 1990 to only 19 per cent in 2000. Even with non-conventional oils added in, total oil supplies and demand are projected at only some 25 million b/d in 2000 compared with over 40 million b/d in 1979.

Solid fuels are projected to be the pre-eminent energy form, accounting for 33 per cent of the total balance in 2000, and for the whole of the 10 million b/d oil equivalent net rise in energy consumption through the 1990s. Nuclear energy is expected to double from about 9 million b/d oil equivalent in 1990 to 18 million b/d oil equivalent in 2000 to offset in large measure the decline in conventional oil consumption foreseen over this period. The volume of natural gas supply and demand is expected to show a negligible change during the 1990s.

Neutral Scenario

The growth of energy demand is likely to be about 0.6 per cent in the ten years up to 2000. With continuing improvements in the efficiency of use expected, implying an energy/GNP growth coefficient of 0.2:1, the economies of the industrialized countries could grow by an average 3 per cent during the 1990s. This is marginally better than the average rate achieved from 1973 to 1979 which was associated with average growth of 1 per cent each year in energy consumption. Economic conditions in the OECD countries should be tolerable as per capita incomes continue to rise significantly in real terms, though the contrast with the much greater rate of progress achieved over the twenty-five years culminating in 1973 would be strong and the comparison unfavourable.

One of the major problems will be in managing the transition away from oil dependence. Even counting in non-conventional oils, supplies are projected at only some 20 million b/d, representing 22 per cent of the energy balance in 2000. In these conditions, oil use will be confined to the premium market sectors and significant further improvements in the efficiency of oil use will necessarily occur even in those in order for this rather dramatic change to be realized.

Among the other energy sources, a fall of some 2 million b/d oil equivalent is foreseen in natural gas use during the 1990s, attributable mainly to a declining reserve base in the USA. Use of solid fuels is projected to grow by 8.5 million b/d oil equivalent over the period, and nuclear energy by 7.5 million b/d oil equivalent, whilst total energy consumption rises by only some 5.5 million b/d oil equivalent.

Confrontation Scenario

This scenario offers a rather bleak prospect through the 1990s. Energy demand grows by a negligible 1.5 million b/d oil equivalent during the decade. Economic growth is unlikely to exceed 1.5 per cent annually on average. Economic and political instability is quite likely to occur in these conditions, particularly if actual experience during the 1980s was following the development of this scenario rather than either of the other two. Economic depression on the scale of that which occurred in the early 1930s is not inevitable but seems quite possible under the conditions hypothesized. There could be a threat to the existing political and economic institutions, and some risk that protectionist policies might be seen as the only solution in an increasing number of countries.

A dramatic fall in oil availability will be accompanied by a further significant rise in its real price in this scenario. The change in the energy mix alone might verge on the unmanageable as total oil use, including non-conventional oils, drops to what may appear to be an impracticably low volume of about 13 million b/d in 2000 from 24 million b/d ten years earlier. However, due to the lack of growth in energy demand this would still imply that oil accounted for over 17 per cent of energy demand in the horizon year. A review of premium market sector requirements and probable structural shifts in the world economy plus substitution possibilities and improvements in efficiency suggests that this scenario would be feasible, though political and social conditions might deteriorate. Almost complete independence from OPEC would have been bought at a very high price.

Solid fuels are projected to account for 39 per cent of the energy balance in industrialized countries in the year 2000 in this scenario. Natural gas would account for 17 per cent and nuclear energy for 16 per cent of the total energy mix, in spite of relatively modest growth of the latter.

COMMUNIST BLOC

Co-operation Scenario

The growth of energy consumption is projected to continue to decelerate during the 1990s to an average of about 3.3 per cent each year in this scenario. This is significantly higher than the rate expected in the non-Communist countries, even though the growth of population is likely to be considerably lower.

Though there is unlikely to be any energy supply constraint for the centrally planned economies as a whole in this scenario, the transition away from oil dependence is rather sharp: from 24 per cent of the energy mix in 1990 to 16 per cent in 2000, though the decline is only 1 million b/d over the ten years. This is projected to reflect a decline in production, so that the Communist bloc as a whole has no net oil exports or imports either in 1990 or in 2000 in this scenario.

Growth of energy consumption comes mainly from a 10 million b/d oil equivalent increment in solid fuels over the ten years and nearly 9 million b/d oil equivalent increase in natural gas usage. Natural gas exports to non-Communist countries should be maintained at about 1.5 million b/d oil equivalent through the 1990s in this scenario.

Neutral Scenario

The outlook is for energy consumption to grow at an average annual rate of 2.5 per cent in the 1990s, down from 3.2 per cent in the preceding five years. Since the average rate of growth of population is likely to be only 1 per cent each year this should make it possible to improve per capita incomes.

Conventional oil consumption is projected to fall 1.7 million b/d over the decade but this decline would be offset in part by a small quantity of non-conventional oil. Oil imports would rise to 1.3 million b/d in 2000 from a projected 1.0 million b/d in 1990 in order to obviate a sharper decline in consumption.

Of the 18 million b/d oil equivalent increase in energy consumption projected for the 1990s, solid fuels would account for some 8.5 million b/d oil equivalent and both natural gas and nuclear for a further 4.5 million b/d oil equivalent each in this scenario. In the year 2000 natural gas is projected to account for 25 per cent of energy demand, up from under 20 per cent in 1979.

Confrontation Scenario

The deceleration of energy consumption growth through time is also a feature of this scenario in the Communist bloc countries. The annual rate of increase is projected at 1.6 per cent in the 1990s after 2 per cent in the preceding five years. It remains significantly higher than the 1 per cent annual average projected for the non-Communist countries over the same ten year period.

Oil production in this scenario is projected to fall to 10 million b/d in 2000 from 12.5 million b/d in 1990. Thus even though oil consumption declines by more than 2 million b/d, net oil imports rise to 1.8 million b/d to balance demand. This will bring further pressure to bear on high priced and scarce world oil supplies which are the predominant feature of this scenario.

The use of solid fuels continues to grow in the 1990s in this scenario, but only at a modest 1.0 per cent annual rate. Natural gas should make the largest single increment to rising energy demand, increasing at an annual average of nearly 3 per cent so that it accounts for 28 per cent of energy consumption in 2000. The share of solid fuels would actually fall to little more than 40 per cent from 48 per cent in 1979.

References

(1) Wilson, Carroll L., (Project Director). Energy: Global Prospects 1985 - 2000. Report on the Workshop on Alternative Energy Strategies (WAES), a project sponsored by the Massachusetts Institute of Technology. McGraw-Hill, May 1977.

(2) United Nations Population Statistics and Forecasts, 1950-2000. Selected Demographic Indicators, 1980.

16
THE US ENERGY SITUATION

THE US ENERGY PROBLEM SUMMARIZED

The US energy problem was clearly identified by the US Federal Energy Administration in its 1975 annual report. The same point was made by President Carter in presenting his national energy programme in April 1977, to which reference is made later in this chapter.(1) Among the fossil fuels, coal accounts for a very high proportion of available US resources, but for a very low proportion of the United States' use of fossil fuels. At the other extreme, oil accounts for a remarkably small proportion of US fossil fuel resources, but for almost half the fossil fuels used in the world's largest economy in 1979. It is hardly over-dramatizing the situation to say that the medium term solution to the world's energy problem might be found largely in an acceleration of the trend away from dependence on oil and gas to meet demand for primary energy in the United States. Together oil and gas still accounted for well over 70 per cent of the total in 1979, a decrease of some 4 percentage points since 1973. For oil alone, the proportion was actually higher in 1979 than in 1973. Oil imports, as the balancing form of energy, rose even more, by 4 percentage points, to account for well over 20 per cent of US energy demand in 1979 (see Table 12). This was in spite of President Nixon's call in November 1973 for the USA to become self-sufficient in energy by 1980, the growth of Alaskan North Slope production and the apparent peaking of oil consumption in 1978.(2)

RESERVE BASE (Table 11)

The United States of America is endowed with vast energy resources.

Estimated proved reserves of conventional fuels are about 132 billion tons oil equivalent - 93.1 per cent coal, 3.6 per cent natural gas and 3.2 per cent oil.(3)

Domestic uranium reserves are estimated to be sufficient to support the fuel requirements of the existing *nuclear power plants and those licensed for construction for about 75 years.(1)

---

*Light water reactor technology.

Among non-conventional energy resources, oil shale proved reserves are estimated to be sufficient to yield some 28 billion tons recoverable oil(3), that is larger than Saudi Arabia's reserves of conventional crude oil. Estimated additional resources are more than eight times as large as US proved reserves of oil shale. This is but one of many nice examples which could be quoted to show that the world generally and the major consuming countries are not short of energy resources, but simply of energy which can be exploited at anywhere near the cost of conventional crude oil in the past.

SUPPLY AND DEMAND (Table 12)

During a period of six years up to and including 1973, the consumption of total primary energy in the US was growing at an average rate of 3.8 per cent per year. Corresponding percentage growth rates for individual energy sources were: oil 5.4; natural gas 2.7; coal 1.4; nuclear 48.9 and hydroelectricity 4.2.(4)

As a result of the economic recession and various energy conservation measures implemented following the OPEC oil price adjustments of 1973-74, the historic growth rate of total primary energy consumption in the USA dropped to an average of only 0.7 per cent per year during the period 1973 - 1979. The corresponding growth rate for oil fell to a level of only 0.9 per cent per year. This resulted in an increase in the oil share to 45.5 per cent of primary energy demand in 1979. Natural gas demand fell by some 1.5 million b/d oil equivalent over the six years to 1979, or at an average rate of 2.3 per cent annually, because of the decline in US production over most of the period. The annual growth rate of coal consumption increased to 2.3 per cent annually, or some 1 million b/d oil equivalent increment over the six years. Growth of nuclear and hydroelectricity decelerated in percentage terms, with respective additions of some 1 million b/d and 90,000 b/d oil equivalent to the US primary energy balance between 1973 and 1979. As a result of the accident at the Three Mile Island plant in March 1979, US nuclear electricity production in 1979 fell behind the level of the previous year for the first time in the industry's history.

Seventy-five years ago the USA depended on coal, the most abundant resource, for over 90 per cent of its energy needs.(5) This dependence was reduced to only 20 per cent in 1979, though this represented an increase from the lowest level of less than 18 per cent reached in 1972. The reasons are well-known. Coal's production and use creates environmental difficulties. Oil and gas are not only clean and versatile but they were conveniently available and low priced. Air pollution restrictions on coal and growing use of automobiles also contributed to constraints on its use. Consequently, dependence on oil and gas, the least abundant resources, increased at a rapid pace, reaching a level of about 77 per cent of energy demand in 1972, but declining subsequently to 72 per cent in 1979. Examined in this way, it may appear that an important turning point occurred in the US energy economy in 1972. In point of fact, though, dependence on oil alone reached a peak of 46.8 per cent of US primary energy demand as recently as 1977, the same year as US gross oil imports peaked at over 8.7 million b/d. The explanation is associated with the decline of US oil and gas production in the Lower 48 states, an inevitable consequence of low and declining reserves to production ratios, and lack of realistic pricing policies at that time.

US oil and gas production has declined from the peak levels achieved in 1970 - 1973. Many factors have contributed to this trend. Some significant factors are:

- production from older fields has peaked

- low, regulated prices encouraged consumption and discouraged the search for new oil and gas

- the availability of inexpensive imported oil until 1973 served as a disincentive to domestic production.

After reaching a peak in 1970 at some 11.3 million barrels per day (including NGLs), domestic oil production in 1979 had fallen to only 10.2 million barrels per day. In the meantime there had been a net increase of some 1.2 million b/d in Alaskan production.

So the first deliveries of oil from the Alaskan North Slope have arrested the decline during recent years. In fact, US oil production in 1979 was 4.9 per cent above the lowest level reached in 1976. But to put the Alaskan North Slope increase into the perspective of the decline in the Lower 48 states, it is necessary to make only two observations. The decline in total US oil production (inclusive of NGLs) was resumed in 1979. Total US oil production in that year was well over 1 million b/d less than in the peak year of 1970. This represented an average annual rate of decline of 1.1 per cent, or more than 2.5 per cent for the Lower 48 states alone.

Six states, Texas, Louisiana, Oklahoma, California, New Mexico and Kansas account for over 90 per cent of domestic natural gas production. Texas and Louisiana alone provide 73 per cent. Overall domestic production peaked in 1973, and has been declining ever since. Except for 26 trillion cubic feet discovered in Alaska in 1970, annual additions to natural gas reserves have failed to equal marketed production for the past several years.(5) Moreover, the Alaskan reserves will not provide significant amounts of gas until after 1985, due to the absence of necessary transportation facilities.

The Clean Air Act of 1977 calls for installation of the best available control technology in all major new coal-consuming units. It will result in somewhat stricter stack effluent standards for new boilers.(6) Regulatory decisions related to these and other provisions of the Act will have a major bearing on the future growth of coal consumption. Other important factors include major developments in mining and transportation facilities and expanded research for the development of various technologies to overcome coal's marketability problems, namely, liquefaction, gasification, stack gas desulphurization etc. Concern about safety and security, the problem of nuclear waste, intricate licensing procedures resulting in long lead times, and high and increasing capital costs continue to hamper the growth prospects of nuclear energy, quite apart from specific shocks exemplified by the accident at the Three Mile Island nuclear plant in March 1979. This served to emphasize the downside risk associated with the nuclear contribution to the energy balance.

Early leases of shale oil reserves in the western part of the USA attracted considerable attention and, as recently as 1976, it was expected that shale oil production might exceed one million barrels per day by 1985. However, more recently, pessimism about the magnitude of the physical and environmental problems (impact of mining, water requirements and waste disposal) associated with this level of production, and upward revisions of investment estimates have called into question the economic viability of this alternative energy source.

Estimates of the cost of shale oil production in 1980 dollars range from about US $15 to $35 per barrel and up, using either strip or deep mining, along with current retorting technology.(7) The combined effects of inflation in capital equipment and a recurring tendency for custom-engineered capital projects to overrun their original cost estimates by sizeable margins have led to rapidly escalating new cost estimates for synthetic fuel projects. A three company consortium project that was to go forward was shelved because of cost escalations. However, there is still optimism that a breakthrough might be achieved in in-situ recovery technology. This might bring shale oil costs down to a level consistent with the expected rise in real prices for oil entering world trade.

Until the late 1960s the USA was essentially independent of foreign oil supplies.(5) The country produced and consumed more oil than any other country. Its domestic supply was considered plentiful. However, the combination of decreasing proven reserves, a decline in domestic oil and gas production and rising demand led to rapid growth in oil imports. These increased from 2.5 million barrels per day in 1967 to a peak of more than 8.7 million barrels per day in 1977, an average annual growth rate of over 13 per cent. Consequently, dependence on oil imports increased sharply from 18 per cent in 1967 to 49 per cent in 1977, before falling to 47 per cent of total oil supplies in 1979 and considerably less in 1980. These particular estimates are for gross rather than net oil imports.

Natural gas imports, in the past, have come almost entirely overland: from Canada and, to a very small extent, from Mexico.(6) These imports constitute an important source of natural gas for regions like the Pacific North West in the USA.

Energy production, consumption and net imports in 1980 show the emergence of some apparently healthy trends. Though there was only a 1.5 per cent increase in production from 1979, a 3.5 per cent decline in consumption resulted in a decline of nearly 28 per cent in net imports of energy into the USA. Oil and coal production went up, but output of all other forms of energy was lower. The decline in oil consumption was more than that for all forms of energy. Only coal consumption went up. Net imports of oil declined 21 per cent and natural gas imports by slightly more than this, while net exports of coal increased by over 40 per cent.(2)

US official estimates put these trends in net trade in perspective. US net energy imports, expressed as a percentage of total energy consumed accounted for only 15.8 per cent in 1980 compared with 23.6 per cent in 1977. But the rise in the two preceding years was relatively sharper, from 16.6 per cent in 1975. Between 1975 and 1977 there was a negligible growth in US energy production while the real growth of the US economy was 5.4 per cent in 1976 and 5.5 per cent in 1977.(2)(8) Net energy imports rose sharply in consequence.

Although a short but intense recession was one reason for the sudden drop in petroleum demand during 1980, the biggest factor was another marked improvement in the energy efficiency of the US economy. All the signs are that this trend is continuing in 1981, following the deregulation of domestic crude oil prices. In 1979 economic growth was about 3 per cent yet energy consumption rose only about 1 per cent. In 1980, with the gross national product flat, energy consumption declined by 3.5 per cent and the share of oil in total energy consumption declined by 2.1 percentage points in just one year.(2)

The real test for the US energy economy and the conservation savings made will come if there is a repetition of the circumstances which prevailed from 1975 to 1977. Such a combination could provide a temporary disturbance to the assumptions underlying many forecasts, namely that there will be continuing improvements in efficiency of energy use bringing declines in oil consumption and imports. The initiative already taken to deregulate domestic crude oil prices and the implementation of an acceleration in the timetable to deregulate natural gas prices by 1985 may contribute much to a healthier US energy balance capable of obviating the sudden US pressure on the world oil balance seen previously.

EVOLUTION OF A COMPREHENSIVE NATIONAL ENERGY POLICY

Price Controls on Natural Gas and Oil

The gas price regulation came into effect as a result of the 1954 Supreme Court interpretation of the 1938 Natural Gas Act. This interpretation stated that the Act required cost-based well-head price regulation. By that decision, the Supreme Court divorced gas pricing from the prices of competing fuels in the market-place. This assured protection of the gas consumers from the escalating world oil prices of the 1970s, unless Congress explicitly deregulated gas prices.(9)

Oil price regulations came later. In 1971 President Nixon introduced temporary wage and price controls, including controls on crude oil and petroleum product prices, as anti-inflation measures. Congress extended the temporary oil controls in its emergency legislation at the time of the oil embargo of 1973 and placed them under the newly-formed Federal Energy Administration. Later, when temporary controls were due to expire, Congress again renewed them.(9)

These restrictive price controls resulted in the generation of a wide range of adverse trends. They held US oil and gas production below their potential, stimulated energy consumption artifically, stifled technological breakthroughs for the development of alternative energy sources and led to increased oil imports.

However, any effort to deregulate oil and gas prices encountered powerful opposition from the consumer advocates. These had the support of a large segment of the American public who had come to take cheap and plentiful energy for granted. Suggestions of an approaching oil or gas shortage were generally believed to be oil industry propaganda in the cause of higher profits.

Environmental Regulations and Health and Safety Controls

The momentum of environmental, health and safety movements was built up in the sixties by the attention given to issues such as the 'smog' problem in Los Angeles, the effect of pesticides on birds (Rachel Carson's 'Silent Spring'), the Santa Barbara Channel oil spill, cigarette smoking, car exhaust emissions and safety equipment, drugs, occupational health and safety, nuclear plant safety and the hazards of waste disposal. Much of the major environmental legislation and health and safety controls was in place by the early 1970s.(9)

This legislation and such controls have now become powerful obstacles to the development of environmentally disruptive or risky sources of energy such as coal, oil shale and nuclear.

Energy Policy - Basic Issues

The major issues involved in the ongoing energy policy debate in the USA since 1973 can be summarized as follows:

- Conservation measures,

- Oil and gas price deregulaton and income distribution,

- Environmental, health and safety risks in energy development,

- The extent of government involvement in energy industry development and control,

- Foreign policy concerns. (President Carter suspended nuclear fuel reprocessing and the breeder reactor out of concern for nuclear proliferation. The supply security issue has led to the development of the strategic petroleum reserve programme.)

Energy Legislation

Since 1973, most of the efforts to devise a national energy policy have required new legislation. It was not until the Carter presidency that the USA was able to achieve some measure of success in this regard.

The United States is a federal democracy with a written constitution which was carefully designed to limit the exercise of power by the executive, the legislative and judicial branches of government through a system of 'checks and balances'. In addition, substantial power is reserved for the states which can often ignore Federal policy-making with impunity. The President of the United States can be powerless when he does not have a clear legislative mandate to act. He can attempt to persuade Congress to provide him with the legislation he needs, but he has no guarantee that Congress will agree with his proposed solutions to problems. And in this effort to get legislative approval for the exercise of Presidential power, the courts always stand ready to judge a particular law or Presidential action unconstitutional, thereby nullifying it.(9) Thus compromise is usually an essential pre-condition for the acceptance and implementation of policy proposals formulated in the White House.

After the previous failures of Republican Presidents with hostile Democratic Congresses President Carter took the initiative of formulating a comprehensive national energy plan. This was at a time when oil imports were increasing rapidly to meet rising US energy demand.

In his speech of 20 April, 1977, presenting a comprehensive national energy programme for the United States, President Jimmy Carter explained his country's energy problem in the following words:

> "The heart of our energy problem is that our demand for fuel keeps rising more quickly than our production, and our primary means of solving this problem is to reduce waste and inefficiency.
>
> Oil and natural gas make up 75 per cent of our consumption in this country, but they represent only about 7 per cent of our reserves. Our demand for oil has been rising by more than 5 per cent each year, but domestic oil production has been falling lately by more than 6 per cent.
>
> Our imports of oil have risen sharply - making us more vulnerable if supplies are interrupted - but early in the 1980s even foreign oil will become increasingly scarce. If it were possible for world demand to continue rising during the 1980s at the present rate of 5 per cent a year, we would use up all the proven reserves of oil in the entire world by the end of the next decade.
>
> Our trade deficits are growing, we imported more than 35 billion dollars worth of oil last year, and we will spend much more than that this year. The time has come to draw the line.
>
> We could continue to ignore this problem - but to do so would subject our people to an impending catastrophe . . ."(1)

President Carter's proposed comprehensive National Energy Program was based on the following stated national energy policy principles, strategies and goals.(10)

Principles

The national energy plan is based on ten fundamental principles:

- The USA can have an effective and comprehensive energy policy only if the government takes responsibility for it and if the people understand the seriousness of the challenge and are willing to make sacrifices.

- Healthy economic growth must continue. Only by saving energy can the USA maintain her standard of living and keep her people working.

- The USA must protect the environment. US energy problems have the same cause as her environmental problems - wasteful use of resources. Conservation helps the USA solve both at once.

- The USA must reduce her vulnerability to potentially devastating embargoes. The USA can protect itself from uncertain supplies by reducing her demand for oil, by making the most of her abundant resources such as coal, and by developing a strategic petroleum reserve.

- The USA must be fair. Her solutions must ask equal sacrifices from every region, every class of people, every interest group. Industry will have to do its part to conserve, just as consumers will. The energy producers deserve fair treatment, but the government will not let the energy companies profiteer.

- The cornerstone of US policy is to reduce demand through conservation. The emphasis on conservation is a clear difference between this plan and others which merely encouraged crash production efforts. Conservation is the quickest, cheapest, most practical source of energy.

- Prices should generally reflect the true replacement cost of energy. The USA would only be cheating herself if it makes energy artificially cheap and uses more than it can really afford.

- Government policies must be predictable and certain. Both consumers and producers need policies they can depend on so that they can plan ahead.

- The USA must conserve the fuels that are scarcest and make the most of those that are more plentiful. The USA cannot continue to use oil and gas for 75 per cent of its consumption when they make up only 7 per cent of her domestic reserves. The USA needs to shift to plentiful coal while taking care to protect the environment, and to apply stricter safety standards to nuclear energy.

- The USA must start now to develop the new, unconventional sources of energy it will rely on in the next century.

Strategy

The objectives of the national energy plan were related to time horizons:

- In the short term, to reduce dependence on foreign oil and to limit supply disruptions.

- In the medium term, to weather the eventual decline in the availability of world oil supplies caused by capacity limitations.

- In the long term, to develop renewable and essentially inexhaustible sources of energy for sustained economic growth.

The major strategies for reaching these objectives were listed as follows:

- The implementation of an effective conservation programme for all sectors of energy use so as to reduce the rate of demand growth to less than 2 per cent per annum, thereby helping to achieve both the short and medium term goals.

- The conversion of industry and utilities using oil and natural gas to the use of coal and other more abundant fuels, reduction of imports, and making natural gas more widely available for household use, thereby helping to achieve both the short and medium term goals.

- A vigorous research and development programme to provide renewable and essentially inexhaustible resources to meet United States' energy needs in the next century, thereby helping to achieve the long term goal.

National Energy Goals

The President proposed, as part of his comprehensive energy legislation, the following energy goals to be achieved between 1977 and 1985. The Congress was requested to support these goals by enacting a joint resolution of the Senate and House committing the nation to:

- Reducing annual growth of United States energy demand to less than two per cent;

- Reducing oil imports from a potential level of 16 million barrels a day to less than 6 million barrels daily, about one eighth of total energy consumption;

- Achieving a 10 per cent reduction in gasoline consumption;

- Insulating 90 per cent of all residences and other buildings;

- Increasing coal production on an annual basis by at least 400 million short tons;

- Using solar energy in more than two and a half million homes.

President Carter's approach to price regulation in the first phase of his National Energy Plan was complex. Oil and gas price regulation (including control of previously unregulated intra-state gas) was to be continued, albeit on more liberal terms. Residential users of gas were to continue to benefit from price controls, but industrial users were subjected to an industrial use tax to force conservation. A Crude Oil Equalization Tax (COET) was designed to capture the economic rent for the Government between controlled prices and world oil price levels.(9)

Congress struggled for a year and a half before it gave him, instead, a long term and complex phased deregulation of natural gas, and no action at all on COET, or for that matter any price legislation on oil. Since the oil price controls were temporary, Carter could in theory have let them expire when he could not get his preferred approach to oil pricing. The message from Congress, however, had been that temporary controls could be continually renewed until something acceptable was proposed.(9)

The National Energy Act was passed by the US Congress on October 15, 1978. President Carter's original proposal was a five part package designed to bring about energy savings equal to 4.5 million barrels of oil equivalent per day by 1985.

After the many changes Congress made in the original proposal, the final Energy Act was estimated to bring about savings of only 2.4 to 3.0 million barrels of oil equivalent per day.

The National Energy Act consists of the following five bills:(11)

- National Energy Conservation Policy Act of 1978.
- Power Plant & Industrial Fuel Use Act of 1978 (Coal Conversion).
- Public Utilities Regulatory Policy Act, 1978.
- Natural Gas Policy Act of 1978.
- Energy Tax Act of 1978.

According to the US Department of Energy, the five bills are estimated to contribute the following energy imports saving in 1985.(11)

Conservation	Energy Imports Savings in Thousands of US barrels/day
Building Conservation Programmes/Appliances	410
Auto and Truck Standards	265 (a)
Utility Rate Reform	0 - 160
Natural Gas Pricing	1000 - 1400 (b)
Coal Conversion	300
Energy Taxes:	
Residential Tax Credits	225
Oil and Gas User Tax	-
Gas Guzzler Tax	80
Business Energy Credits	110
TOTAL	2390 - 2950

(a) Assumes EPCA penalties are increased to full extent permitted and completion of administrative action to implement truck standards.

(b) The range of these estimates depends on the degree of oil displacement which occurs as a result of increased gas supply. Increases in gas supply may displace LNG, propane and butane as well as oil. The price displacement ratios will depend upon world oil prices.

In the second phase of his National Energy Plan President Carter announced that he would use his authority to let crude oil controls expire, but only if Congress gave him a windfall profits tax. This tax was actually an excise tax since it would tax away much of the increment between previously regulated crude prices and the world price. Any efforts to apply the more punitive gasoline taxes common in many other countries ran foul of consumer and Congressional resistance. Carter's complicated 'import fee' proposal during 1980 was overturned by Congress just before extended Court tests of its constitutionality.(9)

Carter's Plan II announced in July 1979 proposed to reduce US oil imports by 4.5 million b/d in the coming 10 years and to finance a massive US $142 billion conservation and synthetic fuels programme with revenues from the proposed excise tax on decontrolled crude oil. The heart of this programme was the creation of an Energy Security Corporation (ESC) with unprecedented power and authority to spend $88 billion during 1980 - 1990 to boost development of 2.5 million b/d of substitute fuels.(12)

Other elements of the programme were:(12)

- To establish a three member Energy Mobilization Board (EMB) empowered to expedite permitting and construction of critical energy facilities.

- To provide new incentives for development of heavy oil, unconventional gas and oil shale.

- To require utilities to cut current oil consumption by 50 per cent saving up to 750,000 b/d of oil.

- To set up a major new residential and commercial conservation programme designed to save 500,000 b/d of oil by 1990.

- To provide $2.4 billion a year in aid to low-income families.

- To provide a total of $16.5 billion during 1980-90 for improvements in the nation's mass transit system and in automobile fuel efficiency.

After Senate clearance in March 1980, Carter signed the US $227.7 billion excise tax on revenues from phased decontrol of US oil production.(13)

Carter wanted the revenues placed in a trust fund for development of a synthetic fuels industry and tax credits for energy production and conservation. But Congress chose to place the tax revenues in the general fund and adopted a guideline which allotted only 15 per cent for production/conservation.(13)

The bill levied taxes only on producers' extra income resulting from crude price decontrol, but when combined with existing royalties and income taxes, it would leave producers with only about 18 per cent of the estimated $1 trillion to be generated from decontrol. The tax was to start phasing out in 1988 or when the $227 billion was raised, whichever was later. If the target amount were not to be reached by January 1991, phase-out would begin anyway.(13)

Later, during 1980, Congress also cleared the bill on the creation of 'The Synthetic Fuels Corporation' but rejected the 'Energy Mobilization Board' bill.

Carter was aware that the obstacles to development were of three types - overly restrictive regulations, unfavourable project economics, and the stalemate created by conflicting approval jurisdictions. He recognized that tighter environmental and safety restrictions on new energy supply projects could be accommodated at a cost; and if Federal financing and R & D support could make the difference, the President with a willing Congress could provide the stimulus to supply development by means of a Federal funding bill. This approach was one of the centre-pieces of his second phase plan. It is also a fact that favourable economics alone - with or without Federal subsidy - may not be enough to permit development if the existing dispersion of authority at local, state or Federal bureaucratic levels obstructs individual projects. Attempts to concentrate power at the Federal level may be needed if development is not to become hopelessly bogged down among conflicting jurisdictions.(9)

Carter's proposed Energy Mobilization Board was an attempt to systematize such a Federal expediting function. However, Congress, in an election year, was not yet prepared to accept the creation of such an authority.(9)

Latest Developments

The immediate decontrol of oil prices in the USA, effective 28 January 1981, only a week after President Reagan took office, underscores his Administration's aim to deregulate the energy industry and let free-market forces determine price, supply and demand. It assumes oil price decontrol will spur greater production, further curb consumption and result in cutting oil imports.

An early assessment was that the United States energy policy was likely to take a more free-market, supply-oriented turn following the landslide victory of Republican Ronald Reagan in the presidential election.(14) Although such a trend began to surface during the Democratic administration, Reagan's current energy positions are dominated by this orientation. In the long campaign, his statements on energy issues emphasized "less government", less regulation, less taxation - and more supply rather than less demand.

The startling Republican resurgence in Congress promises a more favourable reception for a Reagan energy policy than almost anyone expected. Republicans won almost their first majority in the Senate in 26 years and made significant gains in the House, though the Democrats maintained a majority there. On some important issues such as the Budget, Conservative Southern Democrats have voted with the Republicans to secure a majority in the House as well as the Senate. Such a tactic may be an effective route for making progress with energy legislation as well, particularly on contentious issues such as natural gas price deregulation.

The complexities of the energy issues and the need to realize promised tax cuts and higher defence spending are expected to temper Reagan's energy policy ideas with pragmatism. In addition, a conflict may arise between his free-market principles and his desire to increase supplies. A push for supplies from new technologies may lead to greater rather than less government involvement, even though Reagan is a strong proponent of private sector enterprise. A decision to withdraw Federal funds from an international project (with Japan and West Germany) for solvent refined coal suggested that principles and budgetary constraints were still the essential determinants of policy five months into the new Reagan Administration's term of office. Earlier evidence of the same sort was the decision to reduce funding for Carter's nascent Synthetic Fuels Corporation.

On important energy issues, Reagan's public positions have included:

- Decontrol oil prices and eliminate mandatory crude and product allocation rules.

- Speed up decontrol of natural gas. Although many in the new Congress may favour overhauling the current timetable (decontrol by 1985), it may not be easy.

- Modify the 'windfall profits' tax. Although originally against the tax, Reagan may need the revenue for his promised general tax cut. Exemptions may be limited to small oil producers and royalty owners.

- Relax some environmental standards to ease the way for increased energy supplies from coal, refinery expansion, synfuels and other new technologies.

- Reduce, if not dismember, the Department of Energy. Congress is not likely to abolish the Department - only limit its regulatory power.

- Revitalize the nuclear power industry. Congressional sensitivity to widespread opposition to nuclear power expansion, plus decreasing demand from utilities, may, however, continue to restrain its development.

- Expand the leasing of Federal lands, including offshore areas. Reagan would be able to meet this campaign promise through existing law.

The US National Energy Policy has made substantial progress on matters related to prices and conservation. The energy supply uncertainties are still largely unresolved, though.

The Reagan Administration's Energy Policy Plan announced in July 1981 examines the Federal government role in the market. This includes the following statement:

> "Increased reliance on market decisions offers a continuing national referendum which is a far better means of charting the Nation's energy path than stubborn reliance on government dictates or on a combination of subsidies and regulations."(15)

It goes on to mention the President's early action to end oil price controls and to point out that enabling citizens, businesses and state and local governments to make rational energy production and consumption decisions represents a radical departure in approach from the prevailing policy instituted after the first shock of rapid oil price increases in 1973 and 1974.

This latest document recognizes that oil imports have decreased substantially. But even though efficient displacement of imported oil is an important objective, achieving a low level of US oil imports at any cost is not a major criterion for the nation's energy security and economic health. Even at its current high price, imported oil in some cases is substantially less expensive than search for available alternatives, according to the official text.

It proceeds to point out that public spending is appropriate (and will continue) in long term research with high risks, but potentially high payoffs. In most cases, however, using public funds to subsidize either domestic energy production or conservation buys little additional security and only diverts capital, workers, and initiative from uses that contribute more to society and the economy.

The most direct impact of Federal government on the energy future of the USA is identified as arising from its position as steward of the outer Continental shelf and Federal lands comprising one third of the total area of the country. These lands are estimated to contain 35 per cent of US coal, 40 per cent of natural gas and as much as 85 per cent of the nation's oil. The Federal role is said to be to bring these resources into the energy market-place while simultaneously protecting the environment.

The uncertainties of future energy consumption are summarized in the form of levels in the year 2000. The annual average compound rates of growth implied over the next twenty years range from little more than zero up to 2.3 per cent from the depressed level recorded in 1980. US oil imports are projected at 6 million b/d in 1985, 5 million b/d in 1990 and only 1.5 million b/d by the year 2000 in this most recent official document.(15)

ENERGY PROJECTIONS

Co-operation Scenario

To anyone familiar with the vast number of energy and oil forecasts made for the USA every year, the estimates included even in this optimistic scenario may come as a considerable surprise. This is mainly because of the rapidity of the decline in the oil share of the energy consumption mix after 1990. Perhaps it is necessary to inform some readers that some, though not all, long term forecasts of US oil and energy prospects are made for the USA alone, without due regard for the evolution of the oil and energy balance worldwide. A specific feature of this balance is the probable evolution in production and consumption of oil in the oil exporting countries which is discussed in Chapter 19. Another consideration is the pressure on world oil supplies from non-OPEC developing countries reviewed in Chapter 20. Finally, there are the prospects for oil production and consumption in the USA.

On the supply side I have felt bound to take a relatively pessimistic view about indigenous production in spite of the continuing optimism still to be found among well-informed sources and the early initiative of the new Reagan Administration to complete the deregulation of domestic crude oil prices almost immediately on taking office in January 1981. The fact is that although Federal lands and offshore discoveries of a significant size may be made, the geologically interesting parts of the Lower 48 states have been very extensively drilled already; oil production peaked as long ago as 1970, in spite of the additional 1.5 million b/d or so from the Alaskan North Slope; and the reserves to production ratio for conventional crude oil and natural gas liquids together had fallen to less than 9:1 by the end of 1980.(2)(16)

On the demand side very significant improvements have already been made in the pre-eminent oil using sector, namely the US automobile. As smaller and more fuel-efficient cars started to be introduced they began to have an increasingly significant impact on total US gasoline consumption which may have reached an all time peak in 1978. Further progress is being made each year with newly produced automobiles. Even after there is a slow-down in the rate of improvement, there will continue to be some reduction in gasoline use per million miles driven as less efficient automobiles are phased out of use. Significant improvements will be made in oil dependent sectors such as air transportation, while there remains considerable scope for substitution in most other sectors under the impetus of oil product prices derived from conventional crude rising relative to other forms of energy.

This rather long preamble has been given because of the need for and likelihood of reductions in US oil consumption to be effected over the period to 2000 in order to maintain a reasonable global oil market equilibrium. A rather dramatic decline in oil's share of the energy mix occurs in the great majority of countries, so that the world's largest user and importer will necessarily play a large role in achieving a well-balanced global equilibrium.

In 1979 the USA accounted for about 28 per cent of the world's oil consumption but for only 5 per cent of the world's population. In the year 2000 in this scenario the respective figures are likely to be a little more than 12 per cent of oil consumption and a little more than 4 per cent of the world population.(17) This latter proportion is almost the same as the US share of the world's remaining proved oil reserves at the beginning of 1981, according to Oil and Gas Journal estimates.(18) Thus US per capita oil consumption is likely to fall from just over 5.5 times the world average to about 3 times the world average in this scenario.

Table 46 in Part III of this book contains the statistical estimates of the prospective US consumption mix on which this chapter is based. It shows that in this scenario US oil consumption could rise slightly between 1979 and 1985, even though a rather dramatic decline of 8 per cent occurred in 1980 and a further fall seems probable in 1981. Thus a rise to another peak in about 1985 seems likely to depend on a strong and sustained cyclical economic recovery. Another factor influencing US oil consumption over this period will be the extent of the incremental contribution from other forms of energy, particularly coal and nuclear. The third important influence will be the way the energy/GNP growth ratio moves in conditions of strong economic growth. One can notice here that the experience of the 1976 to 1979 period was relatively encouraging, and that is taken into account in the projections. After about 1985, US oil consumption is likely to fall absolutely, even in this scenario, at a rate of some 700,000 b/d annually over the fifteen years to 2000. This rate of decline implies a sharp drop in the conventional oil share of energy consumption, from 43 per cent in 1985 to 33 per cent in 1990, and to what seems a dramatically low 16 per cent in 2000. However, by that time shale oil and synthetic oil from coal, etc, are likely to be making a significant contribution. If these are added in, the total is likely to be about 10.5 millon b/d, or some 22 per cent of US energy consumption in 2000.

Even in this scenario US energy consumption is projected to grow at an average rate of 1.6 per cent annually between 1979 and 1985, falling to about only 0.5 per cent on average between 1985 and 1990, because of supply constraints. If energy consumption grows less than projected in the earlier period, it might be able to grow faster in the late 1980s. Energy consumption will then grow at about 1 per cent on average through the 1990s. In order to sustain this increase, solid fuel use is likely to rise annually by at least 500,000 b/d oil equivalent on average over the twenty-one years from 1979 to 2000: less than this in the earlier years, but more later in the period. Nuclear energy is also projected to supply increasing increments with the passage of time and to account for nearly 8 million b/d oil equivalent by the year 2000. This is likely to be about one sixth of US primary energy consumption.

Neutral Scenario

The main feature of this scenario is that US oil consumption is projected never again to reach the peak of 18.4 million b/d (net of processing gain) attained in 1978. The average rate of decline in conventional oil use is projected to be about 1 per cent annually between 1979 and 1985, rising to about 5 per cent between 1985 and 1990, and then to about 6 per cent on average through the 1990s. Having regard to the fact that these falls are expected to be offset in part by a significant rise in synthetics and shale oil, this rather dramatic rate of decline can be put into the perspective of a similar rate of decline likely to characterize the period from 1978 to 1981. The fall over these three years will probably exceed 5 per cent annually. Moreover, during this period nuclear energy was not making a significant incremental contribution to US energy consumption because of the accident at Three Mile Island.(2) From the relatively low level of nuclear energy input to the US economy in 1979, it is projected to add more than 5 million b/d oil equivalent of increase by the year 2000 in this scenario.

The management of the transition away from US dependence on oil and gas is likely to depend heavily on the performance of coal production. Also important will be a readiness to burn it much more extensively in the industrial and electricity sectors, as well as the development of nascent gasification and liquefaction industries. Coal use is projected to grow at an average rate of 4.3 per cent from 1979 to the year 2000.

US natural gas use is projected to fall slightly between 1979 and 1985 and then to recover marginally by 1990 on the assumption that some incremental supply from Alaska will become available by that year to more than offset an expected decline in the Lower 48 states. There is also some possibility of increased supplies from Mexico and Canada, as well as LNG from such exporters as Algeria and Nigeria. Nevertheless, supply and demand for natural gas is projected to fall by about 2 million b/d oil equivalent through the 1990s on account of a probable sharp fall in production among the existing main gas

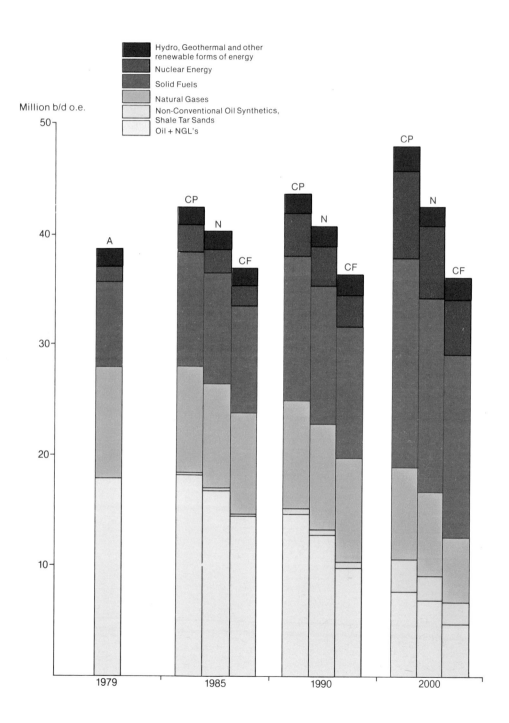

producing states. There must be some prospect that this gas supply scenario will prove to be over pessimistic if the US is able to find and recover natural gas from great depths. The current reserves to production ratio is less than 10:1 and appears to be a formidable constraint on significant increases in US natural gas production in the medium term.

Confrontation Scenario

In this case the US achieves effective energy independence by the year 2000 as oil consumption falls at a faster rate than conventional oil production and catches it up at a volume of barely 5 million b/d at the end of the twentieth century. A few years ago this sort of scenario might have been regarded as implausible. Now it seems more credible and likely, even if such a scenario must be a source of serious concern in the USA because of the implications for economic fluctuations and instability. Some will criticize the very low US oil production estimate used here. The answer to that has been discussed above. Though conventional oil and NGLs would account for little more than 13 per cent of the US energy consumption, the additional supplies of projected shale oil and synthetics would bring the proportion to nearly 19 per cent of the total. The recent decline in use of conventional oil products is certainly consistent with the average annual decline approaching 600,000 b/d to 1985, even allowing for some economic recovery.

Natural gas supply and demand is projected to fall more sharply in this scenario than in those discussed above, while nuclear and solid fuels are likely to grow more modestly. Therefore, the consumption of all forms of primary energy is projected to fall at an average rate of slightly over 0.5 per cent annually from 1979 to 1990. It should be remembered that in the year 1980 US energy consumption actually declined by 3.5 per cent while the level of US economic activity showed virtually no change from the year before. In fact, a trend analysis of the estimates of the relationships between US GNP and energy consumption from 1973 to 1980 suggests that if this is extrapolated through the 1980s then the US economy could grow at an average rate of 2.0 per cent annually from 1980 to 1985 and at 2.5 per cent annually from 1985 to 1990 with no growth in energy consumption at all Since the US population is expected to grow at an average rate of less than 1 per cent annually during the 1980s, this scenario seems to imply a strong need for effective political and economic measures to control a difficult market situation in the energy sector, rather than resort to inappropriate regulation.

Postulating a declining use of energy does suggest a rather uncomfortable period, even though the rate of fall is less than 250,000 b/d oil equivalent on average each year until 1990. This might be accompanied by political and social risks. It is to be expected that the Reagan Administration and its successor in 1985, together with the US Congress, will take national and international political decisions which they are uniquely well placed to do, and thus obviate the onset of the confrontation scenario conditions described in Chapter 14. I have in mind particularly the need for substantive progress in 1981 and 1982 on issues such as US relations with the Arab countries and Israel and the North-South dialogue. The latter is due to be resumed in October 1981. Progress on these matters has been less than satisfactory in recent years, mainly because there has been more partiality and less objectivity than one might have hoped for.

References

(1)   Carter, President Jimmy. A National Energy Plan for the USA, 20 April 1977.

(2)   US Department of Energy. Monthly Energy Review, June 1981.

(3)   World Energy Conference, Survey of Energy Resources, 1980.

(4) BP Statistical Review of the World Oil Industry, 1979 and earlier issues. British Petroleum Company Limited.

(5) US Federal Energy Administration. National Energy Outlook, 1976.

(6) World Energy Outlook. Exxon Background Series, April 1978. Exxon Corporation.

(7) Energy in Profile. Shell Briefing Service, December 1980. Shell International Petroleum Company Limited.

(8) US Department of Commerce. Survey of Current Business, December 1980.

(9) Jensen, James, T. Why US Oil Policy is such a Riddle. Petroleum Intelligence Weekly, 29 September 1980.

(10) US Official Text of President Carter's Energy Program -- A Factual Survey, United States Information Service, 22 April 1977.

(11) United States Explains Its New Energy Act. Petroleum Intelligence Weekly, Special Supplement, 30 October 1978.

(12) Bachman, W.A., Washington Editor. Carter Plan Places Emphasis on Synthetics. Oil and Gas Journal, 23 July 1979.

(13) Senate Clears Massive Crude Excise Tax. Oil and Gas Journal, 31 March 1980.

(14) Reagan Landslide Signals Shift in US Energy Focus. Petroleum Intelligence Weekly, 10 November 1980.

(15) US Department of Energy. Securing America's Energy Future. The National Energy Policy Plan. July 1981.

(16) American Petroleum Institute, 1981.

(17) United Nations Population Statistics and Forecasts, 1950-2000. Selected Demographic Indicators, 1980.

(18) Worldwide Oil and Gas at a Glance. Oil and Gas Journal, 29 December 1980.

17
THE WEST EUROPEAN ENERGY SITUATION

THE WEST EUROPEAN ENERGY PROBLEM SUMMARIZED

Western Europe imports more oil than any other region of the world - more than 13 million barrels a day in 1979 or 37 per cent of inter-regional world trade in oil. It is also relatively more heavily dependent on oil to meet its primary energy needs than is the USA. Oil accounted for 55 per cent of Western Europe's energy demand in 1979. Oil produced in the region represented some 16 per cent of consumption in that year.(1)

However, there have been some considerable changes in the West European oil and energy scene since 1973. In 1979 oil consumption was still some 375,000 barrels a day lower than it had been in 1973. Its net demand on world oil supplies fell by 6.5 percentage points between these two years to 18.6 per cent of the total in 1979. This change was attributable to an increase of 2.9 percentage points in Western Europe's share of world production over the six years combined with a decline of 3.6 percentage points in its share of world consumption. Its use of other forms of primary energy rose by an average of over 3 per cent annually from 1973 to 1979. Growth in use of primary energy per unit of GNP declined by an average of 1.5 per cent each year over this period. Notwithstanding the recession which affected most of Western Europe in 1975, the average rate of economic growth achieved was 2.5 per cent annually from 1973 to 1979, which, with the low population growth in the region, implies a growth of GNP per capita of about 2 per cent each year. The growth in the energy coefficient with respect to GNP averaged about 0.4:1 over these six years.(1)(2)

RESERVE BASE (Tables 13 and 14)

Western European countries differ widely in their energy resource endowments.

Oil and gas reserves are located mainly in the North Sea (the UK and Norway), and the Netherlands (gas reserves). Italy, West Germany, France, Denmark, Greece, Austria and Spain have smaller reserves.

Coal reserves are located mainly in West Germany and the UK. France, Spain, Belgium and Sweden have smaller reserves.

---

* Western Europe is virtually coterminous with OECD Europe (see Glossary).

Uranium reserves are located in France, Denmark, Portugal, Spain, Italy, Sweden and Finland.

Total estimated proved reserves of conventional fuels in Western Europe are estimated to be close to 59 billion tons oil equivalent - 89.7 per cent coal, 5.7 per cent natural gas and 4.6 per cent oil.(3)

Among non-conventional energy resources oil shale reserves (mainly in Scotland) are estimated to be 2.6 billion tons oil equivalent.

SUPPLY AND DEMAND (Table 16)

Over the six years culminating in 1973, consumption of total primary energy was growing at an annual average rate of 5.8 per cent, though consumption of coal was declining at 2.9 per cent each year, on average. Oil alone accounted for 80 per cent of the growth of 7 million barrels a day oil equivalent in energy consumption. Natural gas accounted for another 2 million barrels a day oil equivalent of the increase, so that oil and gas together accounted for more than the total increment in European energy consumption between 1967 and 1973.(1)

The annual average compound rates of change were as follows:

West European Energy Consumption

	1967 to 1973 annual average rate of change per cent
Oil	+8.3
Natural Gas	+29.2
Coal	-2.9
Nuclear	+12.5
Hydroelectricity	+1.3
Total Primary Energy	+5.8

As a result of the economic recession and the various conservation measures adopted in different Western European countries after 1973, the historic growth trend of total primary energy consumption dropped to an average of only 1.0 per cent per year. The rapid growth in oil consumption up to the year 1973 turned into an average annual decline of 0.6 per cent after 1973, though it rose at an annual rate of 2.2 per cent on average after 1975. The exceptionally high growth rate of natural gas consumption dropped to 6.2 per cent per year after 1973. On the other hand, the growth rate of nuclear power increased appreciably and the rapid decline in coal consumption slowed down considerably then increased sharply in 1979.

In 1979, total primary energy consumption in OECD Europe amounted to 1286 million tons oil equivalent - oil 55.1 per cent, natural gas 14.3 per cent, coal 19.5 per cent, nuclear 3.2 per cent and hydroelectricity 7.9 per cent.(1)

After many years of declining coal production in the cheap oil era, local production of energy within Western Europe began to increase with the development of Dutch natural gas, nuclear energy and North Sea oil and gas. These made a substantial contribution to the relief of pressure on world oil supplies in the 1970s. Together with the lower rate of economic growth and the improvements in the efficiency of energy use, these developments led to a fall in West European oil imports which averaged nearly 400,000 b/d annually between 1973 and 1979.(1)

EVOLUTION OF A COMPREHENSIVE ENERGY POLICY

During recent years, the European Economic Community (EEC)* Commission has put forward a number of policy guidelines designed to alleviate the short, medium and long term energy related problems of member countries. The ten countries which are now members of the EEC constitute the major part of Western Europe. The table at the end of this section shows the relative importance of the four largest economies, their energy and oil consumption and average growth rates over the six years to 1979. Similar data are shown for the ten EEC members as a whole in the context of all 19 countries constituting OECD Europe*. The EEC is represented at the annual economic 'summit' conferences of the seven leading industrialized countries. The United States, Japan, the four largest EEC members, West Germany, France, Great Britain and Italy, plus Canada attend in their own right.

The statement issued at the conclusion of the European Council meeting in Bremen in July 1978 affirmed that the Community must apply itself in future to the joint assessment and co-ordination of the energy programmes of the various member states. It also set the objectives of reducing energy import dependence to 50 per cent and cutting to 0.8:1.0 the ratio between the rate of increase for energy consumption and that for gross domestic product. Later during 1978, the Commission proposed the following detailed recommendations in the light of the Bremen statement.(4)

The steps advised were:

- to limit Community oil imports to 500 million tonnes in 1985,
- to adopt existing proposals for hydrocarbon exploration and refining,
- to take steps against a further decline in coal-burn and coal production,
- to recover the slippage in the nuclear programme,
- to provide incentives for energy savings,
- to encourage the development of new energy sources,
- to intensify the co-ordination of member states' energy policies,
- to ensure a favourable climate for energy investment.

In 1980 the revised recommended targets were further revised to reduce Europe's dependence on imported oil to 40 per cent of total energy consumption by 1990 and to reduce to 0.7:1.0 the ratio between the rate of increase for energy consumption and that for gross domestic product. It was also recommended that 77 per cent of electricity produced should come from coal and nuclear energy; that the share of renewable energies should increase and price policies should be on a more realistic basis.(5)

The Energy Department of the European Commission makes long and short term energy forecasts, advises on investment and pricing, analyses and evaluates energy data and

---

* A listing of all member countries can be found in Appendix IV in Part III.

assesses the validity of national programmes which it tries to co-ordinate. The task is enormous. Setting up a policy is obviously extremely complex in an area involving a tangled web of scientific research, technological development, economic growth, social progress, international relations, financial and trade balances - not to mention environmental impacts and the diversity of political and economic philosophies espoused by member governments.(6) Even within countries, policies can change over time: a case in point is the sudden change in emphasis in France from nuclear and towards conservation consequent upon the change in government in May-June 1981.

The areas where the Commission can be said to have achieved some measure of success are the setting up of common targets for energy conservation, for reducing dependence on oil and for developing alternative sources, some of which it admits will not be fulfilled; community financing of new coal-fired and nuclear powered electricity plants and other energy projects; the boosting of research and development programmes dealing with alternative energy sources; and the anti-crisis mechanisms devised. The European Commission has obtained a 106 million European Currency Unit* commitment for 1981 from the European Parliament. The funds are required to finance research and demonstration projects which help to encourage investments in the energy sectors. Indeed, it is not so much the money that helps as the authority to place funds in areas where their impact is greatest. Loans for energy projects (coal technology research, uranium prospecting, etc.) amounted in 1980 to 45 per cent of the total loans granted under the various Community operations. A recent Commission report, however, acknowledges that some member countries will never manage to implement the recommended policy guidelines.

For the Community of the nine member countries (not including Greece, the most recent entrant) the Commission's latest statistics and forecasts show a rather impressive reduction in net oil imports in absolute volume and percentage dependence on them between 1976 and 1981. 64 per cent of the reduction was accounted for by the increase in indigenous production of crude oil, mainly in the British sector of the North Sea, which rose 74 million metric tons or about 1.5 million b/d over the five years.

European Economic Community of Nine Member Countries

	1976	1977	1978	1979	1980	1981
Net oil imports in million metric tons	518	481	472	474	424	403
Dependence on imported oil, per cent of primary energy	55.5	51.9	49.5	47.6	44.2	42.1

Source: European Economy No 8, March 1981.

Less industrialized countries within the Community such as Greece and Ireland whose energy needs will necessarily expand as their industrial sectors grow are the countries most lacking substantial energy resources. Even though planning to increase its coal production threefold, Greece will only manage to reduce its oil dependence from 74 per cent to 54 per cent. It is also starting to produce oil. Ireland will have a great struggle trying to bring its oil share down to less than 65 per cent. Also, Ireland is planning to use gas to increase its electricity production, and is making no effort to develop nuclear energy. It is also weak on conservation.

Even more developed countries such as Italy, Belgium and Denmark are not expected to be able to reduce their oil consumption to the prescribed 40 per cent of total energy use by

---

* when European Currency Unit = $1.20

1990. Italy's very ambitious energy programme over the next ten years includes a drastic reduction in relative oil dependence from 68 per cent in 1979 to 50 per cent of consumption in 1990, a massive switch to coal and natural gas, and considerable development of nuclear energy. As recently as 1979, 83 per cent of energy consumed in Italy was imported. Belgium is planning a massive switch to coal and nuclear energy for electricity production. Denmark is trying to reduce its present 77 per cent dependence on oil - the highest in the Community - by resorting to coal and developing its own natural gas resources.

The three countries with the largest economies - Germany, France and Britain - are all expected to meet the 40 per cent oil target. Indeed, they are well on the way to meeting some of the prescribed objectives. France's nuclear programme is impressive, but is likely to suffer a setback as a result of the change of government in May-June 1981. The lack of emphasis put on coal-fired electric plants is a drawback. Britain already produces 82 per cent of its electricity from coal-fired or nuclear plants - a figure expected to rise to 86 per cent by 1990. Although Germany is expected to reach the 40 per cent oil target, it is rather heavily dependent on gas to produce electricity and has run into difficulties in implementing its nuclear programme.

The Netherlands is, reportedly, making no effort to comply with Community targets. It is planning to increase its oil imports to 50 per cent of its energy consumption despite its considerable natural gas reserves which it plans to deplete more slowly. Its plans for a heavier switch to coal are scheduled only for the late 1980s. An aggravating circumstance is the expected decline (over 25 per cent) of its natural gas production by 1990, so that the natural gas share of Dutch energy consumption is projected to fall throughout the 1980s from the high level of 43 per cent in 1979.

In the matter of energy prices and taxes, all member states now appear to be convinced of the advisability of making the consumer pay realistic prices for energy. Although such prices, inclusive of taxes, are relatively low in Greece, Luxembourg and the Netherlands, the general tendency is to use prices and taxes to encourage conservation. In this area, however, there is scope for improvement in all ten countries. Energy efficiency and savings are a vital part of any effective energy policy. After the very substantial savings made in recent years through elimination of waste, programmes now require more substantial investments and progress. France, Germany, Denmark and the Netherlands are the countries where public spending is highest for energy conservation; Britain, after a good start, has fallen back somewhat. However, as the following table shows, it has achieved the most rapid rate of reduction in oil consumption since 1973. Over this period, the increase in oil production from Britain's North Sea reserves has eased considerably the region's dependence on external oil supplies.

The Community as a whole is planning to invest 500 billion European Currency Units (Ecu) in the energy sector between 1981 and 1990, some 2.1 per cent of GNP, compared with 1.5 per cent in 1968-80. But this overall figure covers wide disparities between countries.

European Economic Energy and Oil Perspective

	Proportion of West European Total in 1979			Annual Average Percentage Rate of Change from 1973 to 1979		
	GDP	Consumption		GDP Growth	Consumption	
		Primary Energy	Oil		Primary Energy	Oil
	%	%	%	%	%	%
West Germany	24.7	22.2	20.7	+2.4	+1.2	-0.3
France	18.5	15.2	16.7	+3.0	+0.8	-1.2
United Kingdom	13.0	17.2	13.3	+1.3	-0.5	-3.0
Italy	10.5	11.5	14.3	+2.6	+1.2	-0.4
Other EEC (6 countries*)	12.3	13.7	14.4	+2.5	+0.7	-0.6
Total EEC (10 countries)	79.0	79.7	79.3	+2.4	+0.7	-1.1
Other OECD Europe (9 countries)	21.0	20.3	20.7	+2.4	+2.2	+1.2
	Billion US Dollars	Million mtoe	Million mt			
Total OECD Europe	3086.9	1286.2	709.2	+2.4	+1.0	-0.6

* Inclusive of Greece not yet a member of the EEC in 1979.

Sources: OECD Main Economic Indicators, various issues.
BP Statistical Review of the World Oil Industry 1979.

ENERGY PROJECTIONS

Co-operation Scenario

Western Europe continues to occupy a critically important position in the world oil import picture, in spite of the build up of production in the North Sea in recent years.

Thus, in a world where oil supplies are likely to be increasingly constrained in relation to demand, even assuming only gradually rising real prices which are a feature of this scenario, the oil share of the energy mix seems bound to decline after 1985. The assumption is made that North Sea production will rise relatively slowly from the levels of early 1981, and that future discoveries will be relatively modest, consisting of small fields. A marginal increase in oil consumption is possible between 1979 and 1985, though after the declines evident in 1980 and the first few months of 1981 it seems more likely

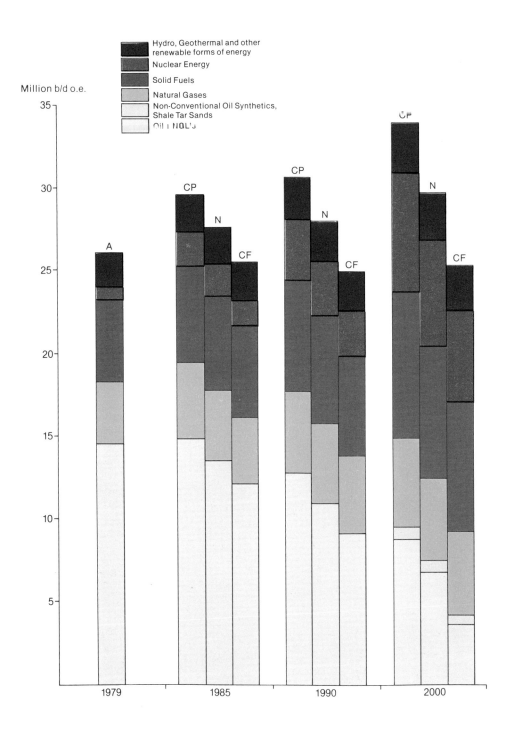

Actual and Prospective Energy Consumption Mix of Western Europe
(Million barrels a day oil/o.e.)

Figure 12

that oil consumption in 1985 will be lower than it was in 1979. This is even allowing for fairly strong economic recovery over the 1982 to 1985 period.

Oil consumption in Western Europe has never regained the 1973 peak, with residual fuel oil consumption dropping especially sharply in subsequent years of recession or low economic growth. If there is no strong recovery from the low level of West European oil consumption likely in 1981, then there is probably a better chance of avoiding the decline in consumption projected in this scenario for the second half of the 1980s. Nevertheless, a fall of some 4 million b/d is projected for the 1990s so that oil consumption of less than 9 million b/d in 2000 would represent about 26 per cent of the West European energy consumption in that year. This is still significantly more than the 16 per cent projected for the United States of America.

In this scenario primary energy consumption is projected to grow at about 2 per cent annually between 1979 and 1985. In mid-1981 this appears likely to turn out to be too high an estimate. Subsequently, significantly lower average growth of only some 0.7 per cent is foreseen on the assumption of supply contraints, especially of oil. These may not materialize fully if consumption grows more slowly than foreseen and the North Sea and other producers elsewhere are prepared to allow output to rise in the late 1980s at the expense of probable falls in the 1990s. Even after allowing for the rather sharp fall in oil consumption, total energy consumption should grow by about 1 per cent on average each year during the 1990s. This assumes a near doubling in the nuclear contribution. This alone is projected to account for more than the whole of the net growth in energy consumption. The attitude of the European public toward nuclear energy seems likely to have to undergo a rather fundamental change over the next few years if 3.5 million b/d of oil equivalent nuclear energy is to be added to the European energy balance during the 1990s. The financial implications are considerable too, while government or EEC initiatives, or lack of them may be of critical importance. Solid fuels, with much of the coal imported, should contribute about 2 million b/d oil equivalent during the course of the 1990s. Natural gas supply and demand should rise too, including larger import volumes, but the proportion of the energy balance which it accounts for is likely to rise less than 2 percentage points over the twenty-one years from 1979 to 2000.

Neutral Scenario

As can be seen from Table 47 in Part III, the growth of energy consumption follows a significantly lower line in this scenario than in the co-operation scenario discussed above. But it follows a similar profile, with growth averaging 1 per cent each year from 1979 to 1985, or the same as between 1973 and 1979. It declines to a negligible 0.2 per cent annually in the 1985 to 1990 period. This reflects the probability of a relative world energy shortage during this period with likely implications for a reduced rate of economic growth. During the 1990s European energy consumption should be able to grow at 0.5 per cent annually, or rather more. Over the period from 1973 to 1979 the European energy growth coefficient relative to GNP averaged 0.4. Since population growth in Western Europe is very low, the average rates of growth in energy consumption should be compatible with increases in per capita real incomes, except possibly in the second half of the 1980s.

The oil share of the energy consumption mix is projected to fall to 49 per cent in 1985, then more precipitately to 39 per cent in 1990, before declining to 23 per cent in 2000. By that time coal is expected to have supplanted oil as Western Europe's leading energy source, accounting for 27 per cent of total consumption in this scenario. Nuclear energy is projected to become the third most important source of energy by the year 2000, though the same qualifications about nuclear apply here as were mentioned in the co-operation scenario above. Supply and demand for natural gas are expected to rise very slightly during the 1990s, assuming an increase in the volume of imports from various potential suppliers.

Confrontation Scenario

The rather dramatic feature of this scenario is the rapid fall in oil consumption after 1985. Conventional oil is projected to account for 37 per cent of energy consumption in 1990, but to fall to only 14 per cent of the total in the year 2000, or fourth place in the European energy mix. In part this reflects relatively large increases in oil prices. Even at the level of some 3.6 million b/d a small volume of net imports is expected to be needed as North Sea production declines through the 1990s. This projection is sensitive to various assumptions concerning future discoveries and depletion policies. These will influence the level and profile of North Sea production. There is the probability that a higher oil price regime will make it possible to maintain a lower plateau of production over a longer period.

Supply and use of solid fuels, nuclear energy and natural gas are all expected to rise throughout the period and to become more important than oil in the European energy balance by the year 2000 in this scenario. In spite of this, supply and demand for primary energy are projected to decline at an average of nearly 0.5 per cent annually during the 1980s and then to rise only marginally during the next decade. Thus total consumption should be slightly lower in 2000 than it was in 1979. Many countries in Western Europe may be hard pressed to achieve any significant and sustained rises in real per capita incomes in these circumstances, particularly during the 1980s.

References

(1) BP Statistical Review of the World Oil Industry 1979. The British Petroleum Company Limited, 1980.

(2) Main Economic Indicators, monthly, various issues. Organisation for Economic Co-operation and Development.

(3) World Energy Conference, Survey of Energy Resources, 1980.

(4) EEC Commission Urges Action on Energy. Petroleum Economist, November 1978.

(5) The European Community and the Energy Problem. European Community, January 1980.

(6) Leblond, Doris. European Energy Report. Petroleum Economist, January and March 1981.

18
THE JAPANESE ENERGY SITUATION

JAPANESE ENERGY PROBLEM SUMMARIZED

In early 1974 Japan seemed to be one of the most vulnerable of the leading economies as a result of the sudden increases in oil prices. In 1973 oil had accounted for 76 per cent of its energy demand and nearly all of that was imported.

By 1979 Japan had achieved a rather remarkable readjustment. It consumed slightly less oil in that year than in 1973, whereas over the six preceding years culminating in 1973 oil consumption had more than doubled. In 1979 oil accounted for just under 70 per cent of energy demand.(1) Japanese demand for other forms of primary energy had risen at an average rate of over 5 per cent annually. The rate of growth of primary energy demand in total averaged little more than 1 per cent each year, associated with economic growth of just over 4 per cent. This compared with over 10 per cent annual growth of energy demand on average before 1974 required to sustain a similar average increase in the Japanese economy.(2)

The slow-down in the rate of growth of the Japanese economy after the 1973 oil price increases was relatively sharper than in the USA or Western Europe. The most notable feature of the Japanese performance was that the use of primary energy per unit of GNP declined remarkably between 1973 and 1979. Over the six year period the energy growth coefficient with respect to GNP averaged less than 0.3:1 and even after the abnormal years of 1974 and 1975 it was below 0.5:1. This change was associated with a reduction in the relative importance of energy intensive industries in the Japanese economy.

In just six years Japan had moved some way from its heavy dependence on imported oil in pursuit of its declared policy of diversifying the types of energy used and the sources of supply. In 1979 alone the supply of and demand for nuclear electricity grew by 18 per cent, natural gas use by 30 per cent, hydroelectricity by 14 per cent, and coal use by 8.5 per cent.

RESERVE BASE (Table 17)

The indigenous energy resources of Japan are negligible. The resources of hydropower and coal have been no more than sufficient to maintain levels of indigenous production during recent years.(3)

SUPPLY AND DEMAND (Table 18)

Prior to 1974, the consumption of total primary energy in Japan was growing at an average rate of 10.1 per cent per year. Corresponding percentage growth rates for the various sources of energy were: oil 14.0; natural gas 17.6; coal 2.3; nuclear 50.2; and hydropower -1.4. 1973 was an abnormally poor year for the production of hydroelectricity. As a result of the economic recession and various energy conservation measures adopted, following OPEC's oil price readjustments of 1973-74, the high growth rate of primary energy consumption dropped to a level of only 1.2 per cent per year. Oil consumption has not returned to the level of 1973. Growth of natural gas consumption increased to an average of 27.0 per cent per year from 1973 to 1979. Apart from oil, coal was the only form of primary energy to account for a smaller share of Japanese primary energy demand in 1979 than in 1973.(1)

In 1979 the percentage composition of Japan's primary energy demand was: oil 69.7; natural gas 5.8; coal 15.4; nuclear 3.9; and hydroelectricity 5.2. Further remarkable progress was achieved in 1980 as Japanese oil consumption fell 9.9 per cent below the 1979 level while the Japanese economy expanded 4.4 per cent.

In the absence of adequate indigenous sources of energy, Japan will remain heavily dependent on oil, gas and coal imports to satisfy her energy demands.

NON-CONVENTIONAL ENERGY RESOURCES

Japan is also actively involved in research and development of non-conventional energy resources such as fast-breeder nuclear reactors, nuclear fusion, solar energy (The Sunshine Project), geothermal energy, liquefaction of coal, utilization of organic waste, and oceanic energy. However, because of the long lead times involved, these sources will not begin to make a worthwhile contribution to Japanese energy supply until after 1990.

ENERGY POLICY

In October 1978, the overall Energy Council, an advisory body to Japan's Ministry of International Trade and Industry (MITI), published a report on Japan's long term energy policy. Its main recommendations were as follows:(4)

- To diversify the sources of crude imports away from the Middle East towards the Pacific region.

- To increase the proportion of imports contracted through deals with national oil companies and producer governments.

- To double the share of imports developed by Japanese companies and expand exploration in Japan and overseas.

- To broaden financial and technical co-operation with oil exporting countries.

- To enlarge the scope of the Japanese National Oil Corporation (JNOC).

- To increase oil stockpiles.

- To consolidate the local industry.

- To step up usage of natural gas.

MITI also announced during late 1979 plans to establish the Alternative Energy Corporation in 1980. The Corporation's planned operations include the development of overseas coal in countries such as Australia and Canada, installation of solar systems in 7.8 million private homes and public housing units by fiscal 1990, commercial operation of coal liquefaction plants with a combined capacity of 25,000 tons/day, and financing the shift of fuels from oil to coal or LNG at ultra low interest. Funds needed for these operations were estimated to amount to about US $10 billion during the ten year period from 1980 to 1990. This sum represented some 1 per cent of Japan's GNP in 1979. The main problems related to the establishment of this Corporation were the vast funds required and the fact that its concept ran counter to the basic government policy of rationalizing the administrative structure.(5)

The schedule for starting nuclear power stations is encountering considerable delays due to environmental concerns. Nevertheless, the latest recommendation is that nuclear power generating capacity be increased to about 76 GW in 1995, implying an annual average growth rate of some 12 per cent. It is based on the premise that nuclear energy is to replace petroleum as Japan's foremost energy source.

Japanese officials published in May 1981 an interim report entitled 'In Search of Japan's Oil Strategy for the 1980s'. The results are reportedly by no means definitive, though they give an interesting insight into Japanese thinking. Japan's oil policies and industrial structures are recognized as being in a transitional phase between the earlier period of "quantitative expansionism" and the emerging period of what is called "qualitative substantialism".(6)

The first chapter evaluates international management and remarks that rapid developments have been made in attaining greater management over petroleum related situations in terms of demand as well as supply. It remarks that this has been responsible for the gradual change-over from confrontation to co-operation. It goes on to discuss the decline in oil dependency in terms of a guarantee of economic security, the international balance of payments implications, bargaining power relative to the oil producing nations, and efficient utilization of oil resources. Other aspects of this first chapter include the normalization of oil deals aiming at supply stability. It mentions differences among OPEC members over the question of oil pricing and the readiness in the oil consuming nations to lend support to the more moderate elements in OPEC. It goes on to consider emergency measures and stockpiles and the promotion of dialogue directly with those involved in devising governmental policy in individual oil producing nations, especially important in view of the oil producing nations' recently expanding interest in downstream operations.

A second main chapter examines the outlook for petroleum demand in the light of the possibilities for alternative energy sources and developments in energy conservation. It reviews the imbalance across the barrel with demand for middle distillates remaining strong relative to residual fuel oil while the crude oil import slate becomes heavier. LPG and LNG are seen as having important roles to play in this context. The remarkable change in the progression of demand growth is the subject of comment, since the extent of change has far exceeded initial expectations. A relatively optimistic view is held regarding oil supply, since the oil price rises are accompanied by advances in secondary and tertiary recovery and the return to profitability of formerly unprofitable oil fields.

Other important developments foreseen include an increased volume of Japanese product imports as oil producing countries increase their refining capacity. Japan is stepping up its efforts to increase crude oil imports from producing countries on a direct deal basis. Increased co-operation with oil producing countries is envisaged, particularly through an extension of economic and technical aid.(6)

ENERGY PROJECTIONS

Co-operation Scenario

Japan has demonstrated a remarkable resilience in responding to the substantial oil price increases of 1973-74 and 1979-80. This has been achieved through a competitive response in export markets, significant structural shifts in the relative importance of different industrial sectors, and limiting growth of domestic demand while the country's external balance of payments was being righted and the value of the Yen kept under discreet control on the foreign exchange markets. All this has been done at the expense of a very significant reduction in the average rate of Japanese economic growth: some 4 per cent annually from 1973 to 1979 compared with the remarkably high 10 per cent or more previously. In 1979 Japanese dependence on oil remained very high at 70 per cent of its energy needs, down six percentage points from the peak six years earlier. Even so, this is a noteworthy achievement.

Over the period to 2000 a sharp acceleration is expected to take place, even in this co-operation scenario, as Japan makes good its declared intention of reducing its dependence away from oil and diversifying its energy consumption mix. One recent manifestation of the earnestness with which this policy is being pursued already is the willingness of some Japanese buyers to pay a crude oil equivalent price for liquefied natural gas (LNG) supplies on an FOB basis, and in spite of higher transport costs, to exporters such as Indonesia and Abu Dhabi. Consumption of liquefied petroleum gas (LPG) is also projected to rise rapidly, according to official estimates, over the five years to 1985.

By 1985 the oil share of the Japanese energy mix should be down to about 56 per cent, even if energy consumption grows by an average of 3.5 per cent annually from 1979. Currently, this appears an improbably high rate, capable of sustaining a much higher rate of economic growth to 1985 than seems probable in mid-1981. The average rate of Japanese GNP growth of 4.1 per cent recorded between 1973 and 1979 was sustained by a rise in energy consumption which averaged only 1.2 per cent. This yields an energy growth coefficient of barely 0.3:1. Thus if growth in the use of other forms of energy rises in the way projected in Table 48 (see Part III), then the chances are that Japanese oil consumption in 1985 will be less than the 5.4 million b/d foreseen in this scenario. However, the latest official projection shows Japanese oil imports at 5.71 million barrels a day in 1985. The Japanese Five Year Supply and Demand Plan shows total crude oil supply and demand rising at an annual average rate of 3.3 per cent to 1985 from the depressed 1980 volume.(7)

The share of oil is further projected to decline to 45 per cent of Japanese energy consumption by 1990. After completing these estimates I was interested to see that this percentage was being actively discussed in Japan as a desirable objective for the same year. After that, though, my projection allows for a further acceleration in the declining Japanese dependence on oil during the 1990s as consumption falls by nearly 50 per cent while energy consumption rises at 1.2 per cent on average each year. Thus Japanese dependence on oil declines to only 20 per cent of its energy needs in 2000, a drop of fifty percentage points in only twenty-one years. If a small quantity of synthetics is added in, it should be about 24 per cent. Nevertheless, this still marks the most remarkable transition away from oil dependence anywhere in the world: as energy consumption rises by some 3.5 million b/d over the twenty-one years to 2000, so oil use is projected to fall by 3.2 million b/d. Coal is likely to be the largest single incremental contributor to rising Japanese energy consumption, just counterbalancing the decline in conventional oil use. I feel confident that Japan, perhaps alone among the major countries, is capable of making such a remarkable transition over the extended period foreseen. This prospect can be seen in a historical perspective. Japanese oil consumption passed through the 2.3 million b/d level as recently as 1967, and rose by nearly 3 million b/d in only six years subsequently.(1)

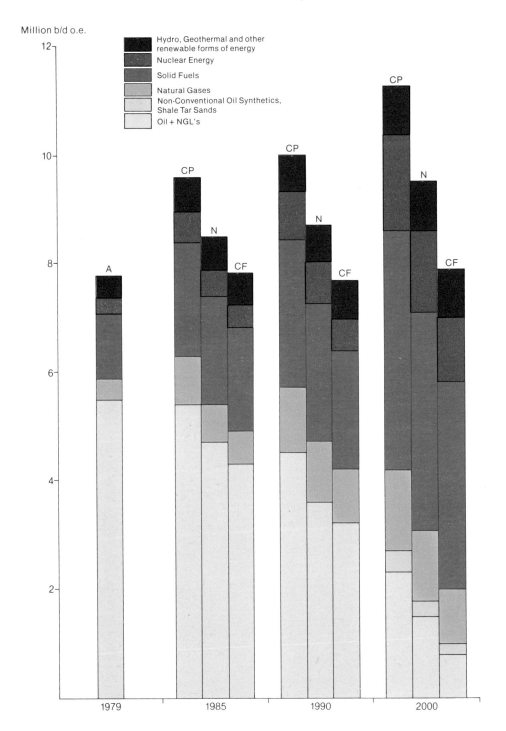

Neutral Scenario

The rate of growth in Japanese energy consumption projected in this scenario looks rather more likely to match the reality of the next few years than the co-operation scenario discussed above. An average increase of 1.4 per cent is shown for the six years from 1979 to 1985, followed by a slower rate of 0.5 per cent on average from 1985 to 1990, reflecting supply constraints. This might rise to about 1 per cent annual average during the 1990s, sufficient to sustain a healthy rate of per capita economic growth. The main problem facing the Japanese economy in this scenario would be the even sharper rundown in oil use with conventional oil falling to 1.5 million b/d out of 9.5 million b/d oil equivalent energy consumption. This would throw up harsher challenges in terms of the demands for technological innovation and sectoral adjustment under less favourable political and economic conditions than those prevailing in the co-operation scenario. Once again, coal looks like becoming the dominant energy form in Japan before 2000. Liquefied coal may be rather more significant in the Japanese energy mix at that time than suggested in Table 48, having regard to the points made above.

Confrontation Scenario

The projection in this scenario probably goes further than any other in this book in straining the credibility of what could happen in a disorderly world. This is not because of the negligible change in Japanese energy consumption over the period to 2000, but because of the phenomenal decline in oil use to only 1 million b/d inclusive of non-conventional material. This reflects a low level of output and export by OPEC countries as well as generally confused political and economic conditions worldwide. Because of Japan's export competitiveness this might include a prolonged period of discrimination against Japanese products in international trade. The transition into this undesirable situation would be difficult to control. There is a strong worldwide incentive to avoid the risk of a damaging downward deflationary spiral likely to accompany such an energy scenario, not least for Japan which could suffer relatively more than most other oil importing countries if such a scenario materialized.

References

(1) BP Statistical Review of the World Oil Industry, 1979. The British Petroleum Company Limited, 1980.

(2) Main Economic Indicators, monthly, various issues. Organisation for Economic Co-operation and Development.

(3) World Energy Conference, Survey of Energy Resources, 1980.

(4) Segal, Jeffrey. Long Term Oil Strategy Succeeding For Japan. Petroleum Economist, September 1979.

(5) Marubeni Petroleum Reports, 1 October and 16 December 1979. Marubeni Corporation, Japan.

(6) In Search of Japan's Oil Strategy for the 1980's. Interim report of Study Group on Basic Oil Industry Issues, Natural Resources and Energy Agency, Ministry of International Trade and Industry (MITI). Translated into English in Japan Petroleum and Energy Weekly, 13 May 1981 and several subsequent issues.

(7) Annual Five Year Oil Supply and Demand Plan for Fiscal Years 1981 - 1985 inclusive. Ministry of International Trade and Industry (MITI). Translated into English in Japan Petroleum and Energy Weekly, 1 and 8 June 1981.

19
THE OPEC DEVELOPING COUNTRIES' ENERGY SITUATION

OPEC ENERGY PROBLEMS SUMMARIZED

Some readers in the industrialized countries may be surprised to learn that OPEC countries have energy problems. These are essentially of two types. The first relates to the high exponential rate of growth in world demand for oil and gas up to 1973, associated with the fact that reserves are limited and non-renewable, yet remain the predominant natural resource of oil exporting countries. The second problem relates to the prospective rapid growth in demand for oil and gas within OPEC countries themselves.

The first problem can be illustrated by reference to what would have happened by 1979 had the actual growth of Middle Eastern oil production continued at the 13.5 per cent annual rate actually achieved during the five years up to 1973. Instead of the Middle East having a reserves to production ratio of 45 years : 1 in 1979, the region's proved oil reserves to annual production ratio would have fallen to only 20 years : 1 by the end of 1979, assuming the same finding rate of new oil reserves as actually occurred. This might well mean the world now having the opportunity to achieve a reasonably orderly transition to increased dependence on alternative energy sources, whereas the continuation of the trend up to 1973 suggested the prospect of approaching the edge of the energy precipice without adequate preparation. Over-rapid depletion of oil reserves might have led, within a few more years, to catastrophic economic disruption in which individual standards of human economic well-being suffered considerable absolute declines over a lengthy period and in very many parts of the world.

The second problem of rapid growth of oil consumption in OPEC member countries is associated with the existing low reserves to production ratios in several of them. While oil and gas prices need to be kept low to promote competitive growth of infant industries, unrealistically low prices could force some oil exporters to become oil importers at a relatively early date. Furthermore the volume of oil entering world trade could decline rather dramatically.

RESERVE BASE (Tables 19 and 20)

The thirteen member developing countries of the Organization of the Petroleum Exporting Countries (OPEC) are Algeria, Ecuador, Gabon, Indonesia, Iran, Iraq, Kuwait, Libya, Nigeria, Qatar, Saudi Arabia, United Arab Emirates (Abu Dhabi, Dubai and Sharjah) and

Venezuela. Most of them are endowed with relatively large oil and gas reserves. About 79 per cent of proved oil reserves and 61 per cent of proved natural gas reserves in the non-Communist world are located in OPEC member countries. Saudi Arabia is in a pre-eminent position as her proved oil reserves are estimated to be about 168.4 billion barrels or 30 per cent of proved oil reserves in the non-Communist world at the end of 1979.(1)(2)

Proved reserves of conventional fuels in all OPEC member countries are estimated to be close to 91 billion tons oil equivalent: 67.9 per cent oil, 28.5 per cent natural gas and only 3.6 per cent coal (coal is located in Indonesia and Venezuela).(3)

Moderate sized uranium reserves are known to exist in Gabon.

In the case of non-conventional sources of energy, Venezuela is known to have large reserves of heavy oil estimated to be around 277 billion tons oil equivalent.

According to various authoritative sources, the best prospects for gross additions to oil and gas reserves exist in OPEC member countries. These include both enhanced recovery and additions to known reserves which may be revealed by further exploration.

As in most other developing countries, there has been a general lack of proper exploratory efforts to discover uranium and unconventional sources of energy in the past. There could be chances for many more new discoveries in these fields with increased exploratory efforts.

Prospects for the development of solar energy in most OPEC member countries, especially in Saudi Arabia, are very promising.

SUPPLY AND DEMAND (Table 22)

The most important source of energy consumed in OPEC countries is oil, though at some 65 per cent of the total in 1979 the oil proportion was actually lower than in many oil importing countries. Also, the 2.2 million barrels a day of oil consumed in that year was still extremely low in relation to crude oil production which exceeded 30.9 million b/d during 1979.(4) This was achieved in spite of the significant fall in output which occurred that year in what had been the second largest producer among OPEC members. By 1979 OPEC exports of crude oil and refined products still amounted to over 93 per cent of crude oil production, only 2 percentage points less than six years earlier.

The average annual rate of oil consumption rose to 14.5 per cent from 1973 to 1979, the volume of consumption having more than doubled over the five years to 1978. Crude oil production and OPEC exports of crude and refined products together were almost the same in 1979 as they had been in 1973. Over the six years culminating in 1973 production, exports and consumption had all been rising at similar average annual compound growth rates of between 10.5 and 11.0 per cent.

As long as the oil and gas resources remained under the virtual control of the traditional concessionaire companies, OPEC member countries failed to obtain the true benefits from their wealth of natural resources. For this reason economic development in these countries advanced less than it might have done over a prolonged period. Following the emergence of national control over their natural resources the rate of development has started to accelerate. This phenomenon has been discussed in detail in Chapters 3 and 4. Consequently the current low levels of domestic energy consumption in OPEC member countries (Table 22) are expected to rise substantially in the coming decades. The prospects are discussed below and elsewhere.(5)

BASIC PROBLEMS AND POLICY OPTIONS

All the thirteen OPEC member countries are basically developing countries faced with the typical problems of developing countries: improving the general standard of living of their populations including food, housing and general living environment; development of social infrastructures such as transport, communication, electrification and irrigation; development of local industries and both domestic and export markets in the developed world for their products. There are also problems related to the transfer of appropriate technology from the developed countries, international trade barriers, and the recent unfavourable developments in the international monetary system, including the erosion in the real value of accumulated OPEC country surpluses.

The economies of OPEC member countries are exceptionally vulnerable in the long term. Most of them are very heavily dependent on their hydrocarbon resources. These are depleting at a rapid pace, owing to the world's heavy dependence on these non-renewable and finite resources.

In order to safeguard their long term economic well-being, these countries must preserve adequate petroleum reserves for their own long term requirements. They must also diversify their economic base before they exhaust their precious petroleum resources. The development of all alternative renewable and non-renewable sources of energy is also imperative for OPEC members, as it is for other countries.

The attitudes of the thirteen OPEC member countries and their dependence on petroleum production also differ greatly, according to each country's special circumstances. Iran, Indonesia, Nigeria, Algeria and Venezuela, for instance, have relatively large populations and need high oil revenues to support economic development programmes. These countries must balance the need for immediate income against the desire to prolong the flow of their oil revenues over the long term by limiting production levels. Their reserves to production ratios are relatively low. Several other less populous OPEC countries are in a similar position. On the other hand, Saudi Arabia, UAE and Kuwait, with relatively small populations, have been producing at levels well in excess of domestic revenue requirements in order to satisfy the international market demand. In this process these countries have accumulated temporary financial surpluses beyond their internal absorptive capacity. Protection of the real value of these financial assets against inflation and currency depreciation is a serious problem. During the Conference on International Economic Co-operation, held in Paris during December 1975 to June 1977, the Group of Nineteen representing the Developing World proposed the following measures to be adopted by the developed countries regarding the investment of the assets of OPEC member countries in their markets:

- Greater access to financial markets and other investment opportunities in developed countries.

- Non-discriminatory treatment of such investments and assets in developed countries.

- Protection of the real value of financial assets against inflation and currency depreciation in developed countries.

- Commitments by the developed countries to safeguard the investments of the oil exporting countries in their markets against:

    - Confiscation, freezing and any other coercive measures to deprive these countries of their investments and income therefrom. In the case of expropriation and nationalization, developed countries should guarantee an appropriate compensation.

- Unwarranted restrictions on prompt conversion and transfer of investment and income therefrom from a host country's currency into freely convertible currencies at agreed rates.

- A larger role for developing countries in international financial institutions and effective participation in the decision-making process concerning the international financial and monetary problems.

ENERGY PROJECTIONS

Co-operation Scenario

There can be no doubts among rational people that co-operation is the preferred scenario for OPEC countries just as much as it is for all other countries in the world. The policies of Saudi Arabia have consistently been based on the principle of international co-operation, and this will continue in the future. The Deputy Secretary General of OPEC discussed interdependence and the need for co-operation in the following terms in early 1981.

> "Although oil producing countries differ in economic and social structures, it is nevertheless, fair to estimate that the time horizon needed for achieving such structural changes in their economies will be generally longer than that envisaged for the energy transition. If the latter is made at a much faster rate than the former, development prospects in the oil producing countries could be compromised before an adequate level of economic structural change has been realized. Consequently, there should be a balance between the pace of economic and social development in the oil producing countries and the pace of transition in the western countries towards a lesser dependence on imported oil. The faster the rate of development, the greater are the incentives for the oil producing countries to cooperate with the consuming countries in smoothly accelerating the energy transition."(6)

If the state of the world economy outside OPEC countries is relatively unhealthy, the oil exporters themselves will be unable to develop satisfactorily in either a quantitative or qualitative sense, however the objectives of human endeavour might be defined. Oil exporting countries need to use the unique opportunities they have been blessed with in the form of oil and gas reserves to put in place an economic and social infrastructure during the next generation which is capable of surviving and advancing as their reserves of hydrocarbons are depleted. This can only be done in an interdependent world.

The population of OPEC member countries grew rapidly, at an average annual rate of nearly 2.8 per cent, from 1960 to 1975. According to the United Nations' medium variant and most likely projection, the total population of OPEC member countries will increase on average at a rate of nearly 2.7 per cent annually from 1975 to 2000. Thus by the year 2000 the population of OPEC countries is projected to be just over 560 million.(7) The present rate of increase is above this average, but it will decline later. This rate is significantly higher than for the Less Developed Regions (as defined by the United Nations)(3) of the world as a whole. The average for all of these over the last quarter of the twentieth century is projected at just under 2.1 per cent annually, though falling rather sharply from 2.21 per cent annually from 1975 to 1980 to only 1.84 per cent from 1995 to 2000.(7)

In these circumstances the energy consumption of OPEC countries is projected to grow rapidly in this scenario in order to sustain high rates of economic growth. An average rate of 12 per cent is foreseen from 1979 to 1985, declining slightly to 11 per cent annually from 1985 to 1990 in this scenario. After that, a rather lower rate of about 9 per cent is anticipated. These high rates are foreseen as being an essential part of the process of putting an industrial infrastructure in place by the end of the twentieth century. There will be some tightening of the international oil market as a consequence of this development, but a significant part of the growth in OPEC energy consumption will come from natural gas and natural gas liquids which were either flared to waste or remained unexploited in the past.

The growth in OPEC oil consumption in this scenario is very similar to the projection made by the OPEC Secretariat in 1980.(5) It assumes, of course, not merely a favourable international environment, but also peaceful conditions within and between member countries. Thus for OPEC oil consumption to rise to as much as 4.2 million b/d in 1985 and then to 6.9 million b/d in 1990 it is likely that there will have to be some important changes in the circumstances which prevailed in 1979 and 1980. Nevertheless there has been a very rapid increase in consumption in most member countries in the recent past. Some oil exporting countries with low reserves to production ratios may need to reconsider their policies for pricing of domestic oil products in order to maintain a reasonable volume of oil for export. Even if there are some changes of this sort, the growth in domestic oil demand is likely to remain very strong in most OPEC member countries in this scenario. Nevertheless, the oil share of energy consumption is likely to decline even in this scenario as several OPEC members with substantial natural gas reserves develop them over the next twenty years.

In consequence, the use of natural gas is projected to increase to account for 30 per cent of total OPEC energy consumption in the year 2000 compared with 61 per cent for oil. This assumes further development of gas resources in Saudi Arabia as well as in countries such as Qatar, Abu Dhabi, Algeria, Nigeria, Indonesia and Iran. At the beginning of 1981 Iran's estimated proved reserves of natural gas were 485 trillion cubic feet, or 48 per cent of the OPEC total and 18 per cent of the world total.

Neutral Scenario

OPEC energy consumption is projected to grow at nearly 10 per cent on average each year from 1979 to 1985 in this scenario, declining to about 9 per cent annually from 1985 to 1990. After that, annual growth is likely to fall to little more than 7 per cent in the 1990s. If these averages seem high to some readers they can be put into the perspective of actual growth of 11 per cent on average from 1973 to 1978, based on UN estimates of consumption in the thirteen member countries.(8) Some further perspective can be gained from the fact that even in 1978, energy consumption per capita in the industrialized countries was nearly ten times as great as in the OPEC countries, on average. Based on the energy consumption projections of this scenario shown in Table 50 and the latest UN population projections, OPEC per capita energy consumption in the year 2000 will still be at only some 35 per cent of the level prevailing in the industrialized countries in 1978. This is surely an effective argument against those who assume that the growth of the OPEC energy consumption over the next twenty years will be sustained at the expense of others.

In this scenario OPEC oil production might fall to 21 million b/d by the year 2000 with Saudi Arabia at 8.5 million b/d. OPEC oil exports are projected at about 11 million b/d in the 2000 horizon year.

The lower energy consumption growth rate of this scenario is likely to require a substantially lower growth of oil consumption which is projected to rise to about 10 million b/d in the year 2000. By that time it would account for about 55 per cent of total energy consumption in OPEC countries, some 10 percentage points lower than in 1979.

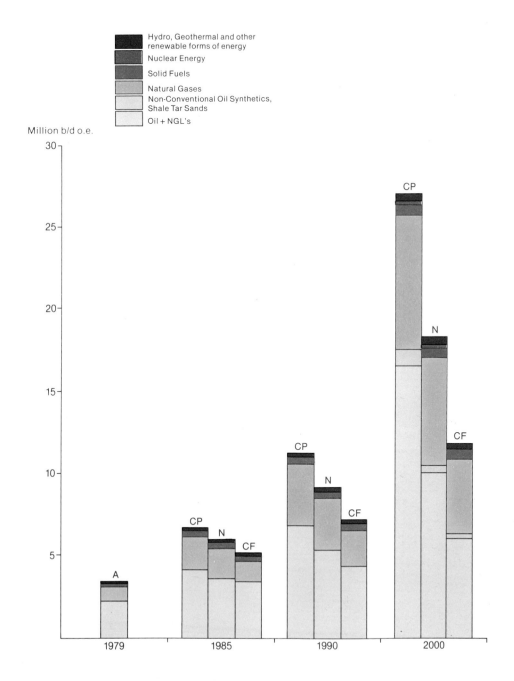

Figure 14

Actual and Prospective Energy Consumption Mix of OPEC Member Countries
(Million barrels a day oil/o.e.)

Confrontation Scenario

An unfavourable world trading and political environment would lead to a big reduction in OPEC oil production in this scenario, and also to a sharp decline in the volume of OPEC oil available for export. A much lower rate of growth in OPEC member countries' consumption of oil and other energy is another feature of this scenario, as OPEC per capita GNP growth rises rather slowly. The large drop in export volume is likely to be offset in part by considerably larger real price rises per barrel for those countries able to maintain significant exports in what are likely to be very disturbed and disorderly international market conditions.

As I pointed out in describing this scenario in Chapter 14, the character of it makes the scenario inherently more unstable than the other two discussed above. This is true of OPEC member countries, just as it is of other parts of the world. In spite of their oil and gas resource endowments, political and economic conditions would inevitably be affected by the deterioration in international relationships. Saudi Arabia's oil production is projected at 5 million b/d in this scenario, and that of OPEC members as a whole might fall from 24 million b/d in 1985 to only 12 million b/d in the year 2000 (see Table 35).

OPEC energy consumption is projected to grow at slightly more than 7 per cent in the early 1980s and at a little less than this rate during the second half of the decade. In the 1990s OPEC energy consumption rises by little more than 5 per cent annually in this scenario, with the oil share of the energy mix falling by ten percentage points during the decade. This implies oil consumption rising to only about 6 million b/d in the year 2000, more than 10 million b/d less than in the co-operation scenario.

For OPEC member countries there would be a remarkable difference between the standards of living prevailing in this scenario and the far better conditions which characterize the year 2000 prospect in the co-operation scenario. It is important that people in OPEC countries come to realize this, just as much as those in other countries, so that appropriate political decisions are taken. A specific disadvantage of this scenario for OPEC member countries is the very large measure of independence from OPEC oil supplies which the industrialized countries will have achieved by the year 2000.

References

(1) Petroleum Statistical Bulletin, 1979. Ministry of Petroleum and Mineral Resources, Kingdom of Saudi Arabia.

(2) Worldwide Oil and Gas at a Glance. Oil and Gas Journal, 31 December 1979.

(3) World Energy Conference, Survey of Energy Resources, 1980.

(4) OPEC Annual Statistical Bulletin, 1979. Organization of the Petroleum Exporting Countries.

(5) Domestic Energy Requirements in OPEC Member Countries. OPEC Papers, Volume I No.1, August 1980.

(6) Problems of World Energy Transition: A Producer's Point of View, Dr. Fadhil J. Al-Chalabi, Deputy Secretary General of OPEC, at Seminar on 'Development Through Co-operation' in Rome, April 1981, sponsored by OAPEC, Italy and the South European Countries.

(7) UN Population Statistics and Forecasts, 1950-2000. Selected Demographic Indicators, 1980.

(8) World Energy Supplies 1973-1978. United Nations, 1979. Statistical Papers Series J No.22.

20
THE NON-OPEC DEVELOPING COUNTRIES' ENERGY SITUATION

NON-OPEC DEVELOPING COUNTRIES' ENERGY PROBLEM SUMMARIZED

The number of countries in this group totals over one hundred. They are highly diverse in many important respects. Some are very populous, others are sparsely inhabited or have small populations. Some are extremely poor with a very low standard of living and a GNP per capita of well under $300 in 1977, while others are newly industrializing, dynamic economies with rapidly expanding exports and a GNP per capita of from $1000 to about $3000 in 1977.(1)(2) Similarly, primary energy demand varies considerably, in terms of per capita consumption, in terms of relative dependence on oil, and in terms of the degree to which financing of oil imports represents a greater or lesser burden.(3)(4) Furthermore, there is a high degree of concentration of large indigenous energy reserves in only a few developing countries.(5)(6)

In order to examine this diversity and reduce it to manageable proportions it is useful to recognize that the ten largest economies account for some 66 per cent of the total GNP of all non-OPEC developing countries. The next ten in rank order account for a further 15 per cent, and those ranking from twenty-first to thirtieth account for another 7 per cent. The top of the rank order for primary energy demand and oil consumption contains more or less the same developing countries. A listing of these ten largest economies is shown in the table on the next page, together with the composition of their primary energy demand in 1978 and the extent of the change in their dependence on oil in the five years since 1973. Also shown in aggregate are the next two groups, ranking eleventh to twentieth and twenty-first to thirtieth in order of GNP. Of course, there are special cases. Mexico increased its oil production significantly during this period and has continued to do so since. Mexico is not merely atypical of the countries considered in this chapter: it is exceptional in having abundant oil reserves.

In 1978 the top ten developing countries alone accounted for 63 per cent of the population and 72 per cent of the commercial energy consumption of all the countries considered in this chapter. In aggregate the thirty largest economies accounted for 88 per cent of GNP of all non-OPEC developing countries, 87 per cent of the total population in all of them and 90 per cent of the commercial energy consumption in 1978. There is also a considerable diversity between these 30 countries, all of which had a GNP of at least $5 billion in 1978.

Among the ten largest economies only India is in the poorest category in terms of GNP per capita. The same can be said for its populous neighbours Pakistan and Bangladesh

Country	Population In 1978 Millions	GNP in 1978 Total $ Billions	GNP in 1978 Per Capita $	Total Million tons oil equivalent	Per Capita tons oil equivalent	Primary Energy in 1978 — Oil %	Natural Gas %	Solid Fuels %	Hydro + Nuclear Electricity %	Change in oil share of Energy Mix from 1973 to 1978 percentage points	Annual Average Growth Rates from 1973 to 1978 Primary Energy %	Annual Average Growth Rates from 1973 to 1978 GNP %	Energy/ GNP Ratio
Brazil	119.5	180.0	1,510	62.4	0.5	74.9	1.8	9.5	13.8	-6.1	7.9	6.8	1.2
India	643.9	117.5	180	77.3	0.1	28.2	1.6	65.6	4.6	-2.7	4.8	4.8	1.0
Mexico	65.4	91.9	1,400	63.0	0.9	68.2	20.4	8.5	2.9	-0.4	6.9	4.1	1.7
Argentina	26.4	53.4	2,030	33.6	1.3	69.5	25.4	3.1	1.9	-5.7	1.4	1.2	1.2
South Korea	36.6	48.0	1,310	34.2	0.9	61.0	NIL	37.8	1.1	+7.1	9.1	10.6	0.9
Yugoslavia	22.0	46.1	2,100	30.3	1.4	46.2	5.9	41.1	6.8	+5.3	5.2	6.6	0.8
South Africa	27.7	43.8	1,580	65.6	2.2	18.0	NIL	80.9	1.0	-1.8	5.0	2.8	1.8
Republic of China (Taiwan)	17.1	26.7	1,560	18.9	1.1	75.9	7.4	13.4	3.4	+9.8	10.9	8.4	1.3
Philippines	45.6	24.4	530	10.7	0.2	94.1	NIL	1.8	4.1	-2.0	4.5	6.1	0.7
Thailand	44.5	23.4	530	10.0	0.2	94.7	NIL	1.4	3.9	-1.9	5.1	8.0	0.6
Total Above top 10 Countries	1048.7	655.2	625	406.0	0.4	53.1	6.6	35.6	4.7	+0.4	5.9	5.5	1.1
Total, next 10 Countries ranking 11-20	213.0	152.4	715	79.4	0.4	76.7	11.6	7.5	4.3	-1.1	4.5	5.4	0.8
Total, next 10 Countries ranking 21-30	177.4	65.0	365	23.1	0.1	86.6	7.2	2.6	3.7	-1.1	7.3	5.8	1.3

Sources: 1980 & 1979 World Bank Atlas and United Nations World Energy Supplies 1973-1978

which are respectively in the second and third groups of ten developing countries by rank order of GNP. Most of the others among the top ten developing country economies had an average GNP per capita of about $1500 or more in 1978.

There is also a great diversity of energy consumption per capita, the relative dependence on oil and the change in oil's place in the consumption mix since 1973 even among the ten largest economies, as can be seen from the table on the preceding page. There are also big differences in the rates of GNP growth achieved since 1973, but rather less difference in the energy/GNP growth coefficient. For the thirty countries combined, the growth of energy with respect to GNP amounted to 1.04:1.

NON-COMMERCIAL ENERGY

Much of the energy consumed in the non-OPEC developing countries is local and non-commercial, that is firewood, animal dung and vegetable waste. In India, for example, non-commercial energy has been estimated to constitute nearly 60 per cent of total energy consumption in 1960 and 48 per cent in 1970. It is expected to remain a significant though decreasing percentage in future.(7)

Although non-commercial sources are important and much of the world depends on these fuels for some part of their energy needs, it is impossible to find comprehensive reliable statistical information on them. Therefore, the following detailed discussion of the energy situation in non-OPEC developing countries is restricted to commercial sources of energy only.

RESERVE BASE (Tables 23 to 27)

Non-OPEC developing countries account for 40 per cent of the world's population, but possess only 8.5 per cent of the world's proved reserves of the main fossil fuels that is 44.7 billion tons oil equivalent - 76.6 per cent coal, 14.3 per cent oil and 9.1 per cent natural gas.(5).

The distribution of these conventional fuel reserves is highly uneven. In the case of oil, over 90 per cent of proved reserves of nearly 70 billion barrels are located in only the top ten countries of those listed on the next page. Mexico alone accounts for about 63 per cent of these reserves.(6) Natural gas is only slightly less concentrated: over 83 per cent of proved reserves are located in ten countries only: namely, Mexico, Argentina, Malaysia, Pakistan, Bangladesh, Brunei, Trinidad & Tobago, Bahrain, India and Thailand. Mexico, Malaysia and Pakistan alone account for about 46 per cent of these reserves.(6) Five of the top ten countries in rank order of gas reserves are also among the top ten developing countries by rank order of oil reserves. In the case of coal, 88 per cent of economically recoverable reserves are located in only five countries: namely, India, South Africa, Botswana, Swaziland and Yugoslavia (mainly lignite/subbituminous coal). South Africa alone accounts for 42 per cent of these reserves and India for another 22 per cent.

In the case of non-conventional sources of energy, oil shale reserves are estimated to be about 2.4 billion tons oil equivalent in Morocco and Jordan.(5)

Uranium resources are concentrated in South Africa, Namibia, Niger, Brazil, Argentina, India and the Central African Republic.

Annual potential generation of electricity from the installed and potential capacity of hydro-resources is estimated at 4.4 million gigawatt hours which is almost 44.5 per cent of total world hydro-potential. However, by 1974 only 4 per cent of the potential resources had been developed.

There has been generally a lack of concerted exploratory efforts in non-OPEC developing countries of both the conventional and non-conventional sources of energy. Increased efforts are expected to realize potential new discoveries.

OIL RESERVES, PRODUCTION AND CONSUMPTION BY COUNTRY

The lack of proved oil reserves and established oil production in the non-OPEC developing countries can be illustrated most forcefully by observing that, including Mexico, there were only fourteen countries among them which possessed both proved oil reserves at the end of 1980 and oil production averages for the year 1980 which exceeded 0.1 per cent of the world totals for proved reserves and production. Of these fourteen countries only seven of them had proved reserves to production ratios of at least 20:1. All these countries are listed below in rank order of proved oil reserves at the beginning of 1981.

Non-OPEC Developing Country Oil Producers

(All countries accounting for at least 0.1 per cent of world proved reserves, i.e. 650 million barrels or more, and at least 0.1 per cent of world crude oil production, i.e. 60,000 barrels a day or more).(6)(3)

Country	Rank Order in terms of Proved Reserves	Proved Reserves millions of barrels Jan 1, 1981	Production in thousands of barrels daily in 1980	Proved Reserves/ Production Ratio in years*	Oil Consumption as % of Production in 1978
Mexico	1	44000	1960	61.5	70
Malaysia	2	3000	280	29.4	59
Egypt	3	2900	585	13.6	46
India	4	2580	185	37.9	195
Argentina	5	2457	490	13.7	101
Oman	6	2340	280	22.9	7
Syria	7	1940	165	32.3	47
Brunei	8	1710	230	20.4	1
Tunisia	9	1652	100	44.6	45
Brazil	10	1300	190	18.6	565
Angola-Cabinda	11	1200	150	21.8	9
Colombia	12	800	125	17.4	95
Trinidad and Tobago	13	700	215	8.9	13
Peru	14	650	190	9.3	89
Total of the above		67229	5145	35.8	83
Total excluding Mexico		23229	3185	20.0	88

* The proved reserves to production ratio expressed in years is derived by multiplying the preceding column, production at a daily rate, by 365/1000 to estimate annual production in millions of barrels, then dividing this estimate into the preceding column, proved reserves. Thus Mexico's production in 1980 at 1.96 million b/d equalled 715.4 million barrels a year. The reserves of 44,000 million barrels, divided by 715.4 equal 61.5 years.

Three of these countries, Brazil, India and Argentina, are net oil importers, the first two being significant ones. Brazil's oil imports increased by more than 200,000 b/d between 1973 and 1978. Along with Mexico these countries are in fact the four largest economies among all developing countries. They all have considerable scope for increased domestic consumption of oil if the rate of new finds is encouraging and production increases. Because of the dominance of Brazil as a consumer among this group of countries their oil consumption represented a high proportion of their oil production. Even with Mexico included, the surplus of production over consumption rose by only some 540,000 barrels a day between 1973 and 1978, though Mexican oil production rose considerably and has risen much more since.

SUPPLY AND DEMAND (Table 28)

The most important single source of energy consumed in non-OPEC developing countries is oil. Total oil consumption in these countries in the year 1979 amounted to 390 million tons compared with indigenous oil production of 264 million tons. Net oil imports to meet demand amounted to 126 million tons. In other words, 32 per cent of oil consumed in this group of countries depended on imports from countries outside the group. In the year 1967, this oil dependence on imports was as high as 56 per cent. Owing to increased oil prices since 1973-74 the annual growth rate of oil consumption has dropped considerably, resulting in an overall decline in the volume of net oil imports associated with a rise in production in countries such as Mexico, Malaysia and Egypt.(8)(6)

Natural gas resources have largely remained undeveloped. The lack of established markets has discouraged the construction of gathering and distributing facilities. Except for a few countries where energy consumption is relatively diversified or where it is the main or single source of indigenous energy, natural gas in non-OPEC developing countries has been produced to meet export demand or for use in maintaining pressure in oil fields. Nevertheless, use of natural gas in this group of countries has been increasing. Growing domestic markets are expected to require large production increases in countries like Pakistan and Egypt. Exports from Afghanistan to the USSR, from Bolivia to Argentina and from Brunei to Japan are also expected to grow in the future.

With the exception of India, Korea, Zimbabwe and South Africa, the coal mining industry in the majority of non-OPEC developing countries is generally very small - a one-mine or one-field activity. Use of coal in these countries is also much less extensive and depends on the availability of domestic supplies, the quality of the resources, and the relative prices of coal and oil. In exceptional cases, namely India and South Africa, coal is the single largest source of energy. Both of these countries meet their coal requirements entirely from domestic sources.(3)

The potential for the development of hydro-resources in non-OPEC developing countries is considerable, but as yet only a small fraction of this potential has been utilized. As most hydroelectric projects are massive undertakings, their capital costs are generally very high and difficult for developing countries to finance. In the context of the North-South dialogue and the Brandt Commission report, this sector is one of the most important among those meriting international co-operation.

Development of nuclear and other sources of energy in most of the non-OPEC developing countries is still in the very early stages.

BASIC PROBLEMS AND POLICY OPTIONS

A World Bank report entitled 'Energy In The Developing Countries', August 1980(9) describes these problems and policy options as follows:

"The appropriate energy policies differ among developing countries. Among those that import oil, the middle income countries*, especially those that are already semi-industrialized, share many of the energy problems of the developed countries. The switch from traditional to commercial forms of energy has already largely taken place, although there remain areas or pockets in which traditional forms of energy still dominate. These are the countries with fast rates of economic growth, whose energy requirements oblige them to buy large quantities of oil on the world market. Most of their commercial energy is used in industry, power generation and transport. These countries have been able to finance their imports by a combination of expanded exports and large foreign borrowings, and some of them have taken appropriate price measures to reduce the growth of demand for oil. They have been able to maintain reasonable, albeit reduced, rates of growth while beginning to make essential adjustments in the structure of their economies. Uncertainties about future markets for their exports, the availability and price of oil, and the extent to which further net foreign borrowings will be prudent, emphasize the need for these countries to exploit domestic resources more fully and to formulate policies and programmes to maximize the efficiency with which commercial energy is used.

Low income countries* derive a half or more of their total energy from wood and agricultural or animal wastes. But many of them depend heavily on petroleum for commercial energy, and are short of the resources needed to develop their own energy supplies. They too must be concerned with the energy efficiency of their development since investments made today will determine their energy requirements as modernization accelerates. One of the difficult choices they face is how to stop the rapid depletion of forests and soil fertility without unduly stimulating the use, and therefore the import, of petroleum. Essential elements of the solution are afforestation, small hydroelectric installations, the more efficient design of cooking stoves and more use of coal. In the longer term, local applications of solar, biomass and other renewable forms of energy hold promise of more abundant energy in rural societies at lower economic and environmental costs.

Some poor countries such as India have a large industrial base and therefore share some of the problems of the middle income countries*, while some of the latter have considerable populations still largely dependent on traditional fuels. The degree of dependence on imported oil also varies widely among countries at similar income levels. Other differences among countries have a bearing on their ability to plan and administer rational energy policies. Some, for example, have been producing coal, oil or natural gas for years and possess national companies with considerable experience in the industry. Others have weak or inexperienced institutions which need strengthening if they are to be capable of planning an energy strategy, advising government about the choice of investments and the management of demand, and informing and educating the public about the importance of conserving energy."

The economies of most of the non-OPEC developing countries, especially the energy deficient countries, do not possess the autonomous strength to adjust easily to the new energy situation. This has been a fundamental constraint on the achievement of maximum progress in the development programmes of these countries.

---

* Low income developing countries are those with a gross national product (GNP) per person below US$360 in 1978; middle income developing countries are those with a GNP per person of US$360 and above in 1978.

The use of energy in these countries will continue to increase as they modernize and industrialize their economies. At the same time, unlike the developed industrialized countries, many of them lack the ability to reduce oil use significantly without hindering development. Their per capita energy consumption is still very low and what little energy they consume is for essential economic purposes such as agriculture, industry and transport. The average per capita commercial energy consumption is less than one tenth the average prevailing in the industrialized countries.(3)

In order to expand, develop and diversify their indigenous and non-conventional sources of energy and build up the related infrastructure, the non-OPEC developing countries will require technology from developed countries and investment funds on a substantial scale.

The problems confronting the developing countries are well recognized in Saudi Arabia. In June 1981 HRH Crown Prince Fahd ibn 'Abd al-Aziz stated that:

> "Our prime concern is to help the Third World countries which are in dire need of the piaster before the dollar .... our stand in favour of moderation on prices derives from our strong ties of friendship with the Third World countries, for which we have no desire to create economic problems."(10)

ENERGY PROJECTIONS

Co-operation Scenario

At the beginning of this chapter I made an analysis of the oil and energy data along with the estimates of population and economic activity for the leading countries considered here. Now they account for about 43 per cent of the globe's inhabitants, but will grow to about 47 per cent in 2000.(11)

I wish to emphasize here my appeal to all countries in the other groups analysed in this book to start working more effectively along the lines of the co-operation scenario discussed in Chapter 14 in order to advance significantly the living standards of the great majority of people in non-OPEC developing countries. Their real incomes are but a small fraction of the average prevailing in the industrialized countries and those now enjoyed by a minority of the populations in the OPEC member countries. Sizeable increases in per capita energy consumption are seen as an indispensable pre-condition for significant advances in living standards for the non-OPEC developing countries over the next twenty years. It is already evident that energy supply constraints could be a major obstacle to the achievement of this objective if appropriate and effective worldwide initiatives are not taken soon.

Both the World Bank and the OPEC Long Term Strategy Committee worked out methods of classifying the large number and diverse assortment of countries included in the developing category. There is also the United Nations categorization. Before making brief reference to these, I wish to discuss the parameters I consider significant in the context of the current situation and the scenario prospects for the next twenty years. The rank order analysis mentioned in the earlier part of this chapter is just a useful and convenient way of identifying the largest countries in the developing world category and examining the differences among them for illustrative purposes. The list given below comprises a qualitative, multi-variate approach covering all the important characteristics in the present context.

- Population, age distribution and rate of growth.

- GNP per capita and growth rate.

- Export percentage of GNP and change in it since 1973.

- Cost of oil imports as a percentage of total export value and change in it since 1973.

- Oil imports as a percentage of total primary energy consumption.

- Abundance or scarcity of energy resources; proved reserves and prospects of future discoveries.

- Political institutions: stability and resourcefulness in developing effective policies.

The World Bank study(9) includes the OPEC countries. This study adopted the fifth of these variables for its basic classification, but also looked at GNP per capita and the extent to which developing countries did or did not have access to fuelwood or other traditional sources of energy. The OPEC Long Term Strategy Committee proposed grouping oil importing developing countries according to their per capita income, their level of economic development, and the size of their payments for oil imports as determined from time to time. Furthermore, it reviewed the potentialities for aid in the development of indigenous energy resources and concluded that the largest group of developing countries, at least 36 in number, has favourable commercial energy prospects. It decided that the scope and intensity of assistance for the development of indigenous commercial energy resources would have to be further elaborated in co-operation with the industrialized countries and international institutions, such as the World Bank and the United Nations, in addition to the developing countries concerned; but it was clear that a separate organization would have to be established, jointly with the industrialized countries, in order to channel the financial aid effectively, and in co-ordination with the technical expertise necessary. The role of the OPEC Special Fund was also considered in this context.

Special Case Variant of the Co-operation Scenario

In 1978 these non-OPEC developing countries had a total population of some 1660 million people with a per capita energy consumption of less than one tenth of the average level prevailing in the industrialized countries.(1)(3) Even the general features of the co-operation scenario yield a rather unsatisfactory advance in energy production and consumption for the non-OPEC developing countries. On the basis of the projections discussed below these imply growth of energy consumption per capita averaging 2.2 per cent annually between 1979 and 2000, but decelerating over time.

This leads me to propose a special variant case for the non-OPEC developing countries. Its main feature is that growth of energy production and consumption during the 1990s is one per cent higher each year than in the co-operation scenario. The realization of this special case will depend on bigger increases in energy production among these countries, but, because of the uneven distribution of energy reserves between them it will certainly involve a very significant stimulus in energy trade among the non-OPEC developing countries as well.

The special case assumes that OPEC countries, the industrialized countries and the centrally planned economies of the Communist bloc all make specific and significant efforts to develop the energy sector, especially oil exploration and production, in the developing countries through direct investment, technology transfer, aid and trade. Its important feature is that a high proportion of the investment, aid and technology will need to be specifically related to the energy sector in order to ensure the requisite growth of production. The transnational oil companies also need to be encouraged to make bigger efforts than they have done so far if the incremental growth of energy production and consumption is to be realized. This special case variant of the co-operation scenario assumes an extra 1.9 million b/d oil production and consumption in the year 2000 as compared with the basic co-operation scenario, and an extra production and consumption of 3.8 million b/d oil equivalent of all energy forms, inclusive of the oil increment. The other 1.9 million b/d oil equivalent will consist mainly of solid fuels, hydroelectricity and natural gas, though nuclear, solar and shale oil should be able to make some incremental contributions in various countries.

The fundamental problem for the non-OPEC developing countries is the low proved oil reserves to production ratio. If Mexico, with 63 per cent of the proved reserves among all these countries by the beginning of 1981 and a reserves to production ratio of 61.5:1 year is excluded, then the proved reserves to production ratio of all the others is just under 20:1 year on the basis of 1980 production volumes and the proved reserves estimates. Since production inclusive of Mexico is projected almost to double between the estimated volume of 5.2 million b/d in 1979 and 10 million b/d in 1990 in this scenario, it is a matter of considerable urgency that immediate steps be taken to increase the rate of exploration very significantly in the 1980s if there is to be any hope of limiting the decline of oil production in most non-OPEC developing countries, other than Mexico and just a very small number of other producing countries.

Oil consumption in this scenario is projected to rise from 8.4 million b/d in 1979 to 11.5 million b/d in 1985, an annual growth rate of over 5 per cent. Subsequently, it will grow more slowly to just over 13 million b/d in 1990, but then there follows the prospect of an absolute decline during the 1990s, unless the special case conditions discussed above are realized. Even in this case the prospect is that oil consumption might fall marginally, unless oil exploration and drilling activities in the 1980s are both very intensive and successful.

However, total primary energy consumption is projected to continue to rise rapidly. An average annual growth of 5.8 per cent is projected from 1979 to 1985, and then 5.0 per cent from 1985 to 1990. Between 1990 and 2000 the rate of growth is projected to fall to 3.6 per cent in the basic co-operation scenario, but it should be possible to raise this to about 4.6 per cent annually in the special case discussed above. On the basis of these two alternatives, the per capita consumption in the year 2000 is compared with the estimate for 1979. There is a very limited closing of the gap between the developing and the industrialized countries, but the difference remains very large in all scenarios in the year 2000.

	Primary Energy Consumption per capita in non-OPEC Developing Countries	Expressed as a Percentage of per capita consumption in industrialized OECD countries
	barrels of oil equivalent	%
1979	3.4	9.2
2000		
Special Case	5.8	13.6*
Co-operation Scenario	5.3	12.4
Neutral Scenario	4.2	11.2
Confrontation Scenario	3.4	10.7

* as percentage of co-operation case in OECD countries.

Neutral Scenario

Oil consumption is projected to peak at a lower level of about 11.4 million b/d in 1990 and then to decline in the 1990s.

Solid fuels, natural gas, nuclear and hydroelectricity are all projected to rise rapidly, especially in the 1990s, even in this scenario. In consequence, the oil share of the energy mix drops sharply from the still high level of 48 per cent in 1990 to 25 per cent in the year 2000. This proportion is high compared with that projected for the industrialized countries in this scenario (18 per cent) and for the centrally planned economies of the Communist bloc (17 per cent) at the same point in time.

Growth of primary energy consumption is projected to grow at 4.2 per cent on average each year from 1979 to 1985, then at 4.0 per cent over the following five years. During the 1990s it is projected to rise more slowly at 2.8 per cent each year, as the prospective fall in oil consumption starts to make itself felt. If, however, in spite of the generally less favourable international political and economic conditions, it is still possible to secure special efforts for developing energy production in these countries during the next ten years, then it might be possible to maintain consumption growth at nearly 4 per cent annually through the 1990s.

Confrontation Scenario

The outlook in this scenario is very bleak for the developing countries. Oil consumption is projected to rise only by some 1.3 million b/d between 1979 and 1990, and then to fall considerably more sharply afterwards, so that the oil share of the energy mix falls to 23 per cent in the year 2000.

Growth in the production and consumption of other energy forms is foreseen as being significantly slower than in the neutral scenario, both as a cause and consequence of the unstable political and economic conditions. Growth of total primary energy consumption is projected at 2.5 per cent average each year through the 1980s, falling to 2.2 per cent in the 1990s (see Table 51). In the unlikely event of it being possible to mount the type of special case effort discussed above under the conditions prevailing in this scenario, then it should be possible to increase growth of energy consumption to an annual average of about 3 per cent in the 1990s from the rise in indigenous production which should follow from the big increase in international investment in the energy sector of developing countries. Increased energy trade is also essential among these countries, though this could be difficult to achieve under the postulated conditions.

In the absence of the special variant case, the growth of energy consumption through the period is only marginally higher than the growth of population in the developing countries which averages about 2.2 per cent from 1980 to 1990 and then some 1.9 per cent from 1990 to 2000. With limited scope for improvements in the energy growth coefficient with respect to GNP growth in these countries, as far as one can see, the confrontation scenario implies negligible growth of per capita GNP in non-OPEC developing countries. This prospect alone should be sufficient to convince the world's political leaders to take immediate and effective measures to ensure that the conditions of the confrontation scenario do not prevail. Worldwide political instability could have unforeseen and highly disruptive consequences, especially for the disadvantaged developing countries. In my view, positive international action in the energy sector, particularly in the non-OPEC developing countries, is likely to prove a prerequisite for avoidance of this confrontation scenario.

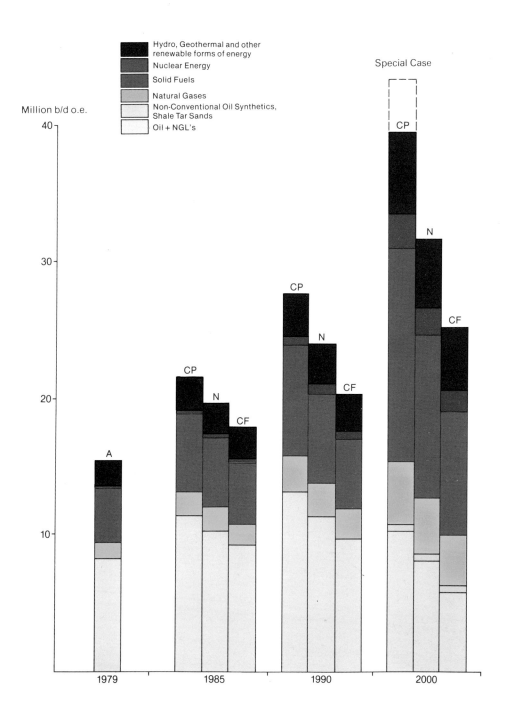

References

(1) World Bank Atlas 1980 and earlier issues. The World Bank, Washington DC.

(2) World Development Report. The World Bank, August 1980.

(3) World Energy Supplies 1973-1978. United Nations, 1979. Statistical Papers, Series J No.22.

(4) International Financial Statistics, Yearbook 1980 and various monthly issues. International Monetary Fund.

(5) World Energy Conference, Survey of Energy Resources, 1980.

(6) Oil and Gas Journal, 29 December 1980.

(7) Report of the Fuel Policy Committee, Govt. of India, 1975.

(8) BP Statistical Review of the World Oil Industry, 1979 and earlier issues. British Petroleum Company Limited.

(9) Energy in the Developing Countries. World Bank 1980.

(10) HRH Crown Prince Fahd ibn 'Abd al-Aziz, Interview with the Beirut daily al-Bayrak published in Saudi Press Agency of 28 June 1981 and reported in English in Middle East Economic Survey Vol. XXIV No.38 dated 6 July 1981.

(11) United Nations Population Statistics and Forecasts, 1950-2000. Selected Demographic Indicators, 1980.

21
THE USSR ENERGY SITUATION

THE USSR ENERGY PROBLEM SUMMARIZED

Several problems but also some opportunities confront the Soviet Union. The declining oil reserves to production ratio, doubts about the future supply of Western technology, and the falling rate of growth in oil production all seem to point to an inevitable decline in net oil exports. Natural gas exports should continue to rise.

Economically, this prospect of the falling volume of oil exports might not cause Soviet planners too many problems because of the higher unit prices which should be realized. In addition, considerably higher prices for gold, arms and timber exported outside the Council for Mutual Economic Assistance (CMEA) have made substantial contributions to the Soviet external current account balance of payments and hard currency earnings.

The problem is mainly a political one. There is a potential loss of control over Eastern Europe as that region's energy demands grow and the USSR is unable to meet them in spite of its own relatively abundant resources, particularly of coal and gas. The record of the last few years suggests that of all the primary energy resources in the USSR, only production of natural gas has consistently met the targets set for it. If recent experience continues for another five years or so, the energy self-sufficiency of the CMEA is in jeopardy. In the medium term, natural gas exports might substitute for oil exports to the West to some extent, although it is doubtful whether production will grow fast enough for this purpose and prove sufficient to make up for the shortfalls in growth of oil, coal and nuclear power production.

RESERVE BASE (Table 29)

The Soviet Union has ample energy resources. Estimated proved reserves of conventional fuels are estimated to be over 135 billion tons oil equivalent. This total is composed of coal 78.8 per cent, natural gas 14.0 per cent and oil 7.2 per cent.(1)

The Soviet Union's recoverable hard coal reserves are marginally less than those of the USA. But in the case of oil and gas the Soviet Union's proven reserves are almost three and a half times those of the US reserves. Its proved reserves to production ratio is only 15:1 for oil, but 60:1 for natural gas.(2)(3)

Among the non-conventional resources, oil shale reserves are estimated to be 6.8 billion tons oil equivalent.(1)

Uranium low cost reserves are estimated at 160,000 tons, or about 30 per cent of those in the USA.(1)

The USSR also has a concentration of the world's largest hydro-generating plants, though total output of hydroelectricity accounts for less than 4 per cent of apparent primary energy demand.

SUPPLY AND DEMAND (Table 30)

Continuing rapid industrial growth has meant that demand for primary energy in the Soviet Union has been rising at a rate well in excess of that in the industrialized countries of the West since 1973.

During the period 1967 to 1973 the average annual growth rate of total primary energy demand was 4.7 per cent. During the period 1973 to 1979 it was 4.6 per cent. By 1979, the Soviet Union's total primary energy demand had reached 1148 million metric tons oil equivalent. This was more than 60 per cent of the US level in that year. The percentage share of the different sources of energy used to satisfy this demand were: oil 38.4; natural gas 26.8; coal 29.8; nuclear energy 1.1 and hydroelectricity 3.9.(4) Ten years earlier oil accounted for only 33.5 per cent of Soviet primary energy demand, and natural gas for only 20.3 per cent. Coal was still the predominant form of energy used in the USSR in 1969, accounting for 41.9 per cent of the total demand. Coal, the most abundant fossil fuel, continued its relative decline in the Soviet Union after 1973, just as it had been doing previously. Moreover, in 1979 and 1980 there has been a rather more alarming absolute decline.

Up to 1975 indigenous oil production was increasing at a consistent average annual growth rate of nearly 7 per cent, reaching a level of 459 million tons in 1975. Since then the percentage rate of growth of Soviet oil production has declined to 5.9 per cent in 1976, 5.0 per cent in 1977, 4.9 per cent in 1978, and more dramatically to 2.4 per cent in 1979, then 2.9 per cent in 1980. Over 25 per cent of this oil production has traditionally been exported: to the CMEA member countries, plus North Korea, to some developing countries, Yugoslavia, Western Europe and Japan. The value of Soviet oil and gas exports is shown in the table below.(5)

USSR Exports in 1978 in Millions of US Dollars

	Oil	Gas
Western Europe	6,619	1,182
USA	159	–
Japan	113	–
Eastern Europe, Cuba North Korea & Vietnam	7,306	1,048
Developing Countries (including Yugoslavia)	1,203	21
TOTAL	15,400	2,251

These exports constitute a very important aspect of the Soviet Union's foreign trade and global policies.

Some observers believe that the Soviet Union wants to preserve its significant role in East European oil markets in order to have an instrument available for the application of political pressure and economic leverage. The Soviets realize that their political dominance in Eastern Europe is based in part on the economically integrated structure of CMEA and that a loosening of economic ties (and the lessening of the energy dependence of most of Eastern Europe on the USSR) could lead to greater political autonomy on the part of East European states. Albania and Romania took advantage of their lack of oil dependence on the USSR to successfully challenge Soviet control mechanisms. The Soviets also wish to make sure that any inroads made by Western oil companies into the oil-deficient East European states will not threaten the dominant Soviet position. Oil exports to Western Europe are used to acquire valuable hard currency and to help buy pipe for the gas grid. With the world oil prices increasing more rapidly than Soviet production costs, the USSR has all the economic incentives to preserve its sales to the West European oil market, but it may well reduce the volume or growth in the volume of these oil exports as it appears to have done in 1979. The modest level of Soviet oil exports to Japan is of political significance in the context of the Chinese oil exports to Japan which were growing until 1979. Soviet oil sales to the developing countries, such as India, are comparatively small in aggregate, but increasing. They are important in particular cases. These exports do not earn hard currency payments, but they can be used to counter potential Chinese influence in these countries.

The Soviet coal industry slipped progressively further behind the targets set for it during the 1976 to 1980 Five Year Plan. By 1980 it was more than 11 per cent or some 1.2 million b/d oil equivalent lower than the original target for that year. The annual rate of change in coal production looks rather ominous in the table below.

Annual Rate of Change in USSR Coal Production from preceding year

	%
1976	+1.5
1977	+1.5
1978	+0.2
1979	-0.6
1980	-0.4

Perhaps one of the reasons is the relatively small increase in funds allocated to investment in the coal industry, a rise of 33 per cent between 1970 and 1978. The Soviet oil industry benefited from a much larger increase of 112 per cent over the same period.

Natural gas is the one Soviet energy sector that has performed relatively well in recent years. The increase in production had made it possible to increase the quantity of gas exported, in part to pay for the vast quantities of steel pipe imported to develop the Soviet gas grid and make exports possible. Consumption of natural gas in the USSR has increased at an annual average rate of 7.8 per cent, representing an increase of nearly 2 million barrels a day oil equivalent from 1974 to 1979.

BASIC PROBLEMS

In spite of the fact that the Soviet Union has vast untapped resources of fossil fuels, its energy planners are facing mounting problems in satisfying the rapid growth of domestic energy demand while serving the vital interest of maintaining revenues from energy exports. The basic problems can be briefly assessed as follows:

- Roughly 70 per cent of the energy in the USSR is consumed on the Western side of the Urals, whereas 70 per cent of the known resources are located east of the Urals, in Siberia and the far north.

- The ageing Urals-Volga oil fields, long the Soviet Union's major source of petroleum, have passed their peak.

- The coal mines of the Moscow Basin are showing increasing evidence of exhaustion.

- Exploitation of Siberian and Arctic fuels is a high priority, but this involves enormous transportation and environmental challenges.

- There has been a move to more marginal relatively high cost production associated with a rather low rate of exploration in new areas.

- There have been no encouraging finds of giant new oil fields in recent years.

- Soviet oil technology is less advanced than in some other industrialized countries. The exploitation of the reserves in Siberia will depend heavily on the extent of Western trade and co-operation in providing pumps, pipe and other oil industry equipment.

In view of the above, the Soviet leadership has been assigning high priority to developing commercial nuclear power. However, construction of new power reactor installations appears to have lagged far behind the ambitious schedule set out in the Five Year Plan that ended in 1980. According to a report in Pravda Soviet nuclear generating capacity amounted to 13 GW by 1980 or some 33 per cent below the original Plan target for that year. It accounted for only 5.6 per cent of the total power output. The Soviet Union's nuclear lag has resulted not from public opposition or regulatory conflicts, as in the Western countries, but can be attributed to pervasive inefficiencies in its machinery industry. Apart from shortages of skilled labour, there is a scarcity of the necessary plant and equipment, such as large casting furnaces, heavy lathes and other machines needed to turn out the massive precision-tooled steel parts for nuclear reactors.

ENERGY PROJECTIONS

Co-operation Scenario

This scenario presupposes both a relatively favourable evolution in the relationship between the Soviet Union and the USA and peaceful coexistence and improved trading links between Communist and non-Communist countries generally. It also presupposes good conditions within the USSR which will favour the growth of the energy sector, especially natural gas.

At global level the effective development of Soviet natural gas seems likely to be one of the most critical elements in the evolution of a healthy world energy balance over the next twenty years. It is capable of achieving incremental export volumes and earnings from West European countries and Japan, as well as contributing to a harmonious transition away from oil dependence in the USSR and Eastern Europe. At the beginning of 1981 Soviet proved reserves of natural gas represented 35 per cent of the world total estimated by the Oil and Gas Journal. The proved reserves to production ratio stood at the relatively high level of 60:1 in terms of the 1980 level of production.

Soviet oil consumption is projected to continue to rise before peaking at about 11.2 million b/d in 1990 and then declining very slightly to about 10.7 million b/d in the year 2000. At that time its share of the energy balance will have fallen to about 22 per cent, after 31 per cent in 1990 and 39 per cent in 1979. Meanwhile consumption of natural gas is projected to grow rapidly from 6.1 million b/d oil equivalent in 1979 to 12.2 million b/d oil equivalent in 1990. By this time it is likely to become the most important contributor to the Soviet energy balance. By the year 2000 Soviet natural gas consumption is projected to grow by another 5 million b/d oil equivalent so that by the millenium it is expected to account for about 35 per cent of total energy consumption.

Soviet energy consumption is projected to rise at 4.7 per cent annually between 1979 and 1985 in this scenario. This is similar to the average rate of growth recorded over the last decade. In the second half of the 1980s, Soviet energy consumption is projected to grow at 3.9 per cent on average each year, falling to 3.1 per cent in the 1990s.

Neutral Scenario

Energy consumption grows more slowly in this scenario and follows the historical trend of a declining rate of increase. After rising at an average 4.4 per cent from 1974 to 1979, it slows to 4.0 per cent on average from 1979 to 1985 and then to 3.2 per cent in the second half of the 1980s. A further slow-down to a growth rate of 2.4 per cent annually occurs in the 1990s as the extraction of new reserves proves more costly and the increase in resources devoted to energy production proves inadequate to sustain a higher rate of growth.

The oil share of the energy consumption mix falls slightly less in this scenario as the growth in other forms of energy is significantly slower than in the co-operation case discussed above. Nevertheless, natural gas is expected to become the most important source of energy for Soviet consumers by 1990. In the year 2000 oil is projected to rank only third, behind both natural gas and solid fuels in the USSR. Oil consumption of 10.2 million b/d compares with a projection of only 6.8 million b/d in the USA in the year 2000, with total energy consumption almost the same in both countries at near 43 million b/d oil equivalent.

This comparison can be taken further by noting that in the USSR oil and gas together are seen as declining from 66 per cent of the energy balance in 1979 to 57 per cent in 2000, whereas in the USA the prospective change in this scenario is much more dramatic: from 72 per cent in 1979 to 33 per cent in 2000.

Confrontation Scenario

Soviet energy consumption rises at a much slower rate than has been customary in the past, effectively reflecting supply constraints. The rate of increase slows from 3.2 per cent on average over the years from 1979 to 1985 and then to 2.1 per cent during the second half of the 1980s. After that, energy consumption grows at only some 1.7 per cent annually as the energy sector suffers relative neglect, new reserves prove difficult to exploit, and there are minimal imports of technology and minimal co-operation and trade with the USA, Western Europe or Japan.

Soviet oil consumption declines by 1.5 million b/d over the ten years from 1990 to 2000. Natural gas provides well over half the net increase in energy consumption, followed by nuclear energy, with solid fuels and hydroelectricity providing modest increments.

References

(1)   World Energy Conference, Survey of Energy Resources, 1980.

(2)   Oil and Gas Journal, 29 December 1980.

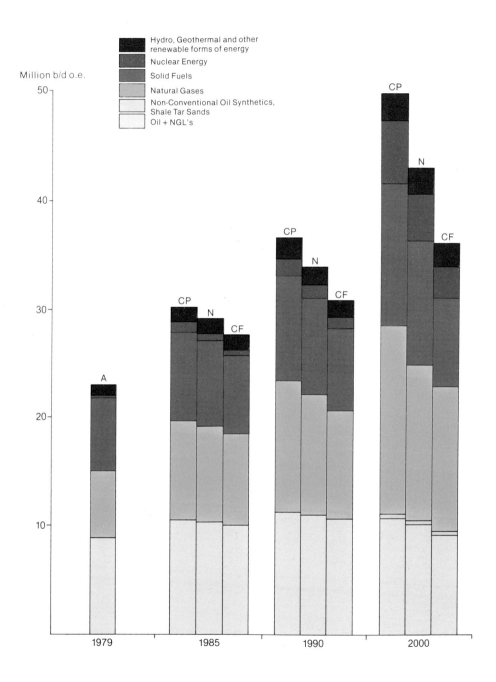

(3) Estimates of Soviet production of oil and gas in 1980, Tass radio broadcast, 23 January 1981.

(4) BP Statistical Review of the World Oil Industry, 1979 and earlier issues. British Petroleum Company Limited.

(5) Soviet Oil, Gas Exports Value Hit High. Oil and Gas Journal, 30 July 1979.

22
THE EAST EUROPEAN ENERGY SITUATION

THE EAST EUROPEAN ENERGY PROBLEM SUMMARIZED

The East European energy problem appears more grave than that in some other areas of the world, though it is a good position in the sense that less than one quarter of its present energy demand is met by oil. Some appreciation of the differences between individual East European countries can be obtained from an examination of the table below. This shows the percentage distribution by type of primary energy consumption and the extent of import dependence. It can be seen that Bulgaria is the most vulnerable country in having both the greatest dependence on imports and reliance on liquid fuels. Oil imports accounted for 18.4 per cent of the region's primary energy demand in 1979, compared with only 10.2 per cent in 1969.

Commercial Energy Consumption Mix of East European Countries and Relative Dependence on Imports in 1978

	Bulgaria	Czechoslovakia	East Germany	Hungary	Poland	Romania
	%	%	%	%	%	%
Solid Fuels	43.0	70.8	72.6	36.0	82.6	21.7
Liquid Fuels	43.6	21.4	20.2	38.0	11.3	26.0
Natural Gas	9.9	6.8	6.5	24.3	6.0	51.0
Hydro and Nuclear Electricity	3.5	0.9	0.7	1.7	0.2	1.3
Total Commercial Energy in million metric tons oil equivalent	30.1	77.6	81.2	25.1	133.2	59.7
Net Energy Imports as % of Consumption	71.7	30.4	32.5	48.4	(6.7) Net Exporter	10.7

Source: United Nations, World Energy Supplies 1973-78 Series J No.22

Part of the solution to this adverse trend could be greater exploitation and use of indigenous solid fuels, such as East German lignite and Polish hard coal. Nuclear co-operation with the Soviet Union holds out some hopes for Czechoslovakia. In Romania, the depletion of the old established oil fields represents a special local problem. Increased gas imports and, perhaps, more electricity imports from the Soviet Union via high voltage lines might contribute to an alleviation of the prospective East European energy problem. The future nuclear contribution remains somewhat problematical. The USSR might also decide to divert oil supplies from other export markets in favour of Eastern Europe for political reasons, in spite of pressure to earn hard currency elsewhere.

Conservation of oil use and improved efficiency of energy use seem likely to be an important way of resolving the prospective shortages in Eastern Europe. However, not too much progress appears to have been made so far.

RESERVE BASE (Table 31)

The East European countries vary widely in energy resources.

Overall, estimated proved reserves of conventional recoverable fuels are just over 30 billion tons oil equivalent. This total is composed of coal 98.1 per cent, oil 0.8 per cent and natural gas 1.1 per cent.(1) Comparatively, Romania is the richest in oil reserves, but even she is increasingly dependent on oil imports nowadays.

Among the unconventional resources, oil shale reserves are estimated to be close to 0.2 billion tons oil equivalent.

Uranium reserves are estimated at 135,000 tons.(1)

SUPPLY AND DEMAND (Table 32)

The demand for primary energy in the East European countries has been rising at a steady average annual growth rate of about 3.6 per cent since 1973, having fallen from an average of 4.5 per cent before that.

By 1979, the size of East European countries' total primary energy demand had reached 441.8 million metric tons oil equivalent.

The percentage shares of the different sources of energy used to satisfy this demand in 1979 were: oil 22.9; natural gas 13.9; coal 61.1; nuclear energy 0.9 and hydroelectricity 1.2. Ten years earlier coal had accounted for 73.6 per cent of the total and oil for only 15.8 per cent.(2)

The average annual growth rate of domestic oil production has declined from 1.9 per cent during 1967 to 1973 to only 0.7 per cent from 1973 to 1979. Total oil production in 1980 was 14.7 million metric tons compared with a planned target of 19.6 million tons. On the other hand demand for oil has been growing at very much higher annual average rates (+ 11.5 per cent from 1967 to 1973 and + 5.1 per cent during 1973 to 1979). Consequently, oil import dependency has grown from 56 per cent in 1967 to 80 per cent of oil consumption in 1979. A significant part of these oil imports is supplied by the USSR. The remainder comes from oil exporting countries of the non-Communist world.

THE BASIC PROBLEM

The basic factors responsible for creating the energy problems in the East European countries are limited indigenous energy resources and a rising preference for oil and gas rather than coal. This results in growing dependence on imported oil, and increasingly on non-Soviet oil from OPEC countries and Mexico which involves hard currency or barter trade. A hard currency shortage and declining demand for East European machinery and equipment in the oil producing states could impose severe limitations on the volume of oil imports from OPEC states. Iran and Iraq have supplied increased quantities of oil to East European countries, but the conflict between these two oil exporters has had an adverse effect on this trade.

ENERGY PROJECTIONS

Co-operation Scenario

The average level of dependence on oil has never been high among the centrally planned economies of Eastern Europe. What is important is the apparent lack of flexibility in the East European energy system compared with that of the USSR. In Romania, which has a long history of oil production, the level of output has been declining for some years. In 1979, oil accounted for 24 per cent of Eastern Europe's oil consumption. Growth in oil consumption of about half a million b/d is projected through to 1990, but, notwithstanding that, the oil share of the energy balance is likely to decline by about five percentage points over this period. Much more significant quantities of indigenous solid fuel and natural gas from the USSR are expected to come into use. Thus all forms of energy consumption are seen as growing at an annual average rate of over 4 per cent up to 1990 in this scenario, or marginally more than over the five years up to 1979.

During the 1990s the increase of energy consumption is projected to decline to an average of just under 3 per cent each year. An absolute fall in oil consumption seems inevitable, so that oil's share of the energy mix declines to only 11 per cent in 2000. Even though nuclear should be capable of starting to make a significant contribution, there are limits to what can be expected from all energy forms other than oil even in this co-operation scenario.

Neutral Scenario

As can be seen in Table 53 in Part III oil consumption continues to rise slowly in this scenario up to about 1990 but then declines more rapidly, so that by the year 2000 it is lower than it was in 1979. Nevertheless, the oil share of the energy mix follows a similar profile through time as it does in the co-operation scenario discussed above.

Use of other forms of energy grows more rapidly than oil through the 1980s. In total, energy consumption rose at an annual average of 3.8 per cent during the period from 1974 to 1979. In this scenario it is projected to increase 3.4 per cent annually from 1979 to 1985, then at less than 3 per cent each year from 1985 to 1990. During the 1990s it is likely to fall further, to about 2.4 per cent each year, reflecting supply constraints, faster rising real energy prices, but also very slow population growth. UN population projections show the most likely rate of increase in the East European population is less than 0.5 per cent each year from 1990 to 2000.(3) Solid fuels remain the dominant form of energy, though their share of the consumption mix falls from 61 per cent in 1979 to 54 per cent in 2000.

Confrontation Scenario

In this scenario energy consumption is projected to rise at less than 2 per cent each year up to 2000. Oil consumption remains relatively static at about 2 million b/d during the 1980s before falling to about 1.3 million b/d in 2000 because of supply constraints. This will represent only 10 per cent of total energy consumption at that time, even though

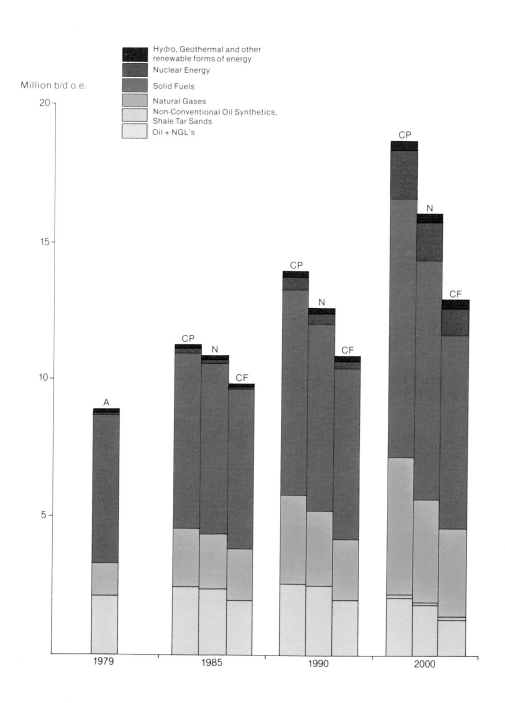

Figure 17. Actual and Prospective Energy Consumption Mix of Eastern Europe (Million barrels a day oil/o.e.)

other forms of energy are projected to grow rather slowly. Natural gas should provide the largest increment, some 2 million b/d oil equivalent between 1979 and the year 2000, even in this scenario.

References

(1) World Energy Conference, <u>Survey of Energy Resources</u>, 1980.

(2) <u>BP Statistical Review of the World Oil Industry, 1979</u>. British Petroleum Company Limited, 1980.

(3) <u>United Nations Population Statistics and Forecasts, 1950-2000. Selected Demographic Indicators, 1980.</u>

23
THE ENERGY SITUATION OF THE PEOPLE'S REPUBLIC OF CHINA

CHINESE ENERGY PROBLEM SUMMARIZED

An impression of the low level of energy consumption in China can be gained from the fact that in 1979 the USA used nearly 14 times as much energy per person as did China, and the USSR used 7 times as much per person.(1)(2)(3)

As in so many parts of the world, coal accounts for a high proportion of China's hydrocarbon resource base. But the relationship of China's coal reserves to its population compares disadvantageously with similar relationships where coal is abundant. The USA has 9.4 times more coal reserves per person than China, and the USSR has 5.3 times as much. At the 1979 rate of coal use in China economically proved reserves are sufficient to meet demand for 130 years, but it is relevant to point out that Chinese coal consumption is estimated to have increased by 80 per cent between 1969 and 1979. For comparison, proved reserves to consumption ratios for coal in 1979 were 309 years to 1 in the USA and 233 years to 1 in the USSR.(1) To complete this perspective, one can observe that, in terms of GNP per capita in 1978, the average US national was 42 times better off than the average Chinese, and the average Soviet citizen was 16 times better off, based on World Bank estimates.(3)

Looked at in these terms, the probability is that by the end of the twentieth century the Chinese need for renewable energy resources is likely to be as great as that of any other major nation or region, if not greater, particularly if the country proceeds with rapid and sustained industrialization in the meantime.

The fact that the Chinese proved reserves to production ratio for oil is little more than 25:1 is significant given the very low level of oil consumption per capita.(4) Forecasts of Chinese oil production rising to 4 to 6 million b/d in 1985 and 6 to 8 million b/d in 1990 are very optimistic. China has to resolve the problem of potential conflict between its intention to industrialize and the need to conserve scarce oil reserves. In addition there is the desire to earn foreign exchange in order to import some technology from Japan and other industrialized countries. The supply of increasing volumes of Chinese oil would provide an attractive means of financing such trade from the point of view of prospective technology supplying oil importers.

Thus China appears to face a short term energy dilemma related to the rate of production, consumption and export of oil. In the longer term it could face a similar dilemma in formulating policies for depleting its coal reserves, unless vast additional reserves are found.

One of the more promising areas worldwide where large oil-bearing reservoirs may be located is offshore China. If such potential finds were realized during the next few years the medium term prospects for the Chinese oil industry and the economy would be transformed.

RESERVE BASE (Table 33)

Estimated proved reserves of conventional hydrocarbon resources in the People's Republic of China are close to 67.2 billion tons oil equivalent. This total is composed of coal 95.1 per cent, oil 4.0 per cent, and natural gas 0.9 per cent.(5)

Among the unconventional resources, oil shale reserves are estimated to be close to 20.7 billion tons oil equivalent.

Uranium reserves are estimated at 166,000 tons.(5)

SUPPLY AND DEMAND (Table 34)

During the period 1967 to 1973 the average annual growth rate of total primary energy demand is estimated at 12.4 per cent, but from a very low base. During the period 1973 to 1979 this figure had come down to little more than 6 per cent annually.

In 1979, the size of China's total primary energy demand had reached some 523 million tons oil equivalent. The percentage shares of the different sources of energy used to satisfy this demand were: coal 78.5; oil 17.4; natural gas 2.4 and hydroelectricity 1.7.(1)(2)

During the past decade indigenous oil production has been increasing from a very low level to more than 2 million barrels a day in 1979. In that year production rose less than 2 per cent, followed by stagnation in 1980. About 15 per cent of this oil production was exported to the non-Communist world, mainly Japan. Exports were rising at a rapid pace until 1978. In 1973 exports were only 2 per cent of production.

POLICY OPTIONS

China has built new pipeline capacity and deep-water port facilities at Dairen, significantly expanded its tanker fleet, and achieved faster growth in productive capacity than in refining capacity. All of these developments suggest an interest in trading some of its oil on world markets. However, unless large new oil reserves are found soon, the combination of high unit prices for exports, rising domestic demand, and a reserves to production ratio which is none too high, will be likely to act as a considerable constraint on growth of oil exports.

Growth in coal output was rapid until 1978, but there has been little increase between 1978 and 1980. It has lagged behind the demand arising from the rapid expansion of thermal-fired electric power plants. This has required the modification of many existing coal-fired plants to burn oil as well. It has also probably triggered the decision to put much new generating capacity (being expanded at 10 per cent annually) on a dual-fired basis.

Potential oil exports suffer from constraints of ideology and high capital costs involved in their development. Until fairly recently the Chinese were reluctant to purchase the equipment and advanced technology from abroad needed to explore offshore provinces.

Furthermore, the virtual absence of external debt relieves China of the need to export more oil than required to finance either its limited purchases of advanced equipment from the industrialized countries or the promotion of its political objectives. Should these revenue needs increase, pressure would mount to increase the volume of oil exports and perhaps coal exports too.

ENERGY PROJECTIONS

Co-operation Scenario

In the last year or two the previous strong growth of Chinese oil production has not been sustained, with consequences both for exports to Japan and domestic consumption. In this scenario a relatively optimistic view has been taken about short to medium term prospects so that Chinese consumption is projected to rise to 2.2 million b/d in 1985. In the absence of further significant oil discoveries offshore or onshore it seems unreasonable to assume any further growth in Chinese consumption, which is therefore projected to remain unchanged over the fifteen years until 2000. This could prove too pessimistic a view because of the favourable prospects for new discoveries offshore. The Chinese have already signed agreements with many foreign oil companies and countries and a programme of exploration and drilling for oil is going ahead.

The great need and potential for increases in Chinese oil consumption can perhaps best be grasped by means of a comparison with the USA. In 1979, Chinese oil consumption was less than three quarters of a barrel per capita per year, whereas in the USA it was about 30 barrels per capita: more than forty times as much. Oil's share of Chinese energy consumption in 1979 was the lowest among the major countries and regions analysed in this book. This can be seen from Table 2 in Part III. The 1979 proportion of oil in the Chinese energy balance is similar to that projected in many other countries and regions for the horizon year 2000. In China, the oil share of the energy mix is projected to decline even from the low level of the recent past to about 9 per cent in 2000 in this scenario. This assumes Chinese oil discoveries do not reveal a further ten or twenty billion barrels at least during the next ten years. Proved reserves in China of 20.5 billion barrels estimated at the beginning of 1981 would be sufficient to support a level of only 2.2 million b/d production for 15 years before hitting a reserves to production ratio of 10:1.

Coal is the dominant form of energy used in China, accounting for over 78 per cent of the total in 1979. This was a higher proportion than anywhere else among the major countries or regions of the world. As it seems to have more potential for growth in China than any other form of energy, it is projected to account for 67 per cent of the rise in total energy consumption between 1979 and 2000 in this scenario. At the horizon year solid fuels are expected to represent over 70 per cent of Chinese energy consumption. Total primary energy consumption is projected to rise at an annual rate of at least 4.5 per cent up to 1990, declining to about 3.3 per cent in the 1990s. This will permit a significant rate of growth in per capita consumption.

Neutral Scenario

In this scenario oil consumption starts to decline from the plateau level of 2.1 million b/d foreseen for the 1985 to 1990 period, so that by 2000 it is back at the level of 1.8 million b/d.

Other forms of energy grow more slowly than in the co-operation scenario, so that total energy consumption is projected to rise at just under 4 per cent annually to 1985, declining to 3.5 per cent each year from 1985 to 1990 and then to only 2.7 per cent in the 1990s.

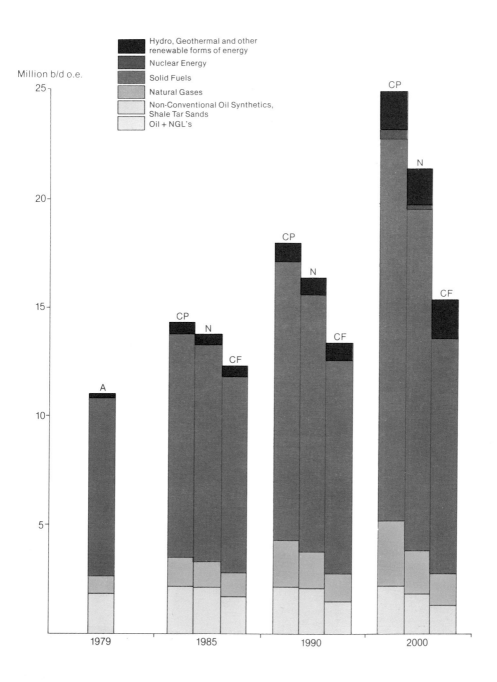

Figure 18

Actual and Prospective Energy Consumption Mix of China
(Million barrels a day oil/o.e.)

Confrontation Scenario

The features of this scenario dictate that foreign oil technologists and equipment play a negligible role. Both oil production and consumption decline slightly through the 1980s, and then at a faster rate in the 1990s. By 2000 oil consumption is projected at only 1.3 million b/d, or 8 per cent of Chinese energy consumption.

Growth of other forms of energy production are also projected to be slower in this scenario, as it is assumed that a less stable economic and political situation will militate against adequate resources being invested in the Chinese energy sector, as is the case elsewhere. Thus total energy consumption is projected to rise at barely 2 per cent annually through the 1980s, falling to 1.3 per cent on average through the 1990s. This can be put into the perspective of population growth, expected to be 1.2 per cent in the current decade and 1.0 per cent annually in the next decade.(6) This allows a slow rate of increase in per capita consumption through the period, even in this pessimistic scenario.

References

(1) BP Statistical Review of the World Oil Industry, 1979. British Petroleum Company Limited, 1980.

(2) Chinese Official Statement of Natural Gas Production in 1979. July, 1980.

(3) 1980 World Bank Atlas. The World Bank.

(4) Oil and Gas Journal. 29 December 1980.

(5) World Energy Conference, Survey of Energy Resources, 1980.

(6) United Nations Population Statistics and Forecasts, 1950-2000. Selected Demographic Indicators, 1980.

Part III
Statistical Data and Appendices

Table 1

WORLD PRIMARY ENERGY CONSUMPTION - Year 1979

Million Metric Tons Oil Equivalent Per Year

REGION	OIL	NATURAL GAS	COAL	HYDRO ELECTRICITY	NUCLEAR	TOTAL
USA	862.9	498.8	384.1	80.1	72.2	1898.1
OECD Europe*	709.2	183.9	250.1	101.7	41.3	1286.2
Japan	265.4	22.1	58.6	19.9	14.7	380.7
Other Developed Countries	127.9	57.6	50.6	63.6	8.5	308.2
All Developed Countries	1965.4	762.4	743.4	265.3	136.7	3874.7
All Developing Countries	521.0	98.9	210.7	87.1	2.8	920.5
Total Non-Communist World	2486.4	861.3	954.1	352.4	139.5	4793.7
Communist Bloc Countries	633.2	380.9	1022.5	59.4	16.3	2112.3
Total World	3119.6	1242.2	1976.6	411.8	155.8	6906.0

*Excludes Yugoslavia which is included in Developing Countries

Sources: BP Statistical Review of the World Oil Industry 1979. Chinese official estimate for natural gas based on Communique from Chinese State Statistical Bureau.

Table 2

COMPOSITION OF WORLD PRIMARY ENERGY CONSUMPTION
PERCENTAGE BREAKDOWN (1979)

COUNTRY/REGION	OIL	NATURAL GAS	COAL	HYDRO ELECTRICITY	NUCLEAR	TOTAL
A - Energy Mix Within Each Country/Group						
USA	45.5	26.3	20.2	4.2	3.8	100.0
OECD Europe*	55.1	14.3	19.5	7.9	3.2	100.0
Japan	69.7	5.8	15.4	5.2	3.9	100.0
Other Developed Countries	41.5	18.7	16.4	20.6	2.8	100.0
All Developed Countries	50.7	19.7	19.2	6.9	3.5	100.0
All Developing Countries	56.6	10.8	22.9	9.5	0.3	100.0
Non-Communist World	51.9	18.0	19.9	7.4	2.9	100.0
USSR	38.4	26.7	29.8	4.0	1.1	100.0
Eastern Europe	22.9	13.9	61.1	1.2	0.9	100.0
China	17.4	2.4	78.5	1.7	-	100.0
Communist World	30.0	18.0	48.4	2.8	0.8	100.0
WORLD	44.8	18.6	28.4	6.0	2.2	100.0
B - Analysis of Each Type of Energy by Country/Group						
USA	34.7	57.9	40.3	22.7	51.8	39.6
OECD Europe*	28.5	21.4	26.2	28.9	29.6	26.8
Japan	10.7	2.5	6.1	5.7	10.5	8.0
Other Developed Countries	5.1	6.7	5.3	18.0	6.1	6.4
All Developed Countries	79.1	88.5	77.9	75.3	98.0	80.8
All Developing Countries	21.0	11.5	22.1	24.7	2.0	19.2
Non-Communist World	100.0	100.0	100.0	100.0	100.0	100.0
USSR	69.6	80.6	33.5	75.8	76.7	54.4
Eastern Europe	16.0	16.1	26.4	9.1	23.3	20.9
China	14.4	3.3	40.1	15.1	-	24.7
Communist World	100.0	100.0	100.0	100.0	100.0	100.0
Non-Communist World	79.7	69.3	48.3	85.5	89.5	69.4
Communist World	20.3	30.7	51.7	14.5	10.5	30.6
WORLD	100.0	100.0	100.0	100.0	100.0	100.0

* Excludes Yugoslavia which is included in the Developing Countries

Sources: BP Statistical Review of the World Oil Industry 1979
Chinese official estimate for natural gas based on
Communique from Chinese State Statistical Bureau.

Table 3

WORLD PRIMARY ENERGY CONSUMPTION IN 1979, PERCENTAGE SHARES AND GROWTH RATES

REGION	OIL	NATURAL GAS	COAL	HYDRO ELECTRICITY	NUCLEAR	TOTAL
(Million Metric Tons Oil Equivalent/Year)						
Non-Communist World	2486.4	861.3	954.1	352.4	139.3	4793.7
Communist Bloc Countries	633.2	380.9	1022.5	59.4	16.3	2112.3
TOTAL WORLD	3119.6	1242.2	1976.6	411.8	155.8	6906.0
(Percentage Shares)						
Non-Communist World: 1979	51.9	18.0	19.9	7.3	2.9	100.0
1973	54.0	19.1	19.2	6.6	1.1	100.0
1967	47.8	18.5	26.1	7.3	0.3	100.0
Communist Bloc Countries: 1979	30.0	18.0	48.4	2.8	0.8	100.0
1973	28.5	15.9	52.8	2.6	0.2	100.0
1967	23.3	15.2	58.7	2.7	0.1	100.0
TOTAL WORLD: 1979	45.2	18.0	28.6	6.0	2.2	100.0
1973	47.1	18.3	28.2	5.6	0.8	100.0
1967	41.4	17.6	34.6	6.1	0.3	100.0
($\pm$ Percentage Annual Average Growth Rates)						
Non-Communist World:						
1973 - 1979	+1.0	+0.6	+2.3	+3.5	+20.6	+1.7
1967 - 1973	+7.5	+6.0	+0.2	+3.5	+27.8	+5.4
Communist Bloc Countries:						
1973 - 1979	+5.7	+7.0	+3.3	+5.8	+26.4	+4.8
1967 - 1973	+9.7	+6.9	+4.2	+5.5	+37.2	+6.1

Sources: BP Statistical Review of the World Oil Industry 1979 and earlier issues. Chinese official estimate for natural gas based on Communique from Chinese State Statistical Bureau.

Table 4

PRODUCTION AND CONSUMPTION OF OIL AND CONSUMPTION
OF OTHER FORMS OF PRIMARY ENERGY

± AVERAGE ANNUAL RATES OF CHANGE (PERCENTAGE)

	PRODUCTION	CONSUMPTION					
	OIL	OIL	NATURAL GAS	COAL	HYDRO ELECTRICITY	NUCLEAR	TOTAL PRIMARY ENERGY
Developing Countries							
1973 - 1979	+0.9	+4.6	+4.2	+5.5	+8.1	+18.7	+5.1
1967 - 1973	+10.4	+7.9	+10.6	+2.2	+8.2	INF	+6.7
Developed Countries							
1973 - 1979	+1.1	+0.2	+0.2	+1.6	+2.3	+20.6	+1.0
1967 - 1973	+2.7	+7.5	+5.6	-0.3	+2.6	+27.4	+5.2
USA							
1973 - 1979	-1.2	+0.9	-2.3	+2.3	+1.0	+22.1	+0.7
1967 - 1973	+1.0	+5.4	+2.7	-1.4	+4.2	+48.9	+3.8
OECD Europe							
1973 - 1979	+34.0	-0.6	+6.2	+0.5	+2.6	+16.4	+1.0
1967 - 1973	-0.8	+8.4	+28.6	-3.0	+1.3	+12.5	+5.8
Japan							
1973 - 1979	-5.5	-0.2	+26.9	-0.6	+36.2	+2.4	+1.2
1967 - 1973	NC	+14.0	+17.6	+2.3	-1.4	+50.2	+10.3
WORLD							
1973 - 1979	+1.9	+1.9	+2.3	+2.8	+3.8	+21.1	+2.6
1967 - 1973	+7.9	+7.9	+6.2	+2.1	+3.8	+28.4	+5.6

NC = No Change
INF = Infinity (Starts from zero)

Sources: BP Statistical Review of the World Oil Industry 1979 and earlier issues.
Chinese official estimate for natural gas based on communique from
Chinese State Statistical Bureau, July 1980

Table 5

ESTIMATES OF TOTAL WORLD ULTIMATELY RECOVERABLE RESERVES OF CRUDE OIL FOR CONVENTIONAL SOURCES AND NGLs

Year	Source	In Billion Barrels	
1920	Anon	43	
1942	Pratt, Weeks & Stebinger	600	
1946	Duce	400	
1946	Pogue	555	
1948	Weeks	610	
1949	Levorsen	1500	
1949	Weeks	1010	
1953	MacNaughton	1000	
1956	Hubbert	1250	
1958	Weeks	1500	
1959	Weeks	2000	
1965	Hendricks (USGS)	2480	
1967	Ryman (Exxon)	2090	
1968	Shell	1800	
1968	Weeks	2200	
1969	Hubbert	1350-2100	
1970	Moody (Mobil)	1800	
1971	Warman (BP)	1200-2000	
1971	Weeks	2290	
1971	* US National Petroleum Council	2657	
1972	* Linden	2935	
1972	* Weeks	3635	
1972	* Moody, Emerick (Mobil)	1796-1891	(1843)
1972	Jodry	1952	
1973	* WEC (USGS)	1343-13432	
1973	Odell	4000	
1974	Bonillas	2000	
1975	* Adams and Kirby (BP)	1993	
1975	* BGR	2453	
1975	Moody (Mobil)	2000	
1975	Odell	3575-4233	
1976	* Klemme (Weeks)	1891	
1977	* World Energy Conference	1760-1900	
1978	* World Energy Conference	927-6935	(mean 2263)
1979	* Halbouty (Moody)	2219	
1979	* Meyerhoff	2190	
1979	* Roorda	2409	
1980	* World Energy Conference	2584	

*   Converted into barrels at rate of 7.3 barrels = 1 ton (WEC).

(1) The estimates may include reserves in the deep sea and Antarctica - such reserves will only be produced at high cost.

(2) Estimates by various geologists rose consistently during the period 1940 - 1960. Since then estimates have tended to converge around 2,000 billion barrels (274 billion metric tons) a convenient figure perhaps.

Table 5 cont.

(3) According to William Stannage, it is not only impossible, but also harmful for geologists and statisticians to establish estimates of the ultimately produceable volume of petroleum to be found in the world. Such views, of course, are controversial.

(4) The World Energy Conference Survey of Energy Resources 1980 gives an estimate of 353.94 billion tons for ultimately recoverable reserves. Of this total only 89.14 billion tons (or little more than 25 per cent), equivalent to some 650 billion barrels, were classified as being proved recoverable reserves at the beginning of the year 1979. It is this part of the total which is the subject of discussion in Parts I and II of this book. Of the remainder, 52.8 billion tons, or nearly 15 per cent of the total, was accounted for by cumulative production to January 1, 1979. The other 212 billion tons, or nearly 60 per cent of the total, is represented by estimated additions to recoverable resources.

Sources: Energy: Global Prospects 1985-2000, WAES Report, 1977.
William Stannage - Resource Estimates and Politics Don't Mix - World Oil, October 1979.
World Energy Conference, Survey of Energy Resources, 1980.

Table 6

ESTIMATES OF WORLD ULTIMATELY RECOVERABLE RESERVES OF NATURAL GAS

Year Made	Source of Estimate	Reserves in Trillion Cu Ft	Reserves in Billion Barrels Oil Equivalent
1956	US Department of the Interior	5000	860
1958	Weeks	5000-6000	860-1035
1959	Weeks	6000	1035
1965	Weeks	7200	1240
1965	Hendricks (USGS)	15300	2640
1967	Ryman (ESSO)	12000	2070
1967	Shell	10200	1760
1968	Weeks	6900	1200
1969	Hubbert	8000-12000	1380-2070
1971	Weeks	7200	1240
1973	Coppack	7500	1300
1973	Hubbert	12000	2070
1973	Linden	10400	1800
1975	Kirkby and Adams	6000	1030
1975	Moody and Geiger	8150	1400*

Source: Kirkby and Adams, Presentation at World Petroleum Congress, Tokyo, May 1975.

* World Ultimately Recoverable Natural Gas Reserves by Region

	Reserves in Billion Barrels Oil Equivalent	% Share of Total
North America	280	20.0
West Europe	77	5.5
Middle East	270	19.3
Rest of World Outside Communist Area	348	24.9
Communist Area	425	30.3
TOTAL World	1400	100.0

Source: Petroleum Resources: How much Oil and Where? Moody and Geiger, Technology Review, March 1975.

NB Estimates of ultimately recoverable world natural gas reserves are subject to even more uncertainty than oil reserve estimates. This is partly because, in the past, much of the gas found was associated with oil and was treated as an unwanted by-product of oil production. Estimates made since 1965 have ranged from 1030 to 2640 billion barrels oil equivalent (140.52 billion to 360.16 billion metric tons oil equivalent).

Source: Energy: Global Prospects 1985-2000, WAES Report, 1977.

Table 7

ESTIMATED PROVED RESERVES OF OIL AND NATURAL GAS

(As of 1/1/1980)

REGIONS	OIL		NATURAL GAS	
	Billion Barrels	% Share	Trillion Cu Ft	% Share
Developed Countries	59.2	9.2	452.0	17.6
OPEC Developing Countries	437.5	68.0	995.9	38.7
Other Developing Countries	56.8	8.8	190.3	7.4
Non-Communist World	553.5	86.0	1638.2	63.7
Communist World	90.0	14.0	935.0	36.3
Total World	643.5	100.0	2573.2	100.0

Sources: Petroleum Statistical Bulletin 1979, The Ministry of Petroleum and Mineral Resources, Kingdom of Saudi Arabia.
The Oil and Gas Journal, Dec 31, 1979.

Table 8

RELATIONSHIP BETWEEN PROVED RECOVERABLE RESERVES AND ESTIMATED ADDITIONAL RESOURCES OF OIL AND NATURAL GAS BY GEO-POLICITAL GROUPINGS

	OIL				NATURAL GAS			
	Proved Recoverable Reserves at 1.1.79		Estimated Additional Resources		Proved Recoverable Reserves at 1.1.79		Estimated Additional Resources	
	Million t.	%	Million t.	%	Billion m^3	%	Billion m^3	%
OECD Countries	7,480	9	35,000	17	11,800	16	51,000	27
OPEC Countries	61,780	69	78,000	36	29,800	40	29,000	15
*Other Countries	7,180	8	35,000	17	5,600	8	48,000	25
Countries with Centrally Planned Economies	12,700	14	64,000	30	26,900	36	64,000	33
Total World	89,140	100	212,000	100	74,100	100	192,000	100

* Other countries comprise non-OPEC developing countries discussed elsewhere in this book.

Million t = Million metric tons

Billion m^3 = Billion cubic metres

Source: World Energy Conference, Survey of Energy Resources, 1980

Table 9

WORLD SOLID FOSSIL FUEL RESOURCES AND RESERVES

TYPE	ESTIMATED TOTAL RESOURCES	of which ESTIMATED TOTAL RESERVES btce	of which ESTIMATED ECONOMICALLY RECOVERABLE RESERVES
Bituminous coal/anthracite	6,936	775	488
Subbituminous coal	3,164	173	111
Lignite	961	113	88
	11,061	1,061	687*

* The World Energy Conference of 1974 estimated economically recoverable reserves at 473 billion tons coal equivalent. The World Energy Conference 1980 estimate of 687 billion tons coal equivalent represents an increase of 45 per cent. Of the 488 billion tons of bituminous coal/anthracite, at least 138 billion tons was estimated to be recoverable at US$30 a ton or less. This part of the economically recoverable coal reserves alone is similar in terms of heat value to the total world proved oil reserves.

DISTRIBUTION OF IDENTIFIED LOW COST RESERVES OF BITUMINOUS COAL/ANTHRACITE (UP TO $30 A TON PRODUCTION COSTS)

USA	73.2
AUSTRALIA	25.4
SOUTHERN AFRICA	25.3
INDIA	12.6
OTHERS	1.7
	138.2 btce

The USSR, China and other Communist countries do not feature in this analysis, though collectively they account for more than 233 billion tons of the world total of 488 billion tons of economically recoverable proved reserves of bituminous coal/anthracite, nearly 48 per cent of the total. Of this higher calorific value category of solid fuel reserves, OECD countries account for nearly 206 billion tons or over 42 per cent of the world total, while OPEC countries account for an estimated 400 million tons or only some 0.1 per cent of the world total.

btce = billion metric tons coal equivalent.

Source: World Energy Conference, Survey of Energy Resources, 1980.

Table 10

WORLD ENERGY RESOURCES
PROVEN REMAINING RECOVERABLE RESERVES

RESOURCE	Non-Communist World	ESTIMATES Communist Bloc Countries	Total World	SOURCE OF ESTIMATE
CRUDE OIL AND NGLs (as of 1/1/1980)	74.5 bt	12.2 bt	86.7 bt	The Oil and Gas Journal Dec 31, 1979
NATURAL GAS (as of 1/1/1980)	38.1 btoe	21.7 btoe	59.8 btoe	The Oil and Gas Journal Dec 31, 1979
COAL (as of 1/1/1979)	242.7 btoe	201.0 btoe	443.7 btoe	World Energy Conference Survey of Energy Resources, 1980
OIL SHALE & TAR SANDS (as at 1/1/1979)	72.7 btoe	6.8 btoe	79.5 btoe	World Energy Conference Survey of Energy Resources, 1980
URANIUM:				
I Reasonably Assured Estimated Additional	1860,000 t 1580,000 t	NA NA	NA NA	World Energy Conference Survey of Energy Resources, 1980
- At price range of 1979 up to US$ 80 per kg of U				
II Reasonably Assured Estimated Additional	737,000 t 980,000 t	NA NA	NA NA	
- At price range of 1979 US$ 80–130 per kg of U				

bt = Billion metric tons
btoe = Billion metric tons oil equivalent
t = Metric ton
NA = Not available

Based on estimates in the Survey of Energy Resources 1980 by the World Energy Conference the percentage distribution of proved recoverable reserves by type is: coal 60.8 per cent; hydrocarbons 33.5 per cent; uranium 5.7 per cent.

Table 11

THE USA
INDIGENOUS ENERGY RESOURCES

RESOURCE	ESTIMATED PROVED RESERVES		SOURCE OF INFORMATION
OIL (as of 1/1/1980)	26.5 bb (3.6 bt)		The Oil and Gas Journal, Dec 31, 1979
NATURAL GAS (as of 1/1/1980)	194 tcf (4.8 btoe)		The Oil and Gas Journal, Dec 31, 1979
COAL			
Economically Recoverable	123.2 btoe[1]		World Energy Conference, Survey of Energy Resources, 1980
Economically Recoverable	248 bt		Energy: Global Prospects 1985-2000, WAES Report, 1977
Known (Measured)	396 bt		Energy: Global Prospects 1985-2000, WAES Report, 1977
OIL SHALE	28 btoe[1]		World Energy Conference, Survey of Energy Resources, 1980
URANIUM	Reasonably Assured	Estimated Additional	
At Price[2] range of up to US$ 80 per kg. U	530,000t	780,000t	World Energy Conference, Survey of Energy Resources, 1980
At Price[2] range of US$ 80-130 per kg. U	178,000t	380,000t	World Energy Conference, Survey of Energy Resources, 1980

1   Conversion factor used: one toe = 45.4×10^9 J
2   1979 US Dollars

bb = Billion US barrels
bt = Billion metric tons
btoe = Billion metric tons oil equivalent
tcf = Trillion standard cubic feet (thousand billion)
toe = Metric tons oil equivalent
t = Metric tons

Energy: A Global Outlook

Table 12

US PRIMARY ENERGY HISTORICAL SUPPLY/DEMAND TRENDS

	Million Metric Tons Oil Equivalent			Average Annual Percentage Growth Rates		Percentage Shares		
	1979	1973	1967	1973-79	1967-73	1979	1973	1967
I CONSUMPTION:								
OIL*	862.9	818.0	595.8	+0.9	+5.4	45.5	44.9	41.0
NATURAL GAS	498.8	572.3	488.1	-2.3	+2.7	26.3	31.4	33.6
COAL	384.1	335.0	308.3	+2.3	+1.4	20.2	18.4	21.2
NUCLEAR	72.2	21.8	2.0	+22.1	+48.9	3.8	1.2	0.1
HYDROELECTRICITY	80.1	75.6	59.1	+1.0	+4.2	4.2	4.1	4.1
TOTAL	1898.1	1822.7	1453.3	+0.7	+3.8	100.0	100.0	100.0
II OIL PRODUCTION	483.1	519.0	487.9	-1.2	+1.0			
III NET OIL IMPORTS	394.0	303.0	115.0	+4.5	+17.5			
IV OIL IMPORTS AS PERCENTAGE OF TOTAL OIL CONSUMPTION	45.7	37.0	19.3					

*Including NGLs

Source: BP Statistical Reviews of World Oil Industry 1967-1979

Table 13

WESTERN EUROPE
INDIGENOUS ENERGY RESOURCES

RESOURCE	ESTIMATED PROVED RESERVES		SOURCE OF INFORMATION
OIL (as of 1/1/1980)	23.6 bb (3.19 bt)		The Oil and Gas Journal, Dec 31, 1979
NATURAL GAS (as of 1/1/1980)	134.9 tcf (3.14 btoe)		The Oil and Gas Journal, Dec 31, 1979
COAL Economically Recoverable	53.3 btoe[1]		World Energy Conference, Survey of Energy Resources, 1980
Economically Recoverable	41 bt		Energy: Global Prospects 1985-2000, WAES Report, 1977
Known (Measured)	225 bt		Energy: Global Prospects 1985-2000, WAES Report, 1977
OIL SHALE	2.6 btoe[1]		World Energy Conference, Survey of Energy Resources, 1974*
URANIUM	Reasonably Assured	Estimated Additional	
At Price[2] range of up to US$ 80 per kg.U	66,300t	47,200t	World Energy Conference, Survey of Energy Resources, 1980
At Price[2] range of US$ 80-130 per kg.U	349,100t	46,100t	World Energy Conference, Survey of Energy Resources, 1980

[1] Conversion factor used: one toe = 45.4×10^9 J
[2] 1979 US Dollars

bb = Billion US barrels
bt = Billion metric tons
btoe = Billion metric tons oil equivalent
tcf = Trillion standard cubic feet (thousand billion)
toe = Metric tons oil equivalent
t = Metric tons

* World Energy Conference, Survey of Energy Resources 1980 shows a total of only 262 million metric tons of recoverable oil. Most of this is located in West Germany.

Table 14

WEST EUROPEAN COUNTRIES

INDIGENOUS RESOURCES OF OIL AND NATURAL GAS

COUNTRY	ESTIMATED PROVED RESERVES (as of 1/1/1980)	
	OIL (bb)	NATURAL GAS (tcf)
Austria	0.141	0.41
Denmark	0.375	2.82
France	0.050	6.30
West Germany	0.480	6.35
Greece	0.150	4.00
Ireland	-	1.00
Italy	0.645	3.50
Netherlands	0.060	59.50
Norway	5.750	23.50
Spain	0.150	0.20
Turkey	0.125	0.50
United Kingdom	15.400	25.00
TOTAL WESTERN EUROPE	23.626	135.78

bb = Billion US barrels
tcf = Trillion standard cubic feet (thousand billion)

Source: The Oil and Gas Journal, Dec 31, 1979.

Table 15

WEST EUROPEAN ESTIMATED OIL PRODUCTION

	1979	1980
Million Barrels Per Day		
NORTH SEA	2.0	2.2
OTHERS	0.3	0.3
TOTAL WESTERN EUROPE	2.3	2.5
Million Metric Tons Per Year		
NORTH SEA	97.0	106.0
OTHERS	15.0	16.0
TOTAL WESTERN EUROPE	112.0	122.0

North Sea = Includes UK and Norway

Others = Austria, Denmark, France, West Germany, Italy, Netherlands, Spain

Source: BP Statistical Review of The World Oil Industry 1980.

Table 16

OECD EUROPEAN COUNTRIES* PRIMARY ENERGY HISTORICAL SUPPLY/DEMAND TRENDS

	Million Metric Tons Oil Equivalent			Average Annual Percentage Growth Rates		Percentage Shares		
	1979	1973	1967	1973-79	1967-73	1979	1973	1967
I CONSUMPTION:								
OIL*	709.2	736.2	455.4	-0.6	+8.3	55.1	60.7	52.8
NATURAL GAS	183.9	128.2	27.6	+6.2	+29.2	14.3	10.6	3.2
COAL	250.1	243.4	291.0	+0.5	-2.9	19.5	20.1	33.7
NUCLEAR	41.3	16.6	8.2	+16.4	+12.5	3.2	1.4	0.9
HYDROELECTRICITY	101.7	87.4	80.8	+2.6	+1.3	7.9	7.2	9.4
TOTAL	1286.2	1211.8	863.0	+1.0	+5.8	100.0	100.0	100.0
II OIL PRODUCTION	111.8	19.3	20.3	+34.0	-0.8			
III OIL IMPORTS*	647.1	760.0	458.0	-2.6	+8.8			
IV OIL IMPORTS AS PERCENTAGE OF TOTAL OIL CONSUMPTION	91.2	103.2	100.6					

*Excluding Yugoslavia which is included in the Developing Countries, except for oil imports

Source: BP Statistical Reviews of World Oil Industry 1967-1979.

Table 17

JAPAN
INDIGENOUS ENERGY RESOURCES

RESOURCE	ESTIMATED PROVED RESERVES		SOURCE OF INFORMATION
OIL (as of 1/1/1980)	0.055 bb (7.432 mt)		The Oil and Gas Journal, Dec 31, 1979
NATURAL GAS (as of 1/1/1980)	0.6 tcf (14.0 mtoe)		The Oil and Gas Journal, Dec 31, 1979
COAL			
Economically Recoverable	0.7 btoe[1]		World Energy Conference, Survey of Energy Resources, 1980
Economically Recoverable	1 bt		Energy: Global Prospects 1985-2000, WAES Report, 1977
Known (Measured)	3 bt		Energy: Global Prospects 1985-2000, WAES Report, 1977
URANIUM	Reasonably Assured	Estimated Additional	
At Price[2] range of up to US$ 80 per kg.U	7,700 t	-	World Energy Conference, Survey of Energy Resources, 1980
At Price[2] range of US$ 80-130 per kg.U	-	-	

[1] Conversion factor used: one toe = 45.4×10^9 J
[2] 1979 US Dollars

bb = Billion US barrels
mt = Million metric tons
bt = Billion metric tons
mtoe = Million metric tons oil equivalent
btoe = Billion metric tons oil equivalent
tcf = Trillion standard cubic feet (thousand billion)
toe = Metric tons oil equivalent
t = Metric tons.

Table 18

JAPANESE PRIMARY ENERGY HISTORICAL SUPPLY/DEMAND TRENDS

	Million Metric Tons Oil Equivalent			Average Annual Percentage Growth Rates		Percentage Shares		
	1979	1973	1967	1973-79	1967-73	1979	1973	1967
I CONSUMPTION:								
OIL	265.4	269.1	122.9	-0.2	+14.0	69.7	75.8	62.4
NATURAL GAS	22.1	5.3	2.0	+26.9	+17.6	5.8	1.5	1.0
COAL	58.6	60.9	53.0	-0.6	+2.3	15.4	17.2	26.9
NUCLEAR	14.7	2.3	0.2	+36.2	+50.2	3.9	0.6	0.1
HYDROELECTRICITY	19.9	17.3	18.8	+2.4	-1.4	5.2	4.9	9.6
TOTAL	380.7	354.9	196.9	+1.2	+10.3	100.0	100.0	100.0
II OIL PRODUCTION	0.5	0.7	0.7	-5.5	NC			
III OIL IMPORTS	275.6	270.0	122.2	+0.3	+14.1			
IV OIL IMPORTS AS PERCENTAGE OF TOTAL OIL CONSUMPTION	103.8	100.3	99.4					

NC = No change

Source: BP Statistical Reviews of World Oil Industry 1967-1979.

Table 19

OPEC DEVELOPING COUNTRIES

INDIGENOUS ENERGY RESOURCES

RESOURCE	ESTIMATED PROVED RESERVES		SOURCE OF INFORMATION
			Petroleum Statistical Bulletin, 1979, The Ministry of Petroleum and Mineral Resources, Kingdom of Saudi Arabia.
OIL (as of 1/1/1980)	437.5 bb (59.9 bt)		The Oil and Gas Journal, Dec 31, 1979
NATURAL GAS (as of 1/1/1980)	995.9 tcf (23.2 btoe)		The Oil and Gas Journal, Dec 31, 1979
COAL Economically Recoverable	3.3 btoe[3]		World Energy Conference, Survey of Energy Resources, 1980
TAR SANDS[1]	20.0 btoe		World Energy Conference, Survey of Energy Resources, 1980
URANIUM[2]	Reasonably Assured	Estimated Additional	
At Price[4] range of US$ 1-10 per kg.U	37,000t	–	World Energy Conference, Survey of Energy Resources, 1980

1   In Venezuela
2   In Gabon
3   Conversion factor used: one toe = 45.4×10^9 J
4   1979 US dollars

bb = Billion US barrels
bt = Billion metric tons
btoe = Billion metric tons oil equivalent
tcf = Trillion standard cubic feet (thousand billion)
t = Metric tons
toe = Metric tons oil equivalent

Table 20

OPEC DEVELOPING COUNTRIES

INDIGENOUS RESOURCES OF OIL AND NATURAL GAS

COUNTRY	ESTIMATED PROVED RESERVES (as of 1/1/1980)	
	OIL (bb)	NATURAL GAS (tcf)
Saudi Arabia*	168.39	95.73
Iran	58.00	490.00
Iraq	31.00	27.50
Kuwait*	68.53	33.50
UAE (Abu Dhabi, Dubai, Sharjah)	29.41	20.50
Qatar	3.76	60.00
Algeria	8.44	132.00
Libya	23.50	24.00
Nigeria	17.40	41.40
Gabon	0.50	0.50
Indonesia	9.60	24.00
Venezuela	17.87	42.80
Ecuador	1.10	4.00
OPEC (Excluding Saudi Arabia)	269.11	900.20
TOTAL OPEC	437.50	995.93
NON-COMMUNIST WORLD	553.53	1638.24
TOTAL WORLD	643.53	2573.24

*Including 50 per cent of Divided Zone

bb = Billion US barrels
tcf = Trillion standard cubic feet (thousand billion)

Source: Petroleum Statistical Bulletin, 1979, The Ministry of Petroleum and Mineral Resources, Kingdom of Saudi Arabia.
Oil and Gas Journal, 31 December 1979.

Table 21

OPEC - CRUDE OIL PRODUCTION AND OUTPUT CAPACITY
(Million Barrels Per Day)

COUNTRY	ACTUAL PRODUCTION Average 1980	ACTUAL PRODUCTION Average 1979	OUTPUT CAPACITY ESTIMATE (1)
Saudi Arabia*	9.90	9.53	11.3
Iran	1.47	3.11	3.0
Iraq	2.65	3.45	4.0
Kuwait*	1.65	2.49	2.8
UAE (Abu Dhabi, Dubai, Sharjah)	1.71	1.83	2.485
Qatar	0.47	0.51	0.65
Algeria	0.94	1.12	1.2
Libya	1.79	2.07	2.1
Nigeria	2.05	2.30	2.4
Gabon	0.18	0.20	0.25
Indonesia	1.58	1.59	1.6
Venezuela	2.17	2.36	2.4
Ecuador	0.22	0.21	0.25
OPEC (excluding Saudi Arabia)	16.88	21.24	23.135
Total OPEC	26.78	30.77	34.435
(Non-OPEC) Non-Communist World	18.33	17.70	NA
Total Non-Communist World	45.11	48.47	NA
Total World	59.84	62.88	NA

*Including 50 per cent share of Divided Zone

Natural Gas Liquids are not generally included in this table.

(1)   PIW's assessment of maximum production sustainable for several months without regard to government ceilings. This is less than installed capacity and does not necessarily reflect maximum production sustainable for long periods without damage to fields.

NA = Not available.

Sources: Middle East Economic Survey 13 July 1981.
Petroleum Statistical Bulletin 1979, Kingdom of Saudi Arabia, Ministry of Petroleum and Mineral Resources.
Petroleum Intelligence Weekly, February 23 and March 23, 1981 and March 10, 1980.
BP Statistical Review of the World Oil Industry 1980.

Table 22

OPEC DEVELOPING COUNTRIES

OIL - HISTORICAL SUPPLY/DEMAND TRENDS

	Thousand Barrels per Day		
YEAR	OIL CONSUMPTION	OIL PRODUCTION	OIL EXPORTS
1967	538.1	16,849.8	15,907.0
1973	983.8	30,988.5	29,521.6
1979	2,217.0	30,928.2	28,868.2
Percentage Average Annual Growth Rates			
1967-1973	+10.6	+10.7	+10.9
1973-1979	+14.5	NC	-0.4
Oil Exports as Percentage of Total Oil Production	1967	1973	1979
	94.4	95.3	93.3

NC = No change

Source: OPEC Annual Statistical Bulletin, 1979.

NON-OPEC DEVELOPING COUNTRIES
INDIGENOUS RESOURCES OF OIL AND NATURAL GAS

Table 23

ESTIMATED PROVED RESERVES
(as of 1/1/1980)

COUNTRY	OIL (bb)	NATURAL GAS (tcf)
Asia-Pacific:		
Bangladesh	-	8.000
Brunei	1.800	7.700
Burma	0.025	0.135
Republic of China (Taiwan)	0.010	0.650
India	2.600	9.300
Malaysia	2.800	17.000
Pakistan	0.200	15.800
Philippines	0.025	-
Thailand	-	8.000
SUB-TOTAL	7.460	66.585
Europe:		
Yugoslavia	0.275	1.000
Middle East:		
Bahrain	0.240	9.000
Oman	2.400	2.000
Syria	2.000	1.500
Others	0.001	0.100
SUB-TOTAL	4.641	12.600
Africa:		
Angola-Cabinda	1.200	1.175
Cameroon	0.140	-
Congo Republic	0.400	2.200
Egypt	3.100	3.000
Ghana	0.007	-
Morocco	-	0.025
Tunisia	2.250	6.000
Zaire	0.135	0.050
SUB-TOTAL	7.232	12.450
Western Hemisphere:		
Argentina	2.400	15.200
Barbados	0.002	-
Bolivia	0.150	5.400
Brazil	1.220	1.500
Chile	0.400	2.500
Colombia	0.710	5.000
Guatemala	0.016	-
Mexico	31.250	59.000
Peru	0.655	1.100
Trinidad and Tobago	0.700	8.000
SUB-TOTAL	37.503	97.700
TOTAL	57.111	190.335

bb = Billion US barrels
tcf = Trillion standard cubic feet (thousand billion)

Source: The Oil and Gas Journal, Dec 31, 1979.

NON-OPEC DEVELOPING COUNTRIES

Table 24

RESERVES AND RESOURCES OF SOLID FOSSIL FUELS

(In million metric tons)

	RESERVES Economically[1] Recoverable	ADDITIONAL[2] RESOURCES
New Caledonia	2	8
Yugoslavia	8,740	1,987
Argentina	117	3,398
South Africa	25,290	33,762
Brazil	910	11,408
Chile	924	3,517
Republic of China (Taiwan)	109	-
Mexico	1,500	1,690
Korea	116	1,049
Malaysia	-	260
Colombia	1,029	7,726
Peru	-	868
Morocco	50	13
Botswana	3,500	100,000
Swaziland	1,820	3,000
Philippines	64	-
Thailand	34	-
Zambia	24	98
Zimbabwe	734	5,820
Egypt	13	-
Central African Republic	1	-
Haiti	-	9
Mozambique	240	155
Pakistan	394	-
Tanzania	200	1500
Zaire	600	-
India	13,134	91,232
Malawi	12	-
Afghanistan	66	400
Burma	2	146
Mali	-	1
Bangladesh	242	1
TOTAL NODCs	59,867	268,048
TOTAL WORLD	687,489	10,001,027

(1) Economically recoverable reserves are that part of proved reserves that are considered exploitable under present local economic conditions using existing available technology.

(2) Additional resources are total measured quantities plus quantities that may be inferred to exist, but additional to economically recoverable reserves. Countries are arranged in decreasing order of GNP per capita in 1978.

Source: World Energy Conference, Survey of Energy Resources, 1980.

Table 25

NON-OPEC DEVELOPING COUNTRIES

URANIUM RESOURCES

At Price of	REASONABLY ASSURED RESOURCES		ESTIMATED ADDITIONAL RESOURCES	
	Up to $80 per kg.U.	$80 to $130 per kg.U.	Up to $80 per kg.U.	$80 to $130 per kg.U.
		Thousand Tons		
Angola	-	-	12.8	-
Botswana	-	0.4	-	-
Central African Republic	18.0	-	-	-
Egypt	-	-	-	5.0
Madagascar	-	-	-	2.0
Morocco	-	-	-	19.6
Namibia	117.0	16.0	30.0	23.0
Niger	160.0	-	53.0	-
Somalia	-	6.6	-	3.4
South Africa	247.0	144.0	54.0	85.0
Zaire	1.8	-	1.7	-
Argentina	23.0	5.1	3.8	5.1
Bolivia	-	-	-	0.5
Brazil	74.2	-	90.1	-
Chile	-	-	5.1	-
Colombia	-	-	51.0	-
Mexico	8.3	-	2.4	-
India	29.8	-	0.9	22.8
South Korea	-	4.4	-	4.0
Philippines	0.3	-	-	-
Yugoslavia	4.5	2.0	5.0	15.5
Total NODCs	683.9	178.5	309.8	185.9
Total WORLD	2318.9	734.6	3154.1	981.4

Source: World Energy Conference, Survey of Energy Resources, 1980.

Table 26

NON-OPEC DEVELOPING COUNTRIES' HYDRO RESOURCES - INSTALLED AND INSTALLABLE CAPACITY

HIGH INCOME Country	Annual Potential Generation (GWh)	MIDDLE INCOME Country	Annual Potential Generation (GWh)	LOW INCOME Country	Annual Potential Generation (GWh)
Argentina	191,000	Angola	48,320	Afghanistan	18,000
Barbados	-	Bolivia	90,000	Bangladesh	6,535
Bahama Is.	-	Botswana	8,952	Burma	225,000
Bermuda	-	British Solomon Is.	-	Burundi	-
Belize	1,500	Cameroon	114,800	Khmer Republic	-
Brazil	519,277	Congo P.R.	45,200	Central African Republic	44,160
Chile	88,600	Eq. Guinea	12,000	Sri Lanka	4,720
Republic of China	5,283	El Salvador	4,551	Chad	10,320
Colombia	300,000	Ghana	15,551	Zaire	660,000
Costa Rica	36,898	Honduras	2,400	Dahomey	7,168
Cyprus	-	Ivory Coast	10,880	Ethiopia	56,020
Dominican Republic	-	Jordan	-	Gambia	-
Fiji	-	Korean Republic	9,925	Guinea	25,600
Guatemala	5,880	Liberia	30,000	Haiti	-
Guyana	72,000	Mauritius	320	India	280,000
Hong Kong	-	Morocco	3,000	Kenya	53,760
Jamaica	-	Mozambique	45,160	Laos	+
Lebanon	-	Papua New Guinea	121,670	Lesotho	2,600
Malaysia	4,467	Paraguay	30,000	Malagasy Republic	320,000
Malta	-	Philippines	19,595	Malawi	400
Mexico	99,360	Zimbabwe	20,000	Mali	10,560
Netherlands Antilles	-	Senegal	17,600	Mauritania	6,000
Nicaragua	18,000	Swaziland	2,800	Nepal	+
Panama	17,000	Syrian A.R.	+	Niger	28,800
Peru	109,154	Thailand	22,584	Pakistan	105,000
Singapore	-	Egypt A.R.	15,000	Rwanda	-
Surinam	1,626	Vietnam Republic	28,143	Sierra Leone	12,000
Trinidad & Tobago	-			Somalia	720
Tunisia	50			Yemen PDR	-
Uruguay	9,496			Sudan	48,000
Zambia	15,336			Tanzania	83,200
Others	360			Togo	1,920
				Uganda	72,000
				Upper Volta	48,000
				Yemen A.R.	-
Sub total:	1,491,287	Sub total:	740,000	Sub total:	2,130,483
TOTAL ALL COUNTRIES					4,361,770
TOTAL WORLD					9,802,420

Notes: - denotes not reported, + denotes quantity unknown

Source: World Energy Conference, Survey of Energy Resources, 1974.

Table 27

NON-OPEC DEVELOPING COUNTRIES

INDIGENOUS ENERGY RESOURCES

RESOURCE	ESTIMATED PROVED RESERVES		SOURCE OF INFORMATION
OIL (as of 1/1/1980)	56.8 bb (7.68 bt)		The Oil and Gas Journal, Dec 31, 1979
NATURAL GAS (as of 1/1/1980)	190.3 tcf (4.43 btoe)		The Oil and Gas Journal, Dec 31, 1979
COAL Economically Recoverable	38.6 btoe[1]		World Energy Conference, Survey of Energy Resources, 1980
OIL SHALE	4.4 btoe[1]		World Energy Conference, Survey of Energy Resources, 1980
URANIUM	Reasonably Assured	Estimated Additional	
At Price[2] range of up to US$ 80 per kg.U	683,900t	309,800t	World Energy Conference, Survey of Energy Resources, 1980
At Price[2] range of US$ 80-130 per kg.U	178,500t	185,900t	World Energy Conference, Survey of Energy Resources, 1980
ESTIMATED PROBABLE HYDROELECTRIC DEVELOPMENT	4.49×10^{18} J in year 2000		World Energy Conference, Survey of Energy Resources, 1980

[1] Conversion factor used: one toe = 45.4×10^9 J
[2] 1979 US Dollars

bb = Billion US barrels
bt = Billion metric tons
btoe = Billion metric tons oil equivalent
tcf = Trillion standard cubic feet (thousand billion)
t = Metric tons

Table 28

NON-OPEC DEVELOPING COUNTRIES

OIL - HISTORICAL SUPPLY/DEMAND TRENDS

YEAR	Thousand Barrels per Day		
	OIL CONSUMPTION	OIL PRODUCTION	OIL IMPORTS
1967	4552	2005	2547
1973	7096	3220	3876
1979	8413	5230	3183
Percentage Average Annual Growth Rates			
1967-1973	+7.7	+8.2	+7.2
1973-1979	+2.9	+8.4	-3.2
Oil Imports as Percentage of	1967	1973	1979
Total Oil Consumption	56.0	54.6	37.8

Sources: Inferred from OPEC Annual Statistical Bulletin 1979 and BP Statistical Review of the World Oil Industry 1979.

Table 29

THE USSR

INDIGENOUS ENERGY RESOURCES

RESOURCE	ESTIMATED PROVED RESERVES		SOURCE OF INFORMATION
OIL (as of 1/1/1980)	67.0 bb (9.1 bt)		The Oil and Gas Journal, Dec 31, 1979
NATURAL GAS (as of 1/1/1980)	900.0 tcf (20.9 btoe)		The Oil and Gas Journal, Dec 31, 1979
COAL Economically Recoverable	106.8 btoe[1]		World Energy Conference, Survey of Energy Resources, 1980
OIL SHALE	6.8 btoe[1]		World Energy Conference, Survey of Energy Resources, 1980
URANIUM	Reasonably Assured	Estimated Additional	
At Price[2] range of up to US$ 80 per kg.U	160,000t	800,000t	World Energy Conference, Survey of Energy Resources, 1980
At Price[2] range of US$ 80-130 per kg.U	-	-	World Energy Conference, Survey of Energy Resources, 1980

1 Conversion factor used: one toe = 45.4×10^9 J
2 1979 US Dollars

bb = Billion US barrels
bt = Billion metric tons
btoe = Billion metric tons oil equivalent
tcf = Trillion standard cubic feet (thousand billion)
toe = Metric ton oil equivalent
t = Metric tons

Table 30

USSR PRIMARY ENERGY HISTORICAL SUPPLY/DEMAND TRENDS

	Million Metric Tons Oil Equivalent			Average Annual Percentage Growth Rates		Percentage Shares		
	1979	1973	1967	1973-79	1967-73	1979	1973	1967
I CONSUMPTION:								
OIL	441.0	325.7	210.7	+5.2	+7.5	38.4	37.3	31.7
NATURAL GAS	307.0	198.8	145.8	+7.5	+5.3	26.8	22.8	22.0
COAL	342.5	315.0	283.3	+1.4	+1.8	29.8	36.0	42.7
NUCLEAR	12.5	3.0	0.5	+26.9	+34.8	1.1	0.3	0.1
HYDROELECTRICITY	45.0	31.6	23.2	+6.1	+5.3	3.9	3.6	3.5
TOTAL	1148.0	874.1	663.5	+4.6	+4.7	100.0	100.0	100.0
II OIL PRODUCTION	586.0	429.0	288.1	+5.3	+6.9			
III OIL EXPORTS	145.0	103.3	77.4	+5.8	+4.9			
IV OIL EXPORTS AS PERCENTAGE OF TOTAL OIL PRODUCTION	24.7	24.1	26.9					

Source: BP Statistical Reviews of World Oil Industry 1967-1979.

Table 31

EASTERN EUROPE

INDIGENOUS ENERGY RESOURCES

RESOURCE	ESTIMATED PROVED RESERVES		SOURCE OF INFORMATION
OIL (as of 1/1/1980)	3.0 bb (405.4 mt)		The Oil and Gas Journal, Dec 31, 1979
NATURAL GAS (as of 1/1/1980)	10.0 tcf (232.6 mtoe)		The Oil and Gas Journal, Dec 31, 1979
COAL Economically Recoverable	29.9 btoe[1]		World Energy Conference, Survey of Energy Resources, 1980
OIL SHALE	NIL shown		World Energy Conference, Survey of Energy Resources, 1980
URANIUM	Reasonably Assured	Estimated Additional	
At Price[2] range of up to US$ 80 per kg.U	135,000t	750,000t	World Energy Conference, Survey of Energy Resources, 1980
At Price[2] range of US$ 80-130 per kg.U	-	-	World Energy Conference, Survey of Energy Resources, 1980

1 Conversion factor used: one toe = 45.4×10^9 J
2 1979 US Dollars

bb = Billion US barrels
mt = Million metric tons
tcf = Trillion standard cubic feet (thousand billion)
mtoe = Million metric tons oil equivalent
btoe = Billion metric tons oil equivalent
t = Metric tons
toe = Metric tons oil equivalent

Table 32

EASTERN EUROPE'S PRIMARY ENERGY HISTORICAL SUPPLY/DEMAND TRENDS

	Million Metric Tons Oil Equivalent			Average Annual Percentage Growth Rates		Percentage Shares		
	1979	1973	1967	1973-79	1967-73	1979	1973	1967
I CONSUMPTION:								
OIL	101.1	75.1	39.0	+5.1	+11.5	22.9	21.0	14.2
NATURAL GAS	61.5	42.1	23.4	+6.5	+10.3	13.9	11.8	8.6
COAL	270.0	235.0	209.0	+2.3	+2.0	61.1	65.8	76.3
NUCLEAR	3.8	1.0	0.1	+24.9	+46.8	0.9	0.3	-
HYDROELECTRICITY	5.4	4.0	2.5	+5.1	+8.1	1.2	1.1	0.9
TOTAL	441.8	357.2	274.0	+3.6	+4.5	100.0	100.0	100.0
II OIL PRODUCTION	20.0	19.2	17.1	+0.7	+1.9			
III OIL IMPORTS	81.1	55.9	21.9	+6.4	+16.9			
IV OIL IMPORTS AS PERCENTAGE OF TOTAL OIL CONSUMPTION	80.2	74.4	56.2					

Source: BP Statistical Reviews of World Oil Industry 1967-1979

Table 33

PEOPLE'S REPUBLIC OF CHINA

INDIGENOUS ENERGY RESOURCES

RESOURCE	ESTIMATED PROVED RESERVES		SOURCE OF INFORMATION
OIL (as of 1/1/1980)	20.0 bb (2.7 bt)		The Oil and Gas Journal, Dec 31, 1979
NATURAL GAS (as of 1/1/1980)	25.0 tcf (581.4 mtoe)		The Oil and Gas Journal, Dec 31, 1979
COAL Economically Recoverable	63.9 btoe[1]		World Energy Conference, Survey of Energy Resources, 1980
OIL SHALE	NIL shown		World Energy Conference, Survey of Energy Resources, 1980
URANIUM	Reasonably Assured	Estimated Additional	
At Price[2] range of up to US$ 80 per kg.U	166,000t	-	World Energy Conference, Survey of Energy Resources, 1980
At Price[2] range of US$ 80-130 per kg.U	-	-	World Energy Conference, Survey of Energy Resources, 1980

1 Conversion factor used: one toe = 45.4×10^9 J
2 1979 US Dollars

bb   = Billion US barrels
bt   = Billion metric tons
tcf   = Trillion standard cubic feet (thousand billion)
mtoe = Million metric tons oil equivalent
btoe = Billion metric tons oil equivalent
t   = Metric tons
toe = Metric ton oil equivalent

Table 34

PEOPLE'S REPUBLIC OF CHINA'S PRIMARY ENERGY HISTORICAL SUPPLY/DEMAND TRENDS

	Million Metric Tons Oil Equivalent			Average Annual Percentage Growth Rates			Percentage Shares		
	1979	1973	1967	1973-79	1967-73		1979	1973	1967
I CONSUMPTION:									
OIL	91.1	53.8	13.9	+9.2	+25.3		17.4	15.0	7.7
NATURAL GAS	12.4*	6.4	0.9	+11.7	+38.7		2.4	1.7	0.5
COAL	410.0	292.5	161.6	+5.8	+10.4		78.5	81.4	89.1
NUCLEAR	-	-	-	-	-		-	-	-
HYDROELECTRICITY	9.0	6.8	5.0	+4.8	+5.3		1.7	1.9	2.7
TOTAL	522.5	359.5	181.4	+6.1	+12.4		100.0	100.0	100.0
II OIL PRODUCTION	106.1	54.8	13.9	+11.6	+30.7				
III OIL EXPORTS	15.0	1.0	NA	-	-				
IV OIL EXPORTS AS PERCENTAGE OF TOTAL OIL PRODUCTION	14.1	1.8	-						

NA = Not available

* Subsequent to publication of the Review for 1979, it was ascertained that this estimate should be 12.4 million metric tons of oil equivalent, based on an official Chinese estimate of natural gas production.

Source: BP Statistical Reviews of World Oil Industry 1967-1979.

Table 35

ACTUAL AND PROSPECTIVE OIL BALANCES
(Conventional Crude + NGLS)
PRODUCTION, CONSUMPTION AND NET TRADE IMPLICATIONS   Million b/d

Year		1973	1979	1985			1990			2000		
Scenario				CP	N	CF	CP	N	CF	CP	N	CF
Saudi Arabia	P	7.7	9.8	12.0	8.5	5.0	12.0	8.5	5.0	12.0	8.5	5.0
Other OPEC	P	23.6	21.5	22.0	21.0	19.0	20.0	16.5	13.0	18.0	12.5	7.0
Total OPEC	P	31.3	31.3	34.0	29.5	24.0	32.0	25.0	18.0	30.0	21.0	12.0
	C	1.5	2.2	4.2	3.7	3.4	6.9	5.4	4.4	16.6	10.1	6.1
	Net Trade	+29.8	+29.0	+29.8	+25.8	+20.6	+25.1	+19.6	+13.6	+13.4	+10.9	+5.9
Other	P	3.2	5.2	9.0	8.0	7.0	10.0	9.0	7.5	8.0	6.0	4.0
Developing	C	7.1	8.4	11.5	10.4	9.2	13.1	11.4	9.7	10.1	8.0	5.8
Countries	Net Trade	-3.9	-3.2	-2.5	-2.4	-2.2	-3.1	-2.4	-2.3	-2.2	-2.0	-1.8
Industrialized	P	13.8	14.8	14.3	14.5	15.0	13.0	13.2	13.8	9.0	8.5	8.0
Countries	C	39.7	40.7	41.4	37.3	32.6	34.2	28.8	23.1	19.8	15.8	10.2
	Net Trade	-25.9	-25.9	-27.1	-22.8	-17.6	-21.2	-15.6	-9.3	-10.8	-7.3	-2.2
All Non-	P	48.3	51.3	57.3	52.0	46.0	55.0	47.2	39.3	47.0	35.5	24.0
Communist	C	47.8	51.3	57.1	51.4	45.2	54.2	45.6	37.2	46.5	33.9	22.1
Countries	Net Trade	-1.0	-1.6	-0.7	-0.2	+0.2	Nil	+1.0	+1.6	Nil	+1.3	+1.8
Centrally	P	10.2	14.4	16.0	15.0	13.5	16.0	14.5	12.5	15.0	12.5	10.0
Planned Economies	C	9.2	12.8	15.3	14.8	13.7	16.0	15.5	14.1	15.0	13.8	11.8
	Net Trade	+1.0	+1.6	+0.7	+0.2	-0.2	Nil	-1.0	-1.6	Nil	-1.3	-1.8
World	P	58.5	65.7	73.3	67.0	59.5	71.0	61.7	51.8	62.0	48.0	34.0
	C	57.0	64.1	72.4	66.2	58.9	70.2	61.1	51.3	61.5	47.7	33.9

Scenarios CP = Co-operation
N = Neutral
CF = Confrontation
P = Production
C = Consumption

Net Exports +
Net Imports -

Production exceeds consumption to allow
for unallocated demand/losses and stock
build when demand is rising but stock drawdown when demand is falling.

Table 36

ACTUAL AND PROSPECTIVE BALANCES FOR OIL SHALE, TAR SANDS AND SYNTHETICS

PRODUCTION, CONSUMPTION AND NET TRADE IMPLICATIONS Million b/d o.e.

Year		1979	1985			1990			2000		
Scenario			CP	N	CF	CP	N	CF	CP	N	CF
Total OPEC	P								1.0	0.5	0.2
	C								1.0	0.5	0.2
	Net Trade								–	–	–
Other Developing Countries	P								0.5	0.5	0.5
	C								0.5	0.5	0.5
	Net Trade								–	–	–
Industrialized Countries	P	0.1	0.4	0.4	0.4	1.0	1.0	1.0	5.0	4.0	3.0
	C	0.1	0.4	0.4	0.4	1.0	1.0	1.0	5.0	4.0	3.0
	Net Trade	–	–	–	–	–	–	–	–	–	–
All Non-Communist Countries	P	0.1	0.4	0.4	0.4	1.0	1.0	1.0	6.5	5.0	3.7
	C	0.1	0.4	0.4	0.4	1.0	1.0	1.0	6.5	5.0	3.7
	Net Trade	–	–	–	–	–	–	–	–	–	–
Centrally Planned Economies	P								0.5	0.5	0.5
	C								0.5	0.5	0.5
	Net Trade								–	–	–
World	P	0.1	0.4	0.4	0.4	1.0	1.0	1.0	7.0	5.5	4.2
	C	0.1	0.4	0.4	0.4	1.0	1.0	1.0	7.0	5.5	4.2

Scenarios CP = Co-operation
N = Neutral
CF = Confrontation

P = Production
C = Consumption

Table 37

ACTUAL AND PROSPECTIVE NATURAL GAS BALANCES

PRODUCTION, CONSUMPTION AND NET TRADE IMPLICATIONS

Million b/d o.e.

Year		1979	1985			1990			2000		
Scenario			CP	N	CF	CP	N	CF	CP	N	CF
Total OPEC	P	1.8	4.0	3.0	2.2	6.0	5.0	3.7	11.0	9.0	6.7
	C	0.9	2.0	1.8	1.3	3.8	3.2	2.2	8.2	6.6	4.6
	Net Trade	+0.9	+2.0	+1.2	+0.9	+2.2	+1.8	+1.5	+2.8	+2.4	+2.1
Other Developing Countries	P	1.3	2.4	2.3	2.2	3.9	3.7	3.5	6.0	5.5	5.0
	C	1.1	1.8	1.7	1.6	2.6	2.4	2.2	4.7	4.2	3.7
	Net Trade	+0.2	+0.6	+0.6	+0.6	+1.3	+1.3	+1.3	+1.3	+1.3	+1.3
Industrialized Countries	P	15.0	14.3	14.5	14.8	13.7	14.0	14.4	13.0	12.0	10.5
	C	15.2	16.7	15.7	15.1	17.4	17.0	16.7	16.6	15.0	13.1
	Net Trade	-1.3	-3.6	-2.4	-1.5	-5.0	-4.3	-3.6	-5.7	-4.7	-4.1
All Non-Communist Countries	P	18.1	20.7	19.8	19.2	23.6	22.7	21.6	30.0	26.5	22.3
	C	17.2	20.5	19.2	18.0	23.8	22.6	21.1	29.5	25.8	21.4
	Net Trade	-0.2	-1.0	-0.6	Nil	-1.5	-1.2	-0.8	-1.6	-1.0	-0.7
Centrally Planned Economies	P	8.4	14.0	13.0	12.0	19.0	17.5	15.0	28.0	22.0	19.3
	C	8.1	12.5	12.0	11.5	16.5	15.5	13.5	25.3	20.0	18.0
	Net Trade	+0.2	+1.0	+0.6	Nil	+1.5	+1.2	+0.8	+1.6	+1.0	+0.7
World	P	26.4	34.7	32.8	31.2	42.6	40.2	36.6	58.0	48.5	41.6
	C	25.3	33.0	31.2	29.5	40.3	38.1	34.6	54.8	45.8	39.4

P = Production
C = Consumption

Net Exports +
Net Imports −

Production exceeds consumption to allow for unallocated demand/losses and stock build when demand is rising.

Scenarios
CP = Co-operation
N = Neutral
CF = Confrontation

Table 38

ACTUAL AND PROSPECTIVE SOLID FUELS BALANCES INCLUDING BIOMASS
PRODUCTION, CONSUMPTION AND NET TRADE IMPLICATIONS
Million b/d o.e.

Year Scenario		1979	1985 CP	1985 N	1985 CF	1990 CP	1990 N	1990 CF	2000 CP	2000 N	2000 CF
Total OPEC	P	0.2	0.3	0.3	0.3	0.4	0.4	0.4	0.6	0.6	0.6
	C	0.2	0.3	0.3	0.3	0.4	0.4	0.4	0.6	0.6	0.6
	Net Trade	−	−	−	−	−	−	−	−	−	−
Other Developing Countries	P	5.2	7.5	6.7	6.0	10.5	8.9	7.2	19.2	15.0	10.7
	C	4.0	5.7	5.0	4.5	8.0	6.5	5.0	15.7	12.0	9.0
	Net Trade	+0.2	+0.5	+0.5	+0.5	+1.0	+1.0	+1.0	+1.5	+1.0	+1.0
Industrialized Countries	P	15.8	19.5	19.0	18.5	23.5	22.5	21.0	33.3	31.5	30.5
	C	14.9	19.5	19.0	18.5	24.0	23.0	21.3	34.0	31.5	30.0
	Net Trade	−0.8	−1.3	−1.3	−1.3	−2.5	−2.0	−1.5	−4.5	−3.0	−1.5
All Non-Communist Countries	P	21.2	27.3	26.0	24.8	34.4	31.8	28.6	53.1	47.1	41.8
	C	19.1	25.5	24.3	23.3	32.4	29.9	26.7	50.3	44.1	39.6
	Net Trade	−0.6	−0.8	−0.8	−0.8	−1.5	−1.0	−0.5	−3.0	−2.0	−0.5
Centrally Planned Economies	P	22.0	27.0	26.0	24.0	33.0	30.0	25.5	45.0	40.0	29.5
	C	20.4	25.0	24.0	22.0	30.0	27.5	23.5	40.0	36.0	26.0
	Net Trade	+0.6	+0.8	+0.8	+0.8	+1.5	+1.0	+0.5	+3.0	+2.0	+0.5
World	P	43.2	54.3	52.0	48.8	67.4	61.8	54.1	98.1	87.1	71.3
	C	39.5	50.5	48.3	45.3	62.4	57.4	50.2	90.3	80.1	65.6

Scenarios   CP = Co-operation   P = Production   Net Exports +   Production exceeds consumption
            N = Neutral   C = Consumption   Net Imports −   to allow for unallocated demand/
            CF = Confrontation   losses and stock build when demand
                                                                            is rising.

Table 39

ACTUAL AND PROSPECTIVE SUPPLY AND DEMAND FOR NUCLEAR ENERGY

PRODUCTION - LOSSES = CONSUMPTION   NO NET TRADE

Million b/d o.e. at input factor

Year		1979	1985			1990			2000		
Scenario			CP	N	CF	CP	N	CF	CP	N	CF
Total OPEC	P	–	–	–	–	–	–	–	0.2	0.2	–
Other Developing Countries	P	0.1	0.3	0.3	0.3	0.7	0.7	0.6	2.5	2.0	1.6
Industrialized Countries	P	2.7	5.5	5.0	4.0	9.0	8.0	6.5	18.0	15.5	12.5
All Non-Communist Countries	P	2.8	5.8	5.3	4.3	9.7	8.7	7.1	20.7	17.7	14.1
Centrally Planned Economies	P	0.3	1.0	0.8	0.6	2.0	1.6	1.3	8.0	6.0	4.0
World	P	3.1	6.8	6.1	4.9	11.7	10.3	8.4	28.7	23.7	18.1

Scenarios CP = Co-operation P = Production
N = Neutral
CF = Confrontation

Table 40

ACTUAL AND PROSPECTIVE SUPPLY AND DEMAND FOR HYDROELECTRIC AND GEOTHERMAL POWER
INCLUDING SOLAR AND OTHER RENEWABLE FORMS OF ENERGY

PRODUCTION - LOSSES = CONSUMPTION   NO NET TRADE
Million b/d o.e. at input factor

Year		1979	1985			1990			2000		
Scenario			CP	N	CF	CP	N	CF	CP	N	CF
Total OPEC	P	0.1	0.2	0.2	0.2	0.2	0.2	0.2	0.4	0.4	0.4
Other Developing Countries	P	1.8	2.3	2.3	2.3	3.2	3.0	2.8	6.0	5.0	4.6
Industrialized Countries	P	5.1	6.0	6.0	6.0	6.9	6.7	6.5	8.8	8.3	7.8
All Non-Communist Countries	P	7.0	8.5	8.5	8.5	10.3	9.9	9.5	15.2	13.7	12.8
Centrally Planned Economies	P	1.3	2.0	2.0	2.0	3.0	2.7	2.6	4.6	4.2	4.0
World	P	8.3	10.5	10.5	10.5	13.3	12.6	12.1	19.8	17.9	16.8

Scenarios CP = Co-operation P = Production
N = Neutral
CF = Confrontation

Table 41

ACTUAL AND PROSPECTIVE TOTAL PRIMARY ENERGY BALANCES
PRODUCTION, CONSUMPTION AND NET TRADE IMPLICATIONS

Million b/d o.e.

Year Scenario		1979	1985 CP	1985 N	1985 CF	1990 CP	1990 N	1990 CF	2000 CP	2000 N	2000 CF
Total OPEC	P	33.4	38.5	33.0	26.7	38.6	30.6	22.3	43.2	30.7	19.9
	C	3.4	6.7	6.0	5.2	11.3	9.2	7.2	27.0	18.4	11.9
	Net Trade	+29.9	+31.8	+27.0	+21.5	+27.3	+21.4	+15.1	+16.2	+12.4	+8.0
Other Developing Countries	P	13.6	21.5	19.6	17.8	28.3	25.3	21.6	42.3	34.0	26.6
	C	15.4	21.6	19.7	17.9	27.6	24.0	20.3	39.5	31.7	25.2
	Net Trade	+2.8	-1.4	-1.3	-1.1	-0.8	-0.1	Nil	+0.6	+0.3	+0.3
Industrialized Countries	P	53.5	60.0	59.4	58.7	67.1	65.4	63.2	87.1	79.8	72.3
	C	78.7	89.5	83.4	76.6	92.5	84.5	75.1	102.2	90.1	76.6
	Net Trade	-28.0	-32.0	-26.5	-20.4	-28.7	-22.4	-14.4	-21.0	-15.1	-7.8
All Non-Communist Countries	P	100.5	120.0	112.0	103.2	134.0	121.3	107.1	172.6	144.5	118.8
	C	97.5	117.8	109.1	99.7	131.4	117.7	102.6	168.7	140.2	113.7
	Net Trade	-2.4	-2.5	-1.6	-0.6	-3.0	-1.2	+0.3	-4.6	-1.2	+1.3
Centrally Planned Economies	P	46.4	60.0	56.8	52.1	73.0	66.3	56.9	101.1	85.2	67.3
	C	42.9	55.8	53.6	49.8	67.5	62.8	55.0	93.4	80.5	64.3
	Net Trade	+2.4	+2.5	+1.6	+0.6	+3.0	+1.2	-0.3	+4.6	+1.7	-0.6
World	P	146.8	180.0	168.8	155.3	207.0	187.6	164.0	273.7	229.7	186.1
	C	140.4	173.6	162.7	149.5	198.9	180.5	157.6	262.1	220.7	178.0

Net Exports +
Net Imports -

P = Production
C = Consumption

Scenarios CP = Co-operation
N = Neutral
CF = Confrontation

Production exceeds consumption to allow for unallocated demand/losses and stock build when demand is rising.

Table 42

ACTUAL AND PROSPECTIVE ENERGY CONSUMPTION MIX OF THE WORLD

Million barrels a day oil/o.e.

		1979	1985	1990	2000
Oil + NGLs	CP		72.4	70.2	61.5
	N	64.1	66.2	61.1	47.7
	CF		58.9	51.3	33.9
Non-Conventional Oil,	CP		0.4	1.0	7.0
Synthetics, Shale,	N	0.1	0.4	1.0	5.5
Tar Sands	CF		0.4	1.0	4.2
Natural Gas	CP		33.0	40.3	54.8
	N	25.3	31.2	38.1	45.8
	CF		29.5	34.6	39.4
Solid Fuels	CP		50.5	62.4	90.3
	N	39.5	48.3	57.4	80.1
	CF		45.3	50.2	65.6
Nuclear Energy	CP		6.8	11.7	28.7
	N	3.1	6.1	10.3	23.7
	CF		4.9	8.4	18.1
Hydro, Geothermal	CP		10.5	13.3	19.8
and other renewable	N	8.3	10.5	12.6	17.9
forms of energy	CF		10.5	12.1	16.8
Total Primary Energy	CP		173.6	198.9	262.1
(in million barrels a	N	140.4	162.7	180.5	220.7
day of oil equivalent)	CF		149.5	157.6	178.0
Oil % of total	CP		42	35	23
	N	46	40	34	22
	CF		39	32	19
Average annual growth	CP		+ 3.6%	+ 2.8%	+ 2.8%
of Primary Energy	N		+ 2.5%	+ 2.1%	+ 2.0%
Consumption	CF		+ 1.1%	+ 1.1%	+ 1.2%

CP = Co-operation Scenario
N = Neutral Scenario
CF = Confrontation Scenario

Table 43

ACTUAL AND PROSPECTIVE ENERGY CONSUMPTION MIX
OF ALL NON-COMMUNIST COUNTRIES

Million barrels a day oil/o.e.

		1979	1985	1990	2000
Oil + NGLs	CP		57.1	54.2	46.5
	N	51.3	51.4	45.6	33.9
	CF		45.2	37.2	22.1
Non-Conventional Oil,	CP		0.4	1.0	6.5
Synthetics, Shale,	N	0.1	0.4	1.0	5.0
Tar Sands	CF		0.4	1.0	3.7
Natural Gas	CP		20.5	23.8	29.5
	N	17.2	19.2	22.6	25.8
	CF		18.0	21.1	21.4
Solid Fuels	CP		25.5	32.4	50.3
	N	19.1	24.3	29.9	44.1
	CF		23.3	26.7	39.6
Nuclear Energy	CP		5.8	9.7	20.7
	N	2.8	5.3	8.7	17.7
	CF		4.3	7.1	14.1
Hydro, Geothermal	CP		8.5	10.3	15.2
and other renewable	N	7.0	8.5	9.9	13.7
forms of energy	CF		8.5	9.5	12.8
Total Primary Energy	CP		117.8	131.4	168.7
(in million barrels a	N	97.5	109.1	117.7	140.2
day of oil equivalent)	CF		99.7	102.6	113.7
Oil % of total	CP		48	41	28
	N	53	47	39	24
	CF		45	36	19
Average annual growth	CP		+ 3.2%	+ 2.2%	+ 2.5%
of Primary Energy	N		+ 1.9%	+ 1.5%	+ 1.8%
Consumption	CF		+ 0.4%	+ 0.6%	+ 1.0%

CP = Co-operation Scenario
N = Neutral Scenario
CF = Confrontation Scenario

Table 44

ACTUAL AND PROSPECTIVE ENERGY CONSUMPTION MIX OF THE INDUSTRIALIZED COUNTRIES

Million barrels a day oil/o.e.

		1979	1985	1990	2000
Oil + NGLs	CP		41.4	34.2	19.8
	N	40.7	37.3	28.8	15.8
	CF		32.6	23.1	10.2
Non-Conventional Oil,	CP		0.4	1.0	5.0
Synthetics, Shale,	N	0.1	0.4	1.0	4.0
Tar Sands	CF		0.4	1.0	3.0
Natural Gas	CP		16.7	17.4	16.6
	N	15.2	15.7	17.0	15.0
	CF		15.1	16.7	13.1
Solid Fuels	CP		19.5	24.0	34.0
	N	14.9	19.0	23.0	31.5
	CF		18.5	21.3	30.0
Nuclear Energy	CP		5.5	9.0	18.0
	N	2.7	5.0	8.0	15.5
	CF		4.0	6.5	12.5
Hydro, Geothermal	CP		6.0	6.9	8.8
and other renewable	N	5.1	6.0	6.7	8.3
forms of energy	CF		6.0	6.5	7.8
Total Primary Energy	CP		89.5	92.5	102.2
(in million barrels a	N	78.7	83.4	84.5	90.1
day of oil equivalent)	CF		76.6	75.1	76.6
Conventional Oil % of total	CP		46	37	19
	N	52	45	34	18
	CF		43	31	13
Average annual growth	CP		+ 2.2%	+ 0.7%	+ 1.0%
of Primary Energy	N		+ 1.0%	+ 0.3%	+ 0.6%
Consumption	CF		- 0.4%	- 0.4%	+ 0.2%

CP = Co-operation Scenario
N = Neutral Scenario
CF = Confrontation Scenario

Table 45

ACTUAL AND PROSPECTIVE ENERGY CONSUMPTION MIX OF THE CENTRALLY PLANNED ECONOMIES

Million barrels a day oil/o.e.

		1979	1985	1990	2000
Oil + NGLs	CP		15.3	16.0	15.0
	N	12.8	14.8	15.5	13.8
	CF		13.7	14.1	11.8
Non-Conventional Oil, Synthetics, Shale, Tar Sands	CP		–	–	0.5
	N	–			0.5
	CF				0.5
Natural Gas	CP		12.5	16.5	25.3
	N	8.1	12.0	15.5	20.0
	CF		11.5	13.5	18.0
Solid Fuels	CP		25.0	30.0	40.0
	N	20.4	24.0	27.5	36.0
	CF		22.0	23.5	26.0
Nuclear Energy	CP		1.0	2.0	8.0
	N	0.3	0.8	1.6	6.0
	CF		0.6	1.3	4.0
Hydro, Geothermal and other renewable forms of energy	CP		2.0	3.0	4.6
	N	1.3	2.0	2.7	4.2
	CF		2.0	2.6	4.0
Total Primary Energy (in million barrels a day of oil equivalent)	CP		55.8	67.5	93.4
	N	42.9	53.6	62.8	80.5
	CF		49.8	55.0	64.3
Oil % of total	CP		27	24	16
	N	30	28	25	17
	CF		28	26	18
Average annual growth of Primary Energy Consumption	CP		+ 4.5%	+ 3.9%	+ 3.3%
	N		+ 3.8%	+ 3.2%	+ 2.5%
	CF		+ 2.5%	+ 2.0%	+ 1.6%

CP = Co-operation Scenario
N = Neutral Scenario
CF = Confrontation Scenario

230 Energy: A Global Outlook

Table 46

ACTUAL AND PROSPECTIVE ENERGY CONSUMPTION MIX OF THE USA

Million barrels a day oil/o.e.

		1979	1985	1990	2000
*Oil + NGLs	CP		19.0	14.6	7.6
	N	17.9	16.8	12.7	6.8
	CF		14.4	9.6	4.8
Non-Conventional Oil,	CP		0.2	0.6	2.9
Synthetics, Shale,	N	–	0.2	0.6	2.3
Tar Sands	CF		0.2	0.6	1.9
Natural Gas	CP		9.6	9.7	8.3
	N	10.0	9.3	9.5	7.4
	CF		9.1	9.4	5.9
Solid Fuels	CP		10.3	13.0	18.7
	N	7.7	10.0	12.5	17.5
	CF		9.7	11.7	16.4
Nuclear Energy	CP		2.5	3.9	7.9
	N	1.4	2.2	3.4	6.6
	CF		1.8	2.7	5.0
Hydro, Geothermal	CP		1.7	1.8	2.1
and other renewable	N	1.6	1.7	1.8	2.0
forms of energy	CF		1.7	1.8	1.9
Total Primary Energy	CP		42.5	43.6	47.5
(in million barrels a	N	38.6	40.2	40.5	42.6
day of oil equivalent)	CF		36.9	35.8	35.9
Conventional Oil % of total	CP		43	33	16
	N	46	42	31	16
	CF		39	27	13
Average annual growth	CP		+ 1.6%	+ 0.5%	+ 0.9%
of Primary Energy	N		+ 0.7%	+ 0.1%	+ 0.5%
Consumption	CF		– 0.7%	– 0.6%	NC

CP = Co-operation Scenario
N = Neutral Scenario
CF = Confrontation Scenario

NC = No Change

* Net of processing gain

ACTUAL AND PROSPECTIVE ENERGY CONSUMPTION MIX OF WESTERN EUROPE

Table 47

Million barrels a day oil/o.e.

		1979	1985	1990	2000
Oil + NGLs	CP		14.8	12.7	8.7
	N	14.5	13.5	10.9	6.7
	CF		12.1	9.1	3.6
Non-Conventional Oil, Synthetics, Shale, Tar Sands	CP		–	–	0.8
	N	–	–	–	0.7
	CF		–	–	0.6
Natural Gas	CP		4.6	4.9	5.3
	N	3.7	4.2	4.8	5.0
	CF		4.0	4.7	5.0
Solid Fuels	CP		5.8	6.7	8.8
	N	5.0	5.7	6.5	8.0
	CF		5.6	6.0	7.9
Nuclear Energy	CP		2.1	3.7	7.2
	N	0.8	2.0	3.3	6.4
	CF		1.5	2.8	5.5
Hydro, Geothermal and other renewable forms of energy	CP		2.2	2.5	2.9
	N	2.0	2.2	2.4	2.8
	CF		2.2	2.3	2.7
Total Primary Energy (in million barrels a day of oil equivalent)	CP		29.5	30.5	33.7
	N	26.0	27.6	27.9	29.6
	CF		25.4	24.9	25.3
Conventional Oil % of total	CP		50	42	26
	N	56	49	39	23
	CF		48	37	14
Average annual growth of Primary Energy Consumption	CP		+ 2.1%	+ 0.7%	+ 1.0%
	N		+ 1.0%	+ 0.2%	+ 0.6%
	CF		– 0.4%	– 0.4%	+ 0.2%

CP = Co-operation Scenario
N = Neutral Scenario
CF = Confrontation Scenario

Table 48

ACTUAL AND PROSPECTIVE ENERGY CONSUMPTION MIX OF JAPAN

Million barrels a day oil/o.e.

		1979	1985	1990	2000
Oil + NGLs	CP		5.4	4.5	2.3
	N	5.5	4.7	3.6	1.5
	CF		4.3	3.2	0.8
Non-Conventional Oil,	CP				0.4
Synthetics, Shale,	N	-	-	-	0.3
Tar Sands	CF				0.2
Natural Gas	CP		0.9	1.2	1.5
	N	0.4	0.7	1.1	1.3
	CF		0.6	1.0	1.0
Solid Fuels	CP		2.1	2.7	4.4
	N	1.2	2.0	2.5	4.0
	CF		1.9	2.2	3.8
Nuclear Energy	CP		0.6	0.9	1.8
	N	0.3	0.5	0.8	1.5
	CF		0.4	0.6	1.2
Hydro, Geothermal	CP		0.6	0.7	0.9
and other renewable	N	0.4	0.6	0.7	0.9
forms of energy	CF		0.6	0.7	0.9
Total Primary Energy	CP		9.6	10.0	11.3
(in million barrels a	N	7.8	8.5	8.7	9.5
day of oil equivalent)	CF		7.8	7.7	7.9
Conventional Oil % of total	CP		56	45	20
	N	70	55	41	16
	CF		55	42	10
Average annual growth	CP		+ 3.5%	+ 0.8%	+ 1.2%
of Primary Energy	N		+ 1.4%	+ 0.5%	+ 0.9%
Consumption	CF		NC	- 0.3%	+ 0.3%

CP = Co-operation Scenario
N = Neutral Scenario
CF = Confrontation Scenario

NC = No Change

Table 49

ACTUAL AND PROSPECTIVE ENERGY CONSUMPTION MIX
OF CANADA, AUSTRALIA, NEW ZEALAND

Million barrels a day oil/o.e.

		1979	1985	1990	2000
Oil + NGLs	CP		3.0	2.4	1.2
	N	2.7	2.3	1.6	0.8
	CF		1.8	1.2	0.8
Non-Conventional Oil,	CP		0.2	0.4	0.9
Synthetics, Shale,	N	0.1	0.2	0.4	0.7
Tar Sands	CF		0.2	0.4	0.5
Natural Gas	CP		1.6	1.6	1.5
	N	1.2	1.5	1.6	1.3
	CF		1.4	1.6	1.2
Solid Fuels	CP		1.3	1.6	2.1
	N	1.0	1.3	1.5	2.0
	CF		1.3	1.4	1.9
Nuclear Energy	CP		0.3	0.5	1.1
	N	0.2	0.3	0.5	1.0
	CF		0.3	0.4	0.8
Hydro, Geothermal	CP		1.5	1.9	2.9
and other renewable	N	1.2	1.5	1.8	2.6
forms of energy	CF		1.5	1.7	2.3
Total Primary Energy	CP		7.9	8.4	9.7
(in million barrels a	N	6.3	7.1	7.4	8.4
day of oil equivalent)	CF		6.5	6.7	7.5
Conventional Oil % of total	CP		38	29	12
	N	43	32	22	10
	CF		28	18	11
Average annual growth	CP		+ 3.8%	+ 1.2%	+ 1.5%
of Primary Energy	N		+ 2.0%	+ 0.8%	+ 1.3%
Consumption	CF		+ 0.5%	+ 0.6%	+ 1.1%

CP = Co-operation Scenario
N = Neutral Scenario
CF = Confrontation Scenario

Table 50

ACTUAL AND PROSPECTIVE ENERGY CONSUMPTION MIX OF OPEC DEVELOPING COUNTRIES

Million barrels a day oil/o.e.

		1979	1985	1990	2000
Oil + NGLs	CP		4.2	6.9	16.6
	N	2.2	3.7	5.4	10.1
	CF		3.4	4.4	6.1
Non-Conventional Oil, Synthetics, Shale, Tar Sands	CP				1.0
	N	-	-	-	0.5
	CF				0.2
Natural Gas	CP		2.0	3.8	8.2
	N	0.9	1.8	3.2	6.6
	CF		1.3	2.2	4.6
Solid Fuels	CP		0.3	0.4	0.6
	N	0.2	0.3	0.4	0.6
	CF		0.3	0.4	0.6
Nuclear Energy	CP				0.2
	N	-	-	-	0.2
	CF				-
Hydro, Geothermal and other renewable forms of energy	CP		0.2	0.2	0.4
	N	0.1	0.2	0.2	0.4
	CF		0.2	0.2	0.4
Total Primary Energy (in million barrels a day of oil equivalent)	CP		6.7	11.3	27.0
	N	3.4	6.0	9.2	18.4
	CF		5.2	7.2	11.9
Conventional Oil % of total	CP		63	61	61
	N	65	62	59	55
	CF		65	61	51
Average annual growth of Primary Energy Consumption	CP		+12.0%	+11.0%	+9.1%
	N		+9.9%	+8.9%	+7.2%
	CF		+7.3%	+6.8%	+5.2%

CP = Co-operation Scenario
N = Neutral Scenario
CF = Confrontation Scenario

Table 51

ACTUAL AND PROSPECTIVE ENERGY CONSUMPTION MIX OF NON-OPEC DEVELOPING COUNTRIES

Million barrels a day oil/o.e.

		1979	1985	1990	2000
Oil + NGLs	CP		11.5	13.1	10.1
	N	8.4	10.4	11.4	8.0
	CF		9.2	9.7	5.8
Non-Conventional Oil,	CP				0.5
Synthetics, Shale,	N	–	–	–	0.5
Tar Sands	CF				0.5
Natural Gas	CP		1.8	2.6	4.7
	N	1.1	1.7	2.4	4.2
	CF		1.6	2.2	3.7
Solid Fuels	CP		5.7	8.0	15.7
	N	4.0	5.0	6.5	12.0
	CF		4.5	5.0	9.0
Nuclear Energy	CP		0.3	0.7	2.5
	N	0.1	0.3	0.7	2.0
	CF		0.3	0.6	1.6
Hydro, Geothermal	CP		2.3	3.2	6.0
and other renewable	N	1.8	2.3	3.0	5.0
forms of energy	CF		2.3	2.8	4.6
Total Primary Energy	CP		21.6	27.6	39.5
(in million barrels a	N	15.4	19.7	24.0	31.7
day of oil equivalent)	CF		17.9	20.3	25.2
Oil % of total	CP		53	47	26
	N	55	53	48	25
	CF		51	48	23
Average annual growth	CP		+ 5.8%	+ 5.0%	+ 3.6%
of Primary Energy	N		+ 4.2%	+ 4.0%	+ 2.8%
Consumption	CF		+ 2.5%	+ 2.5%	+ 2.2%

CP = Co-operation Scenario
N = Neutral Scenario
CF = Confrontation Scenario

For discussion of the special case variant see Chapter 20 of Part II

Table 52

ACTUAL AND PROSPECTIVE ENERGY CONSUMPTION MIX OF THE USSR

Million barrels a day oil/o.e.

		1979	1985	1990	2000
Oil + NGLs	CP		10.6	11.2	10.7
	N	8.9	10.3	10.9	10.2
	CF		10.0	10.5	9.2
Non-Conventional Oil, Synthetics, Shale, Tar Sands	CP				0.4
	N	–	–	–	0.4
	CF				0.4
Natural Gas	CP		9.1	12.2	17.3
	N	6.1	8.8	11.1	14.3
	CF		8.5	10.0	13.3
Solid Fuels	CP		8.2	9.7	13.0
	N	6.8	7.8	8.9	11.5
	CF		7.2	7.5	8.2
Nuclear Energy	CP		0.8	1.5	5.8
	N	0.2	0.6	1.2	4.4
	CF		0.5	1.0	3.0
Hydro, Geothermal and other renewable forms of energy	CP		1.4	1.9	2.5
	N	0.9	1.4	1.7	2.2
	CF		1.4	1.6	2.1
Total Primary Energy (in million barrels a day of oil equivalent)	CP		30.1	36.5	49.7
	N	22.9	28.9	33.8	43.0
	CF		27.6	30.6	36.2
Oil % of total	CP		35	31	22
	N	39	36	32	24
	CF		36	34	25
Average annual growth of Primary Energy Consumption	CP		+ 4.7%	+ 3.9%	+ 3.1%
	N		+ 4.0%	+ 3.2%	+ 2.4%
	CF		+ 3.2%	+ 2.1%	+ 1.7%

CP = Co-operation Scenario
N = Neutral Scenario
CF = Confrontation Scenario

Table 53

ACTUAL AND PROSPECTIVE ENERGY CONSUMPTION MIX OF EASTERN EUROPE

Million barrels a day oil/o.e.

		1979	1985	1990	2000
Oil + NGLs	CP		2.5	2.6	2.1
	N	2.1	2.4	2.5	1.8
	CF		2.0	2.0	1.3
Non-Conventional Oil, Synthetics, Shale, Tar Sands	CP				0.1
	N	–	–	–	0.1
	CF				0.1
Natural Gas	CP		2.1	3.2	5.0
	N	1.2	2.0	2.7	3.7
	CF		1.9	2.2	3.2
Solid Fuels	CP		6.4	7.5	9.4
	N	5.4	6.2	6.8	8.7
	CF		5.8	6.2	7.0
Nuclear Energy	CP		0.2	0.5	1.8
	N	0.1	0.2	0.4	1.4
	CF		0.1	0.3	1.0
Hydro, Geothermal and other renewable forms of energy	CP		0.1	0.2	0.3
	N	0.1	0.1	0.2	0.3
	CF		0.1	0.2	0.3
Total Primary Energy (in million barrels a day of oil equivalent)	CP		11.3	14.0	18.7
	N	8.9	10.9	12.6	16.0
	CF		9.9	10.9	12.9
Oil % of total	CP		22	19	11
	N	24	22	20	11
	CF		20	18	10
Average annual growth of Primary Energy Consumption	CP		+ 4.1%	+ 4.4%	+ 2.9%
	N		+ 3.4%	+ 2.9%	+ 2.4%
	CF		+ 1.8%	+ 1.9%	+ 1.7%

CP = Co-operation Scenario
N = Neutral Scenario
CF = Confrontation Scenario

238 Energy: A Global Outlook

Table 54

ACTUAL AND PROSPECTIVE ENERGY CONSUMPTION MIX OF CHINA

Million barrels a day oil/o.e.

		1979	1985	1990	2000
Oil + NGLs	CP		2.2	2.2	2.2
	N	1.8	2.1	2.1	1.8
	CF		1.7	1.6	1.3
Non-Conventional Oil, Synthetics, Shale, Tar Sands	CP				
	N	-	-	-	-
	CF				
Natural Gas	CP		1.3	2.1	3.0
	N	0.8	1.2	1.7	2.0
	CF		1.1	1.3	1.5
Solid Fuels	CP		10.4	12.8	17.6
	N	8.2	10.0	11.8	15.8
	CF		9.0	9.8	10.8
Nuclear Energy	CP				0.4
	N	-	-	-	0.2
	CF				-
Hydro, Geothermal and other renewable forms of energy	CP		0.5	0.9	1.8
	N	0.2	0.5	0.8	1.7
	CF		0.5	0.8	1.6
Total Primary Energy (in million barrels a day of oil equivalent)	CP		14.4	18.0	25.0
	N	11.0	13.8	16.4	21.5
	CF		12.3	13.5	15.4
Oil % of total	CP		15	12	9
	N	16	15	13	8
	CF		14	12	8
Average annual growth of Primary Energy Consumption	CP		+ 4.6%	+ 4.6%	+ 3.3%
	N		+ 3.9%	+ 3.5%	+ 2.7%
	CF		+ 1.9%	+ 1.9%	+ 1.3%

CP = Co-operation Scenario
N = Neutral Scenario
CF = Confrontation Scenario

Table 55

WORLD ECONOMIC AND ENERGY CONSUMPTION GROWTH

Average Annual Growth Rates and Coefficients

		1973-79			1979-85			1985-90			1990-2000		
		PE %	GNP %	PE/GNP Coeff.	PE %	GNP %	PE/GNP Coeff.	PE %	GNP %	PE/GNP Coeff.	PE %	GNP %	PE/GNP Coeff.
OPEC Countries	CP				12.0	7.0	1.7	11.0	6.5	1.7	9.1	6.0	1.5
	N	11.1	6.5	1.71	9.9	6.5	1.5	8.9	5.9	1.5	7.2	5.1	1.4
	CF				7.3	5.5	1.3	6.8	4.8	1.4	5.2	4.0	1.3
Other Developing Countries	CP				5.8	5.7	1.0	5.0	5.0	1.0	3.6	4.0	0.9
	N	5.4	5.2	1.04	4.2	4.5	0.9	4.0	4.5	0.9	2.8	3.5	0.8
	CF				2.5	3.3	0.8	2.5	3.3	0.8	2.2	3.0	0.7
Industrialized Countries	CP				2.2	4.4	0.50	0.7	2.8	0.25	1.0	4.0	0.25
	N	1.0	2.9	0.34	1.0	2.5	0.40	0.3	1.5	0.20	0.6	3.0	0.20
	CF				-0.4	0.5	NA	-0.4	0.5	NA	0.2	1.5	0.13
Centrally Planned Economies (CPEs)	CP				4.5	4.5	1.0	3.9	4.3	0.9	3.3	4.1	0.8
	N	5.2	5.2	1.00	3.8	4.2	0.9	3.2	4.0	0.8	2.5	3.5	0.7
	CF				2.5	3.2	0.8	2.0	2.9	0.7	1.6	2.6	0.6
Total World	CP				3.6	4.7	0.8	2.8	3.5	0.8	2.8	4.1	0.7
	N	2.7	3.6	0.75	2.5	3.2	0.8	2.1	2.6	0.8	2.0	3.3	0.6
	CF				1.1	1.6	0.7	1.1	1.6	0.7	1.2	2.1	0.6

CP = Co-operation Scenario
N = Neutral Scenario
CF = Confrontation Scenario

PE = Primary Energy
GNP = Gross National Product

Table 56

A COMPARISON OF PROJECTIONS FOR OIL PRODUCTION

Million Barrels a Day

	OPEC			All Non-Communist Countries			Total World		
Actual 1979	31.3			51.3			65.7		
Projections 1985	Hi	Me	Lo	Hi	Me	Lo	Hi	Me	Lo
This book	34.0	29.5	24.0	57.3	52.0	46.0	73.3	67.0	59.5
US DOE (EIA)[1]	30.0	23.9	21.5	54.6	47.8	45.3	–	–	–
US DOE (NEPP)[2]	27.8	26.0	21.5	53.4	48.5	42.7	–	–	–
US DOE (IA)[3]	–	28.8	–	–	51.2	–	–	–	–
US CBO[4]	–	30.1	–	–	55.5	–	–	–	–
US OTA[5]	35.0	–	28.5	59.5	–	49.0	–	–	–
Exxon (1980)[6]	33.0	–	28.0	–	–	–	–	–	–
Shell (1980)[7]	–	30.0	–	–	53.0	–	–	–	–
Socal (1981)[8]	–	24.4	–	–	45.3	–	–	–	–
Texaco (1981)[9]	–	28.3	–	–	51.3	–	–	–	–
IEA (1980)[10]	–	30.8	–	–	56.4	–	–	–	–
WAES (1977)[11]	39.0	–	33.0	63.7	–	55.0	–	–	–
1990									
This book	32.0	25.0	18.0	55.0	47.2	39.3	71.0	61.7	51.8
US DOE(EIA)[1]	34.5	26.1	23.9	61.0	52.2	45.7	–	–	–
US DOE (NEPP)[2]	28.2	24.8	22.0	55.8	47.9	42.5	–	–	–
US DOE (IA)[3]	–	28.9	–	–	52.4	–	–	–	–
US CBO[4]	–	31.3	–	–	57.6	–	–	–	–
Exxon (1980)[6]	33.0	30.0	28.0	–	53.0	–	–	68.0	–
Shell (1980)[7]	–	30.0	–	–	56.0	–	–	–	–
Socal (1981)[8]	–	24.4	–	–	46.0	–	–	–	–
Texaco (1981)[9]	–	30.2	–	–	53.7	–	–	–	–
IEA (1980)[10]	–	31.6	–	–	59.7	–	–	–	–
WB (1981)[12]	–	–	–	–	57.9	–	–	75.8	–
WB (1980)[13]	–	–	–	–	–	–	–	77.3	–
2000									
This book	30.0	21.0	12.0	47.0	35.5	24.0	62.0	48.0	34.0
US DOE (NEPP)[2]	28.1	25.0	20.7	54.0	48.4	42.0	–	–	–
Exxon (1980)[6]	33.0	29.0	28.0	–	55.0	–	–	71.0	–
Socal (1981)[8]	–	27.5	–	–	48.9	–	–	–	–
Texaco (1981)[9]	–	33.7	–	–	58.9	–	–	–	–
WAES (1977)[11]	45.0	–	33.0	69.6	–	51.5	–	–	–
IIASA (1981)[14]	–	–	–	–	–	–	83.0	–	66.9

Hi = High, Me = Medium, Lo = Low.

Sources:
1. United States Department of Energy, Energy Information Administration (EIA), 1980 Annual Report to Congress, March 1981.

2. United States Department of Energy, Office of Policy, Planning and Analysis, Supplement to the National Energy Policy Plan (NEPP), July 1981.

Table 56 (contd)

3. United States Department of Energy, International Affairs, December 1980.

4. United States Congressional Budget Office, May 1980.

5. United States Congress, Office of Technology Assessment, October 1980.

6. Exxon Corporation, World Energy Outlook, December 1980.

7. Shell Oil Company, Free World Energy Forecasts 1985 and 1990, November 1980.

8. Standard Oil Company of California, World Energy Outlook 1981-2000, September 1981. NGLs are excluded.

9. Texaco, Free World Energy Outlook, July 1981.

10. International Energy Agency, Outlook for the Eighties, 1980.

11. Global Energy Prospects 1985-2000, Workshop on Alternative Energy Strategies, 1977.

12. World Development Report 1981, The World Bank, August 1981.

13. Energy in the Developing Countries, World Bank, August 1980.

14. Energy in a Finite World, the International Institute for Applied Systems Analysis, 1981.

The projections shown in Tables 35 to 55 in this book were prepared in late 1980 without specific cross-reference to other published forecasts. It is thus quite fortuitous that the three scenario sets of numbers appear to be fairly near the averages for other forecasts/projections of oil production in OPEC and other non-Communist countries for the year 1985. By 1990, though, the neutral scenario numbers featured in this book (shown in the Me column in this table) are the lowest or close to the lowest numbers published by others. By the year 2000 they are clearly the lowest by a significant margin, more particularly outside OPEC. This probably reflects more on the poor finding rate of recent years than do some other projections, relatively more pessimism about finds of new giant fields during the next few years, the lack of much hope for significant improvements in recovery factors during the next ten to twenty years than may be assured by some others, and the fact of the constraint of low current reserves to production ratios in so many producing countries with an evident need for increased indigenous oil production either to reduce dependence on imports or to increase export revenue.

Table 57

A COMPARISON OF PROJECTIONS FOR OIL AND ENERGY CONSUMPTION

Million Barrels a Day Oil/Oil Equivalent

	Oil Consumption						Energy Consumption					
	All Non-Communist Countries			Total World			All Non-Communist Countries			Total World		
	Hi	Me	Lo	Hi	Me	Lo	Hi	Me	Lo	Hi	Me	Lo
Actual 1979		51.3			64.1			97.5			140.4	
Projections 1985												
This book	57.1	51.4	45.2	72.4	66.2	58.9	117.8	109.1	99.7	173.6	162.7	149.5
US DOE (EIA)[1]	54.6	47.8	45.3	—	—	—	—	100.4	—	—	—	—
US DOE (NEPP)[2]	49.9	48.0	43.4	—	—	—	115.4	105.4	92.6	—	—	—
US DOE (IA)[3]	—	50.7	—	—	—	—	—	—	—	—	—	—
US CBO[4]	—	58.9	—	—	—	—	—	—	—	—	—	—
Exxon (1980)[5]	—	—	—	—	—	—	—	111.1	—	—	159.7	—
Shell (1980)[6]	—	53.0	—	—	—	—	—	104.0	—	—	—	—
Socal (1981)[7]	—	—	—	—	65.0	—	—	109.0	—	—	162.7	—
Texaco (1981)[8]	—	52.0	—	—	—	—	—	107.2	—	—	—	—
IEA (1980)[9]	—	58.9	—	—	—	—	—	—	—	—	—	—
WAES (1977)[10]	63.0	—	58.0	—	—	—	123.2	—	114.1	—	—	—
1990												
This book	54.2	45.6	37.2	70.2	61.1	51.3	131.4	117.7	102.6	198.9	180.5	157.6
US DOE (EIA)[1]	61.6	52.2	45.7	—	—	—	—	115.4	—	—	—	—
US DOE (NEPP)[2]	50.5	46.9	43.8	—	—	—	130.5	116.1	99.1	—	—	—
US DOE (IA)[3]	—	51.4	—	—	—	—	—	—	—	—	—	—
US CBO[4]	—	66.4	—	—	—	—	—	—	—	—	—	—
Exxon (1980)[5]	—	55.0	—	—	70.0	—	—	125.0	—	—	179.6	—
Shell (1980)[6]	—	56.0	—	—	—	—	—	116.0	—	—	—	—
Socal (1981)[7]	—	—	—	—	68.6	—	—	123.8	—	—	189.3	—
Texaco (1981)[8]	—	54.5	—	—	—	—	—	119.9	—	—	—	—
IEA (1980)[9]	—	64.3	—	—	—	—	—	—	—	—	—	—
Mobil (1981)[11]	—	53.0	—	—	—	—	—	122.0	—	—	—	—
World Bank (1981)[12]	—	—	—	—	77.1	—	—	—	—	—	189.7	—
World Bank (1980)[13]	—	—	—	—	77.3	—	—	—	—	—	201.5	—

Table 57 (contd)

Comparison of Forecasts

	Oil Consumption						Energy Consumption					
	All Non-Communist Countries			Total World			All Non-Communist Countries			Total World		
	Hi	Me	Lo	Hi	Me	Lo	Hi	Me	Lo	Hi	Me	Lo
2000												
This book	46.5	33.9	22.1	61.5	47.7	33.9	168.7	140.2	113.7	262.1	220.7	178.0
US DOE (NEPP)[2]	51.4	45.7	38.7	—	—	—	162.7	141.0	109.3	—	—	—
Exxon (1980)[5]	—	61.0	—	—	77.0	—	—	155.3	—	—	226.5	—
Shell (1980)[6]	58.8	—	52.8	—	—	—	160.0	—	140.0	—	—	—
Socal (1981)[7]	—	—	—	—	76.2	—	—	152.2	—	—	247.0	—
Texaco (1981)[8]	—	59.8	—	—	—	—	—	150.0	—	—	—	—
WAES (1977)[10]	93.0	—	75.0	—	—	—	206.8	154.0	159.9	—	—	—
Mobil (1981)[11]	—	58.5	—	83.0	—	66.9	—	—	—	—	—	—
IIASA (1981)[14]	—	—	—	—	—	—	—	—	—	237.2	—	191.4
OECD (1979)[15]	—	—	—	—	—	—	—	—	—	294.7	—	240.7
SRI (1980)[16]	—	—	—	—	96.4	—	—	—	—	—	275.3	—

Sources:

1. United States Department of Energy, Energy Information Administration (EIA), 1980 Annual Report to Congress, March 1981.

2. United States Department of Energy, Office of Policy, Planning and Analysis, A Supplement to the National Energy Policy Plan (NEPP), July 1981.

3. United States Department of Energy, International Affairs, December 1980.

4. United States Congressional Budget Office, May 1980.

5. Exxon Corporation, World Energy Outlook, December 1980.

6. For 1985 and 1990, Shell Oil Company, Free World Energy Forecasts 1985 and 1990, November 1980. For 2000, Shell World, March/April 1980.

7. Standard Oil Company of California, World Energy Outlook 1981-2000, September 1981.

8. Texaco, Free World Energy Outlook 1980-2000, July 1981.

9. International Energy Agency (IEA/OECD), Outlook for the Eighties (Adjusted Case), 1980.

Table 57 (contd)

10. Global Energy Prospects 1985-2000, Workshop on Alternative Energy Strategies, 1977.

11. Mobil, Long Range Energy Needs, speech by E.S. Checket to International Petroleum Seminar, Institut Francais du Pétrole, Nice, March 1981.

12. World Bank Development Report 1981, The World Bank, August 1981.

13. Energy in the Developing Countries, World Bank, August 1980.

14. Energy in a Finite World, the International Institute for Applied Systems Analysis, 1981.

15. OECD, Interfutures, Facing the Future, 1979, in Ruth Leger Sivard, World Energy Survey, 1981.

16. SRI International, World Energy: A Manageable Dilemma, in Harvard Business Review, May-June 1980 and noted in Ruth Leger Sivard, World Energy Survey, 1981.

The neutral scenario numbers (Me column in this table) are fortuitously near the mean of projected levels of oil demand in 1985 quoted by others, but the spread I have considered it appropriate to show between high (co-operation) and low (confrontation) is wider than that shown by others. Perhaps this is because more explicit recognition has been given to political uncertainties than in some other projections, as well as taking necessary account of supply uncertainties in the oil and other energy industries and market demand factors: income and price elasticity uncertainties. The projections in this book for oil consumption in 1990 are slightly lower than the lowest among all the others, though this is not the case for total primary energy consumption. The assumption is made that given the relative price competitiveness and abundance of coal it will start to make increasingly significant contributions to the growth of energy demand. Some erosion in the resistance to nuclear power in industrialized countries is assumed as this source of energy proves itself in terms of economic and technical criteria. The dramatic feature of the projections in this book compared to others is the extent of the decline in oil consumption between 1990 and 2000, particularly in the confrontation scenario. This reflects both a fall in worldwide availability and rising real prices which will involve a considerable backing out of oil use, especially in the industrialized countries. Synthetics are expected to substitute for some of the reduction in conventional crude. The scenario numbers for total energy consumption included in this book for the various scenarios are really quite similar for non-Communist countries as a whole to those published in July 1981 by the US Department of Energy as a part of the new US National Energy Policy Plan. The NEPP oil projections are higher than those included in this book. The supply and demand for other sources of energy is assumed to be higher in this book. My expectation is that oil market developments over the period 1973 to 1981 will have a profoundly significant influence on a restructuring of the energy sector in the industrialized countries through the 1990s. Perhaps the most remarkable feature of the energy transition discussed in this book is the fact that even in the confrontation scenario the volume of oil consumption of all the non-Communist countries in aggregate is projected to be still more than double the 10 million b/d of oil they are estimated to have consumed in 1950.

Table 58

A COMPARISON OF PROJECTIONS FOR WORLD OIL PRICES

Per Cent per Year Rate of Change (Compound)
Based on Constant US Dollar Prices*

		Low	Medium	High
Actual	1960 to 1970		-3.3	
	1970 to 1980		+20.0	
Projected	1980 to 1990			
	This book*	Up to +3.0	+3.0 to +6.0	more than 6.0
	US DOE (EIA)[1]	+0.3	+2.8	+4.7
	US DOE (NEPP)[2]	+1.0	+3.5	+6.3
	DRI[3]		+3.5	
	Exxon[4]		+2.2	
	Gately[5]		-0.3	
	1990 to 2000			
	US DOE (EIA)[1]	+2.8	+3.9	+6.6
	US DOE (NEPP)[2]	+2.0	+3.0	+3.4
	DRI[3]		+3.3	
	Exxon[4]		+2.0	
	Gately[5]		+6.7	
	2000 to 2020			
	US DOE (EIA)[1]	+1.6	+1.8	+1.8

* Based on an estimated price of oil entering world trade of about $31.50 a barrel FOB in 1980 with the price of Arab Light marker crude averaging $28.67 a barrel FOB Ras Tanura. The weighted average price of oil entering world trade appears likely to be only about 10 per cent higher in nominal US dollars in 1981 than in 1980. The movement through the first eight months of 1981 was almost continuously downwards and in contrast to the almost continuous and sharp upward movement through 1979 and 1980.

Other projections in this table are based on US DOE's estimate of a rounded price of $34 a barrel which was approximately the average US refiner acquisition cost of imported crude oil in 1980.

Sources:
1. US Department of Energy, Energy Information Administration (EIA) 1980 Annual Report to Congress, Volume Three Forecasts, March 1981.

2. US Department of Energy, Office of Policy Planning and Analysis. A Supplement to the National Energy Policy Plan (NEPP), July 1981.

3. DRI Forecast, Energy Review Spring 1981 Data Resources Inc.

4. Exxon Corporation, World Energy Outlook, December 1980.

5. Dermont Gately, Modeling OPEC Behaviour in the World Energy Market, IAEE Conference, Toronto, June 1981.

Table 59

AN ANALYSIS OF CHANGES IN OIL AND ENERGY PROJECTIONS OVER TIME

Million Barrels per Day Oil/Oil Equivalent

Forecasts for 1990 Base Scenario, Mid Price

Annual Reports* Dated	OPEC Oil Production	Free World Oil Consumption	Energy Consumption
1977	61.0	82.9	151.1
1978	39.7	68.4	131.8
1979	26.3	52.3	116.3
1980	26.1	52.2	115.8

This analysis shows that the decline in expected OPEC production is almost as large as the decline in energy consumption.

*Source: The U.S. Department of Energy, Energy Information Administration, Annual Reports to Congress.

Appendix I
OPEC RESOLUTIONS, AGREEMENTS AND DECLARATIONS

Appendix I.1 -   Resolution IV.32, Geneva 5-8 April & 4-8 June 1962

Appendix I.2 -   Resolution XVI. 90, Vienna 24-25 June 1968

Appendix I.3 -   Resolution XXI. 120, Caracas December 1970

Appendix I.4 -   Resolution XXII. 131, Tehran 3-4 February 1971
                 Resolution XXII. 132, Tehran 3-4 February 1971

Appendix I.5 -   Full Text of Tehran Price Agreement, Tehran
                 14 February 1971

Appendix I.6 -   Resolution XXXV. 160, Vienna 15-16 September 1973

Appendix I.7 -   OPEC Communique, 16 October 1973

Appendix I.8 -   Arab Oil Ministers Communique 17 October 1973

Appendix I.9 -   Full Text of Participation Agreement
                 20 December 1972 (Signature date)

Appendix I.10 - Solemn Declaration of Sovereigns and Heads
                 of State of OPEC Member Countries
                 4-6 March 1975

Appendix I.11 - Agreement Establishing the OPEC Special Fund
                 28 January 1976

APPENDIX I.1
FOURTH OPEC CONFERENCE HELD IN GENEVA 5-8 APRIL 1962
                                                     AND 4-8 JUNE 1962

RESOLUTION IV.32

The Conference, considering

1. That the Member Countries, acting in pursuance of Resolution I.1, duly protested against the price reduction effected by the Oil Companies in August 1960;

2. That the Oil Companies have so far taken no steps to restore prices to the pre-August 1960 level;

3. That oil production in all Member Countries is the most prominent source of revenue for financing the implementation of projects of economic development and social progress, and that a fall in crude-oil prices reduces the level of oil revenues, thereby bringing about a setback in the realization of the above objectives and seriously dislocating the economy of the Member Countries;

4. That a fall in crude-oil prices impairs the purchasing power of Member Countries in respect of manufactured goods which are fundamental to the economy of developing nations and the prices of which have been steadily rising while those of crude oil have been falling;

5. That the oil industry having the character of a public utility, Member Countries cannot be indifferent to such a vital element of the industry as the determination of the price of oil;

Recommends

That Member Countries should forthwith enter into negotiations with the Oil Companies concerned and/or any other authority or body deemed appropriate, with a view to ensuring that oil produced in Member Countries shall be paid for on the basis of posted prices not lower than those which applied prior to August 1960. If within a reasonable period after the commencement of the negotiations no satisfactory arrangement is reached, the Member Countries shall consult with each other with a view to taking such steps as they deem appropriate in order to restore crude-oil prices to the level which prevailed prior to August 9, 1960; and in any event a report as to the result of negotiations shall be submitted to the Fifth

Appendix I.1

Conference for decision in the light of this paragraph.

That Member Countries shall jointly formulate a rational price structure to guide their long-term price policy, on which subject the Board of Governors is hereby directed to prepare a comprehensive study at the earliest possible date. An important element of the price structure to be devised will be the linking of crude-oil prices to an index of prices of goods which the Member Countries need to import.

Source: OPEC Offical Resolutions and Press Releases 1960-1980, Pergamon Press, 1980, for OPEC.

APPENDIX I.2
SIXTEENTH OPEC CONFERENCE HELD IN VIENNA 24-25 JUNE 1968

RESOLUTION XVI.90
DECLARATORY STATEMENT OF PETROLEUM POLICY IN MEMBER COUNTRIES

The Conference,

recalling Paragraph 4 of its Resolution 1.2;
recognizing that hydrocarbon resources in Member Countries are one of the principal sources of their revenues and foreign exchange earnings and therefore constitute the main basis for their economic development;

bearing in mind that hydrocarbon resources are limited and exhaustible and that their proper exploitation determines the conditions of the economic development of Member Countries, both at present and in the future;

bearing in mind also that the inalienable right of all countries to exercise permanent sovereignty over their natural resources in the interest of their national development is a universally recognized principle of public law and has been repeatedly reaffirmed by the General Assembly of the United Nations, most notably in its Resolution 2158 of November 25, 1966;

considering also that in order to ensure the exercise of permanent sovereignty over hydrocarbon resources, it is essential that their exploitation should be aimed at securing the greatest possible benefit for Member Countries;

considering further that this aim can better be achieved if Member Countries are in a position to undertake themselves directly the exploitation of their hydrocarbon resources, so that they may exercise their freedom of choice in the utilization of hydrocarbon resources under the most favourable conditions;

taking into account the fact that foreign capital, whether public or private, forthcoming at the request of the Member Countries, can play an important role, inasmuch as it supplements the efforts undertaken by them in the exploitation of their hydrocarbon resources, provided that there is government supervision of the activity of foreign capital to ensure that it is used in the interest of national development and that returns earned by it do not exceed reasonable levels;

bearing in mind that the principal aim of the Organization, as set out in Article 2 of its Statute, "is the coordination and unification of the petroleum policies of Member Countries and the determination of the best means for safeguarding their interest, individually and collectively";

recommends that the following principles shall serve as basis for petroleum policy in Member Countries.

Mode of Development

1. Member Governments shall endeavour, as far as feasible, to explore for and develop their hydrocarbon resources directly. The capital, specialists and the promotion of marketing outlets required for such direct development may be complemented when necessary from alternate sources on a commercial basis.

2. However, when a Member Government is not capable of developing its hydrocarbon resources directly, it may enter into contracts of various types, to be defined in its legislation but subject to the present principles, with outside operators for a reasonable remuneration, taking into account the degree of risk involved. Under such an arrangement, the Government shall seek to retain the greatest measure possible of participation in and control over all aspects of operations.

3. In any event, the terms and conditions of such contracts shall be open to revision at predetermined intervals, as justified by changing circumstances. Such changing circumstances should call for the revision of existing concession agreements.

Participation

Where provision for Governmental participation in the ownership of the concession-holding company under any of the present petroleum contracts has not been made, the Government may acquire a reasonable participation, on the grounds of the principle of changing circumstances.

If such provision has actually been made but avoided by the operators concerned, the rate provided for shall serve as a minimum basis for the participation to be acquired.

Relinquishment

A schedule of progressive and more accelerated relinquishment of acreage of present contract areas shall be introduced. In any event, the Government shall participate in choosing the acreage to be relinquished, including those cases where relinquishment is already provided for but left to the discretion of the operator.

Posted Prices or Tax Reference Prices

All contracts shall require that the assessment of the operator's income, and its taxes or any other payments to the State, be based on a posted or tax reference price for the hydrocarbons produced under the contract. Such price shall be determined by the Government and shall move in such a manner as to prevent any deterioration in its relationship to the prices of manufactured goods traded internationally. However, such price shall be consistent, subject to differences in gravity, quality and geographic location, with the levels of posted or tax reference prices generally prevailing for hydrocarbons in other OPEC Countries and accepted by them as a basis for tax payments.

Limited Guarantee of Fiscal Stability

The Government may, at is discretion, give a guarantee of fiscal stability to operators for a reasonable period of time.

Renegotiation Clause

1. Notwithstanding any guarantee of fiscal stability that may have been granted to the operator, the operator shall not have the right to obtain exessively high net earnings after taxes. The financial provisions of contracts which actually result in such excessively high net earnings shall be open to renegotiation.

2. In deciding whether to initiate such renegotiation, the Government shall take due account of the degree of financial risk undertaken by the operator and the general level of net earnings elsewhere in industry where similar circumstances prevail.

3. In the event the operator declines to negotiate, or that the negotiations do not result in any agreement within a reasonable period of time, the Government shall make its own estimate of the amount by which the operator's net earnings after taxes are excessive, and such amount shall then be paid by the operator to the Government.

4. In the present context, "excessively high net earnings" means net profits after taxes which are significantly in excess, during any twelve-month period, of the level of net earnings the reasonable expectation of which would have been sufficient to induce the operator to take the entrepreneurial risks necessary.

5. In evaluating the "excessively high net earnings" of the new operators, consideration should be given to their overall competitive position vis-a-vis the established opeators.

Accounts and Information

The operator shall be required to keep within the country clear and accurate accounts and records of his operations, which shall at all times be available to Government auditors, upon request.

Such accounts shall be kept in accordance with the Government's written instructions, which shall conform to commonly accepted principles of accounting, and which shall be applicable generally to all operators within its territory.

The operator shall promptly make available, in a meaningful form, such information related to its operations as the Government may reasonably require for the discharge of its functions.

Conservation

Operators shall be required to conduct their operations in accordance with the best conservation practices, bearing in mind the long-term interests of the country. To this end, the Government shall draw up written instructions detailing the conservation rules to be followed generally by all contractors within its territory.

Settlement of Disputes

Except as otherwise provided for in the legal system of a Member Country, all disputes arising between the Government and operators shall fall exclusively within the jurisdiction of the competent national courts or the specialized regional courts, as and when established.

Other Matters

In addition to the foregoing principles, Member Governments shall adopt on all other matters essential to a comprehensive and rational hydrocarbons policy, rules including no less than the best of current practices with respect to the registration and incorporation of operators; assignment and transfer of rights; work obligations; the employment of nationals; training programs; royalty rates; the imposition of taxes generally in force in the country; property of the operator upon expiry of the contract; and other such matters.

Definition

For the purposes of the present Resolution, the term "operator" shall mean any person entering into a contract of any kind with a Member Government or its designated agency including the concessions and contracts currently in effect, providing for the exploration for and/or development of any part of the hydrocarbon resources of the country concerned.

Source: OPEC Official Resolutions and Press Releases 1960-1980, Pergamon Press, 1980, for OPEC.

APPENDIX I.3
TWENTY-FIRST OPEC CONFERENCE HELD IN CARACAS IN DECEMBER 1970

RESOLUTION XXI.120

The Conference,

having heard the statement of the Head of the Libyan Delegation with regard to the outcome of the negotiations carried out by that Member Country with its concessionaire companies to correct the unjustifiable basis on which Libyan posted prices had been calculated since their inception;

having heard the statements of the Heads of the Iranian and Kuwaiti Delegations with regard to the recent increases made in the posted prices of certain crude oils and the adoption of a uniform 55 per cent tax rate in those Member Countries, having also noted the statement of the Head of the Saudi Arabian Delegation that an offer of a similar nature has been made to his country;

having noted the recent 20 cent per barrel upward adjustment published by the concessionaire companies in Iraq and Saudi Arabia for the crude oil shipped from East Mediterranean terminals;

having heard the statement made by the Head of the Algerian Delegation on the negotiations being held with the French Government concerning the revision of the fiscal terms applicable to the French oil companies;

having heard the statement made by the Head of the Venezuelan Delegation on the price situation in that Member Country, where some of the concessionaire exporting companies have failed to adjust their export prices to take into account prevailing market conditions, as established in existing reference price agreements, to the eventual detriment of the Venezuelan fiscal revenue;

recalling Resolutions XIII.80, XIII.81 and XIX.105, where the Organization supported the measures that were being taken by the Libyan, Iraqi and Algerian Governments to safeguard their legitimate interests with respect to the upward revision of posted or reference prices and of fiscal revenue;

pursuant to the principles established in Resolution XVI.90, calling for revision of exisitng agreements as justified by changing circumstances, and that the reference price for the purpose of determining the tax liability of the concessionaire companies should be determined by the Governments of Member Countries;

Appendix I.3

having heard the reports presented by the Secretariat concerning the necessity for an immediate elimination of the disparities as well as an upward adjustment of the existing posted or reference prices in all Member Countries;

considering the general improvement in the economic and market outlook of the international oil industry, as well as in its competitiveness with other sources of energy;

resolves that all Member Countries adopt the following objectives:

1. to establish 55 per cent as the minimum rate of taxation on the net income of the oil companies operating in the Member Countries,

2. to eliminate existing disparities in Posted or Tax-Reference Prices of the crude oils in the Member Countries on the basis of the highest Posted Price applicable in the Member Countries, taking into consideration differences in gravity and geographic location and any appropriate escalation in the future years,

3. to establish a uniform general increase in the Posted or Tax-Reference Prices in all Member Countries to reflect the general improvement in the conditions of the international petroleum market,

4. to adopt a new system for the adjustment of gravity differential of Posted or Tax-Reference Prices on the basis of 0.15 cents/bbl/$0.1°$ API for crude oil of $40°$ API and below, and 0.2 cents/bbl/$0.1°$ API for crude oil of $40.1°$ API and above,

5. to eliminate completely the allowances granted to oil companies as from the first January, 1971.

To this end, all Member Countries shall establish negotiations with the oil companies concerned with a view to achieving the above objectives and, recognizing the similarity of geographical location and other conditions in Abu Dhabi, Iran, Iraq, Kuwait, Qatar and Saudi Arabia, a Committee shall be formed consisting of the representatives of Iran, Iraq and Saudi Arabia who shall negotiate on behalf of Abu Dhabi, Iran, Iraq, Kuwait, Qatar and Saudi Arabia with the representatives of the oil companies operating in said Member Countries.

The Committee shall establish negotiations with the oil companies concerned in Tehran within a period of 31 days from the date of the conclusion of the present Conference and report to all Member Countries through the Secretary General the results of the negotiations not later than 7 days thereafter.

Within 15 days of the submission of the Committee's report to Member Countries, an extraordinary meeting of the Conference shall be convened in order to evaluate the results of the Committee's and of the individual Member Countries' negotiations. In case such negotiations fail to achieve their purpose, the Conference shall determine and set forth a procedure with a view to enforcing and achieving the objectives as outlined in this Resolution through a concerted and simultaneous action by all Member Countries.

Source: OPEC Official Resolutions and Press Releases 1960-1980, Pergamon Press, 1980, for OPEC.

APPENDIX I.4
TWENTY-SECOND OPEC CONFERENCE HELD IN TEHRAN
3-4 FEBRUARY 1971

RESOLUTION XXII.131

The Conference,

having heard the report of the three-member Ministerial Committee representing the Member Countries bordering the Gulf on the outcome of their negotiations with the oil companies' representatives on the implementation of Resolution XXI.120;

having also heard the reports of the Heads of the Delegations of Algeria, Libya and Venezuela on the actions taken by their respective Governments towards the implementation of the objectives of said Resolutions;

bearing in mind the sharp increase and the general firming up of crude oil and product prices in the world market coupled with the staggering growth of demand for petroleum in the main consuming countries;

taking note of the continued erosion in the purchasing power of Member Countries' oil revenues, due to worldwide inflation and the ever-widening gap existing between the prices of capital and manufactured goods essential for their economic development and those of petroleum;

recalling the Declaratory Statement of Petroleum Policy embodied in Resolution XVI.90 which provides, inter alia, that the determination of oil posted and tax-reference prices be made by the Governments of the producing countries;

with a view towards safeguarding Member Countries' rightful and legitimate interests in an equitable manner, and recognizing the benefits that stability in the fiscal obligations of the oil industry represents for the consuming countries as well as for those investing in crude oil production;

resolves that each Member Country exporting oil from Gulf terminals shall introduce on 15th February the necessary legal and/or legislative measures for the implementation of the objectives embodied in Resolution XXI.120. In the event that any oil company concerned fails to comply with these legal and/or legislative measures within seven days from the date of their adoption in all the countries concerned, Member Countries Abu Dhabi, Algeria, Iran, Iraq, Kuwait, Libya, Qatar, Saudi Arabia and Venezuela shall take appropriate measures including total embargo on the shipments of crude oil and petroleum products by such company.

In case the oil companies operating in Member Countries concerned express their willingness to comply with the minimum requirements agreed upon by the six Member Countries bordering the Gulf on the implementation of the objectives of Resolution XXI.120 before the expiry of the time limit set out above, then the Member Countries concerned shall refrain from resorting to the legal and/or legislative measures referred to above,

and with respect to Algeria and Libya the necessary legal and/or legislative measures for the implementation of the objectives embodied in Resolution XXI.120 applicable to them shall be introduced at the convenience of their respective Governments. In the event that any oil company operating in these Member Countries fails to comply, within seven days from the date of their adoption, with the same minimum requirements agreed upon by the Member Countries bordering the Gulf plus an additional premium reflecting a reasonably justified short-haul freight advantage for their crude-oil exports, Member Countries Abu Dhabi, Algeria, Iran, Iraq, Kuwait, Libya, Qatar, Saudi Arabia and Venezuela shall take appropriate measures including total embargo on the shipment of crude oil and petroleum products by such company.

Source: OPEC Official Resolutions and Press Releases 1960-1980, Pergamon Press, 1980, for OPEC.

TWENTY-SECOND OPEC CONFERENCE HELD IN TEHRAN
3-4 FEBRUARY 1971

RESOLUTION XXII.132

The Conference,

having heard the report of the Head of the Libyan Delegation on the implementation of Resolution XXI.120,

gives full support to measures taken or to be taken by the Libyan Government for safeguarding the legitimate interests of the Libyan people against any collective act that might be exercised by oil companies operating in the Libyan Arab Republic.

Source: OPEC Official Resolutions and Press Releases 1960-1980, Pergamon Press, 1980, for OPEC.

APPENDIX I.5

FULL TEXT OF TEHRAN PRICE AGREEMENT

The following is the full text of the five-year oil price agreement signed in Tehran on 14 February 1971 between six Gulf oil exporting countries - Abu Dhabi, Iran, Iraq, Kuwait, Qatar and Saudi Arabia - and a group of international oil companies.

Abu Dhabi, Iran, Iraq, Kuwait, Qatar and Saudi Arabia (the said six States being hereinafter known as "the Gulf States" insofar as their exports from the Gulf are concerned) and the Companies listed in Annexe 1 and their affiliates (hereinafter known as "the Companies"), to establish security of supply and stability in financial arrangements agree:

1. The existing arrangements between each of the Gulf States and each of the Companies to which this Agreement is an overall amendment, will continue to be valid in accordance with their terms.

2. The following provisions constitute a settlement of the terms relating to government take and other financial obligations of the Companies operating in the Gulf States as to the subject matters referred to in OPEC Resolutions and as regards oil exported from the Gulf, for a period from 15th February, 1971 through 31st December, 1975. These provisions shall be binding on both the Gulf States and the Companies for the said period.

3. These provisions are:-

No leap-frogging   (a) During this Agreement no Gulf State will seek any increase in government take or other financial obligations over that now agreed regarding Gulf production, as a result of:-

        (1) The application of different terms in:

           (i) any Gulf State as a Mediterranean exporter; or
           (ii) any Mediterranean producer; or
           (iii) any producer from any other area; or

        (2) The breach of contract through unilateral action by any Government in the Gulf; or

(3) The elimination of existing disparities in the Gulf under paragraph (c) (2) (iv) or any settlement under paragraph (c) (3) THIRDLY; or

(4) The application of different terms to any future agreement in any country bordering on the Gulf.

No Embargo (b) The requirements of the six Member Countries of OPEC bordering the Gulf under OPEC Resolutions XXI.120 and XXII.131 are satisfied by the terms of this Agreement. During the period of this Agreement the Gulf States shall not take any action in the Gulf to support any OPEC member which may demand either any increase in government take above the terms now agreed, or any increase in government take or any other matter not covered by Resolution XXI.120.

Financial Terms to meet OPEC Resolution XXI.120 OPEC 120 Par. 1

(c) (1) Total tax rates on income shall be stabilized in accordance with existing arrangements, except that insofar as present tax laws provide for total rates lower than 55 percent, the Companies concerned will submit to an amendment to the relevant income tax laws raising the total rates to 55 percent.

(2) In satisfaction of the several claims arising out of paragraphs 2 and 3 of OPEC Resolution XXI.120

    (i) Each of the Companies shall uniformly increase as from the effective date its crude posted prices at the Gulf terminals of the Gulf States by 33 cents per barrel.

    (ii) (aa) Each of the Companies shall make further upward adjustments to its crude posted prices to the nearest tenth* of a cent per barrel by increasing on 1st June, 1971 each of such posted prices by an amount equal to $2\frac{1}{2}$% of such posted price on the day following the effective date. On 1st January of each of the years 1973 through 1975 a further increase to the nearest tenth of a cent shall be made in each such posted price equivalent to $2\frac{1}{2}$% of the posted price prevailing on 31st December of the preceding year.

    (ii) (bb) Each of the Companies shall increase its crude posted prices on 1st June, 1971 by 5 cents per barrel and by a further increase of 5 cents per barrel on 1st January in each of the years 1973-1975.

    (ii) (cc) Each of the Companies shall further increase its crude posted prices as from the effective date by 2 cents per barrel which, together with paragraph 3 (d) is in satisfaction of claims related to freight disparities.

---

* For each decimal fraction of a cent of 0.05 cents or above the amount is to be increased to the next higher whole 0.1 cent. For each decimal fraction of a cent below 0.05 cents the amount is decreased by this fraction.

Appendix I.5 261

  (iii) The increases included in (ii) above shall be in satisfaction of claims in respect of freight, escalation and of inflation under both OPEC Resolution XXI.120 and OPEC Resolution XXI.122, and also in satisfaction of certain other economic considerations raised by the Gulf States.

  (iv) Each of the Gulf States having an existing claim under negotiation based on posted price disparity has discussed and resolved such claim with the Companies exporting the crude grade concerned as follows:

    In case of Iranian Heavy, Saudi Arab Medium and Kuwait, the posted prices shall each be increased by the Companies concerned by one cent with effect from the effective date. In the case of Basrah after the adjustment provided for in (3) FIRSTLY the posted price will be $1.805 for $35°$ API.

OPEC 120 Paragraph 4

(3) FIRSTLY For crude oil API gravity $30.0°$ to $39.9°$ with effect from the effective date each posted price shall be further increased by the Companies by $\frac{1}{2}$ cent per barrel for each degree such crude is less than $API°$ 40. A table showing the resulting increases before taking into account the settlement of disparities under (c) (2) (iv) is attached (Annexe 2) and forms part of this Agreement.

SECONDLY Posted prices shall apply to shipments falling within the range of .0 to .09 degrees of any full degree of API gravity and shall be subject to a gravity differential on the basis of 0.15 cents per barrel for each full 0.1 degree API.

THIRDLY In the case of crudes under $30°$ API the Governments and Companies shall agree on a basis for adjusting the posted price. However, if no such agreement is reached the same principles applied in FIRSTLY and SECONDLY above shall apply.

OPEC 120 Paragraph 5

The existing percent allowance, the gravity allowance and the $\frac{1}{2}$ cent per barrel marketing allowance shall be eliminated as from the effective date of this Agreement.

 (d) If Libya is receiving a premium for short haul crude which premium is to fluctuate according to freight conditions in accordance with a freight formula and if in respect of any period the premium applied by any major oil company which has production in Libya and the Gulf States exceeds for any reason the lowest level permitted by such formula for such period the Gulf States shall be entitled to additional payments as set out in Annexe 3.

4. "Affiliate" shall mean in relation to any Company, any company which is wholly or partly owned directly or indirectly by that Company.

5. Each of the Gulf States accepts that the Companies' undertakings hereunder constitute a fair appropriate and final settlement between each of them,

and those of the Companies operating within their respective jurisdictions, of all matters related to the applicable bases of taxation and levels of posted prices up to the effective date.

6. The effective date of this Agreement shall be 15th February, 1971.

Appendix I.5

ANNEXE 1

The British Petroleum Company Limited

Compagnie Francaise des Petroles

Gulf Oil Corporation

Mobil Oil Corporation

The Shell Petroleum Company Limited and Shell Petroleum N.V.

Standard Oil Company of California

Standard Oil Company (New Jersey)

Texaco Inc.

Continental Oil Company

Standard Oil Company (Ohio)

Hispanica de Petroleos S.A.

American Independent Oil Company of Iran

Signal (Iran) Petroleum Company

ANNEXE 2

Crude	°API	Present Posting $pb	½ cent pb.x Degrees of Gravity under 40°	Adjusted Posted Price* $pb
Qatar	40	1.93	0	1.93
Abu Dhabi	39	1.88	.005	1.885
Abu Dhabi Marine	37	1.86	.015	1.875
Qatar Marine	36	1.83	.02	1.85
Basrah	35	1.72	.025	1.745
Arabian Light	34	1.80	.03	1.830
Iran Light	34	1.79	.03	1.820
Iran Heavy	31	1.72	.045	1.765
Kuwait	31	1.68	.045	1.725
Arab Medium	31	1.68	.045	1.725

The crude below API Gravity 30° is not covered by this table.

*Subject to paragraph 3(c)(2)(iv)

Appendix I.5

ANNEXE 3

Short Haul Freight

The following provisions shall apply with respect to the implementation of paragraph 3(d) of the Agreement to which this Annexe 3 is attached.

(1) Any major oil company concerned shall pay to each Gulf State (as a supplemental payment) that proportion of a "balancing amount" as such Company's crude production exported from Gulf terminals (including Arabia/Bahrain pipeline) in such Gulf State bear to the total of such Company's crude exports in such period from all Gulf States in the Gulf.

(2) The "balancing amount" will be equal to the monetary amount by which the Company's payments to Libya for the period exceed the monetary amount which the Company would have paid to Libya for the period if it had effected the full reduction of premium permitted by its agreement with Libya or if it had effected a reduction in premium equal to $21\frac{1}{2}$ cents/B which is agreed with the Gulf States to be the short haul premium, whichever reduction is smaller.

(3) "Major Oil Company" for the above purpose means any of Esso, Texaco, Socal, Gulf, Mobil, BP, Shell and CFP.

(4) Illustrative examples of the implementation of the terms of this annexe are shown in Exhibit A, attached.

EXHIBIT A

ILLUSTRATIVE EXAMPLE OF "BALANCING AMOUNT"

	Cents/BBL			
Short haul Premium agreed with the Gulf States	21.5	21.5	21.5	21.5
Libyan "Premium" for illustrative purposes:	18.0	21.5	24.0	30.0
Lesser of Under-Reduction of Libyan Freight Premium or 21½ cents/B:				
1. Libyan Premium should be reduced by 25% but is not	4.5	5.375	6.0	7.5
2. Libyan Premium should be reduced by 50% but is not	9.0	10.75	12.0	15.0
3. Libyan Premium should be reduced by 100% but is not	18.0	21.5	21.5	21.5
4. Libyan Premium should be reduced by 100% but was only reduced to 50%	9.0	10.75	12.0	15.0
5. Libyan Premium should be reduced by 100% but was only reduced by 25%	13.5	16.125	18.0	21.5

To obtain balancing amount:

(a) Multiply figure given under 1-5 by the total Libyan tax rate on income plus (100 percent minus such rate) applied to the royalty, all as applicable to the producer concerned

(b) Multiply resultant dollar/B figure in (a) by the barrels of the major company's crude production exported from Libya

APPENDIX I.6
THIRTY-FIFTH MEETING (EXTRAORDINARY) OF THE OPEC
CONFERENCE HELD IN VIENNA 15-16 SEPTEMBER 1973

RESOLUTION XXXV.160

The Conference,

having examined the prevailing conditions and expected future trends of the crude oil and oil products markets, as well as the worldwide inflation, especially in the industrialized countries;

having reviewed the terms of the Tehran, Tripoli and Lagos Agreements in the light of the above conditions and trends;

noting that the present level of posted prices as determined by those Agreements is no longer compatible with such prevailing conditions and trends thus requiring an upward adjustment;

having noted that the annual escalations provided for in those Agreements are also no longer in line with the current and expected future trends of world inflation, as well as the crude oil and product prices;

recognizing that the oil companies are reaping high unearned profits owing to developments which have occurred since the conclusion of the Tehran, Tripoli and Lagos Agreements, and that such a situation is detrimental to the Member Countries leading to a further deterioration of the value of their oil;

resolves

1. that the Member Countries concerned shall negotiate, individually or collectively, with the oil companies with a view to revising the Tehran, Tripoli and Lagos Agreements in the light of the prevailing conditions and expected future trends in the crude oil and oil product markets, as well as the world inflation;

2. to this end a Ministerial Committee, composed of the Heads of Delegations of the Member Countries bordering the Gulf, be established in order to negotiate collectively the revision of the terms of the Tehran Agreement with the representatives of the oil companies, on the 8th of October, 1973 in Vienna; and

3. to empower the said Committee to call for an Extraordinary Meeting of the Conference if it is deemed necessary.

Source: OPEC Official Resolutions and Press Releases 1960-1980, Pergamon Press, 1980, for OPEC.

APPENDIX I.7

OPEC, 16 OCTOBER 1973

The following communique was issued on 16 October following the meeting in Kuwait of the Ministerial Committee representing the six OPEC Member Countries bordering on the Gulf:

"In accordance with the decision taken in Vienna on 12 October, the Ministerial Committee met in Kuwait on 16 October and decided to:

"(1) In line with OPEC Resolution No. 90 as well as the practice of other OPEC Member States - Venezuela, Indonesia and Algeria - to establish and announce the posted prices of crudes in the Gulf.

"(2) The new posted prices are based on actual market prices in the Gulf as well as in other areas, corrected for gravity differentials and geographical location.

"(3) From this day on, actual market prices will determine the level of corresponding posted prices, keeping the same relationship between the two prices as existed in 1971 before the Tehran agreement. The correction for changing posted prices upwards or downwards will take place when the actual market prices of crude oil exceed or drop below the corresponding level of the new announced prices by one percent.

"(4) The corresponding market price of the new posted price for Arabian Light crude is hereby established and announced at $3.65/barrel. The prices of other crudes will be established accordingly. This price represents only a 17 percent increase over the actual sale of the same crude recently. Consequently, the posted prices for all crudes shall be increased so as to produce the same result.

"(5) The sulfur premium of various crudes will be determined individually by each Member State on the basis of actual market trends.

"(6) The Geneva agreements shall continue to be in force.

"(7) The effective date of the new arrangements and prices will be 16 October.

"(8) In case the oil companies refuse to take crudes on the basis of these arrangements, the producing countries will make available to any buyer the various crudes at prices computed on the basis of Arabian Light at $3.65/barrel f.o.b. Ras Tanura."

APPENDIX I.8

ARAB OIL MINISTERS, 17 OCTOBER 1973

The following communique was issued by the Conference of the Oil Ministers of 10 Arab countries - Saudi Arabia, Kuwait, Iraq, Libya, Algeria, Egypt, Syria, Abu Dhabi, Bahrain and Qatar - which took place in Kuwait on 17 October:

"The Arab oil exporting countries contribute to the prosperity of the world and the growth of its economy through their exports of this wasting natural resource. And in spite of the fact that the production of many of these countries has exceeded the levels required by their domestic economies and the energy and revenue needs of their future generations, they have continued to increase their production, sacrificing their own interests in the service of international cooperation and the interests of the consumers.

"It is known that huge portions of the territories of three Arab states were forcibly occupied by Israel in the June 1967 war, and it has continued to occupy them in defiance of UN resolutions and various calls for peace from the Arab countries and peace-loving nations.

"And although the international community is under an obligation to implement UN resolutions and to prevent the aggressor from reaping the fruits of his aggression and occupation of the territories of others by force, most of the major industrialized countries which are consumers of Arab oil have failed to take measures or to act in such a way as might indicate their awareness of this public international obligation. Indeed, the actions of some countries have tended to support and reinforce the occupation.

"Before and during the present war, the United States has been active in supplying Israel with all the means of power which have served to exacerbate its arrogance and enable it to challenge the legitimate rights of others and the unequivocal principles of the public international law.

"In 1967, Israel was instrumental in closing the Suez Canal and burdening the European economy with the consequences of this action. In the current war, it hit East Mediterranean oil export terminals, causing Europe another shortfall in supplies. This is the third such occurrence resulting from Israel's disregard of our legitimate rights with US backing and support. The Arabs have therefore been induced to take a decision to discontinue their economic sacrifices in producing quantities of their wasting oil assets in excess of what would be justified by domestic economic considerations, unless the international community hastens to rectify matters by compelling Israel to withdraw from our occupied territory, as well as letting the US know the heavy price which the big industrial countries are having to pay as a result of America's blind and unlimited support for Israel.

"Therefore, the Arab Oil Ministers meeting in Kuwait today have decided to reduce their oil production forthwith by not less than 5 percent of the September (1973) level of output in each Arab oil exporting country, with a similar reduction to be applied each successive month, computed on the basis of the previous month's production, until such time as total evacuation of Israeli forces from all Arab territory occupied during the June 1967 war is completed, and the legitimate rights of the Palestinian people are restored.

"The conferees took care to ensure that reductions in output should not affect any friendly state which has extended or may in the future extend effective concrete assistance to the Arabs. Oil supplies to any such state will be maintained in the same quantities as it was receiving before the reduction. The same exceptional treatment will be extended to any state which takes a significant measure against Israel with a viewing to obliging it to end its occupation of usurped Arab territories.

"The Arab Ministers appeal to all the peoples of the world, and particularly the American people, to support the Arab nation in its struggle against imperialism and Israeli occupation. They reaffirm to them the sincere desire of the Arab nation to cooperate fully with all the peoples of the world and their readiness to supply the world with its oil needs as soon as the world shows its sympathy with us and denounces the aggression against us."

APPENDIX I.9

FULL TEXT OF THE PARTICIPATION AGREEMENT

GENERAL AGREEMENT

This General Agreement ("Agreement") sets forth provisions covering participation and is made between the Gulf States listed in Column 1 of Annex 1 and the Companies listed in Columns 2 and 3 of Annex 1.

Preamble

Whereas the Conference of the Organization of Petroleum Exporting Countries (OPEC) has passed certain resolutions demanding participation in respect of existing crude oil Concessions within such countries; and

Whereas the Companies listed in Column 2 of Annex 1 have previously expressed their agreement in principle to participation by the Gulf States party to this Agreement in the Concessions held by such Companies, subject to mutually satisfactory resolution of certain related issues;

Now, therefore, the parties agree as follows:

Article One

(a) This Agreement applies to each crude oil Concession within the jurisdiction of each Gulf State listed in Column 1 of Annex 1 now held by any one or more of the Companies listed in Column 2 of Annex 1 ("Concession").

(b) The provisions of this Agreement shall take effect so as to create rights and obligations only between the Gulf State grantor of each Concession and the Company or Companies (and their successors and assigns) concerned therein.

Article Two

(a) Promptly after the signing of this Agreement, negotiations shall be undertaken between each Gulf State listed in Column 1 of Annex 1 and appropriate Company or Companies listed in Columns 2 and 3 of Annex 1 to conclude a separate agreement applicable to each Concession ("Implementing Agreement") which shall implement the provisions of this Agreement and cover other matters related to participation, whether reserved in this Agreement for resolution in the Implementing Agreement or not dealt with in this Agreement.

(b) Each such Implementing Agreement shall include provision for the structural, organizational or corporate arrangements for the ownership and operation of the Concession concerned. While this Agreement has been prepared in contemplation of an undivided interest form of concession ownership and operation between the Company or Companies concerned and the Gulf State participant, if the corporate form is adopted, the principles and terms of this Agreement shall be adapted in the applicable Implementing Agreement to the corporate form.

Article Three

(a) Each Gulf State shall have an initial percentage level of participation equal to twenty-five percent (25%) in each Concession as provided in paragraph (b) of this Article; thereafter, it shall have the right to acquire percentage increments and resulting percentage levels of participation in accordance with Annex 2, provided the obligations of the Gulf State under the provisions of Article Four and Annexes 2 and 4 applicable to its then existing percentage level of participation have been currently met.

(b) As a participant in a Concession, each Gulf State shall have an interest, directly or indirectly, as the case may be, in that Concession's crude oil concession rights, in the crude oil produced therefrom, and in the Concession's crude oil production facilities, whether such facilities are tangible or intangible, situated within such Gulf State's jurisdiction, equal to its percentage level of participation from time to time in that Concession. In this Agreement, the term "crude oil production facilities" shall include, without limitation, exploration, development, production, pipelines, storage, delivery and export facilities as shall be defined in the applicable Implementing Agreement. For the purposes of this Article and Article 4, the term "crude oil" shall include both crude oil and natural gas; matters relating to natural gas (other than in connection with crude oil production) shall be dealt with, where necessary, by separate agreement between the Gulf State and the Company or Companies concerned. Unless otherwise provided in this Agreement, the Company or Companies concerned shall be relieved of all related concession obligations to the extent of the Gulf State's percentage level. In respect of each Concession, the Gulf State may call upon the Company or Companies concerned to discuss whether and on what terms such Gulf State will also participate in rights and facilities of the Concession within such Gulf State's jurisdiction other than crude oil concession rights and crude oil production facilities. Such other rights and facilities shall include, without limitation, those relating to refining and gas processing (other than in connection with crude oil production).

Article Four

(a) (1) Consideration for the initial percentage level of participation in each Concession shall be an amount equal to twenty-five percent (25%) of the Book Value of the crude oil production facilities (whether in existence or under construction) and of exploration and intangible development (whether complete or in process) of such Concession on the day before the Effective Date for Participation, as determined from the books as used for fiscal purposes in the Gulf State of the Company or Companies listed in Column 2 of Annex 1 holding such Concession pursuant to paragraph (a) (2) of this Article, such determination to be certified by an internally recognized firm of public accountants to be agreed upon between the Gulf State and the Company or Companies concerned prior to or concurrently with the execution of the applicable Implementing Agreement.

(2) For the purposes of this Article, Book Value shall be computed as follows:

(i) for each year of such Concession calculate the difference (whether positive or negative) between capitalized expenditures (including for this purpose all exploration and intangible development costs not capitalized) made in such year and the amount by which the revenue of the Gulf State concerned in respect of such year was reduced as a result of depreciation and amortization (including exploration and intangible development costs for those years in which costs were fully amortized or written-off as incurred) allowed for such year;

(ii) apply to each of the paragraph (a) (2) (i) calculations for years prior to 1945 a multiplier of 1.00 and for 1945 and subsequent years the appropriate multiplier shown in the Middle East Construction Price Factors table set forth in Annex 5;

(iii) determine the sum of all paragraph (a) (2) (ii) calculations.

(b) Consideration for each percentage increment of participation in each Concession shall be an amount equal to a fraction (of which the numerator shall be the percentage increment being acquired and the denominator the percentage interest of the Company or Companies concerned on the day before the date of acquisition of such increment) of the Book Value of such percentage interest of the Company or Companies concerned in the crude oil production facilities (whether in existence or under construction) and the exploration and intangible development (whether complete or in process) of such Concession on the day before the date of acquisition of such increment, as determined from the books as used for fiscal purposes in the Gulf State of the Company or Companies concerned pursuant to paragraph (a) (2) of this Article, except that all calculations shall be brought forward to the date of acquisition of such increment, and certified by an internationally recognized firm of public accountants to be agreed upon between the Gulf State and the Company or Companies concerned.

(c) For the purposes of computations made under paragraph (a) (1), (a) (2) (i) and (b) of this Article, any amounts originally stated in sterling shall be converted and restated in U.S. dollars in respect of each year at the average rate of exchange used for the purpose of computing tax liabilities for such year. In any year when no tax liabilities arose, the average commercial rate of exchange in such year shall apply.

(d) In respect of each Concession, the amounts computed under paragraphs (a) and (b) of this Article shall be paid by the Gulf State concerned to the Company or Companies concerned in accordance with Annex 4.

(e) For the purposes of calculations to be made under paragraph (b) of this Article, continuation of the Middle East Construction Price Factors table set forth in Annex 5 will be prepared by the firm of Haskins & Sells, unless by mutual agreement between the Gulf States and the Companies such preparation is assigned to a different firm of international reputation and with appropriate competence.

Appendix I.9

Article Five

(a) During each full calendar year each Gulf State as a participant shall have a Basic Right to a percentage of each grade of crude oil available at each specified offtake point equal to its percentage level of participation for such year. During each such year the Company or Companies (considered as a group for administrative purposes) concerned in a Concession shall retain the Basic Right to a percentage of each grade of crude oil available at each specified offtake point equal to the difference between one hundred percent (100%) of such crude oil available at such point and the Gulf State's Basic Right percentage thereof.

(b) "Crude oil available" means, in respect of each grade at each specified offtake point in each calendar year, the quantity of that grade of crude oil which installed facilities are capable of producing and delivering during such year at such offtake point. In computing said quantity, relevant operating factors, including force majeure, which apply during such year shall be taken into account.

(c) Each Implementing Agreement shall contain detailed procedures appropriate to the particular circumstances in the Gulf State or relating to the Concession concerned and consistent with those outlined in Annex 3, governing the exercise of Basic Rights. Such procedures, among other things, shall specify offtake points for the total of each grade of crude oil produced in the relevant Concession and shall provide in detail for the exercise of Basic Rights at substantially even rates during each full calendar year.

(d) Under contracts to be entered into in respect of each Concession between each Gulf State and the Company or Companies concerned or their designated subsidiaries (acting either individually or collectively as they may elect in the applicable Implementing Agreement), such Company or Companies or their designated subsidiaries shall purchase and the Gulf State shall sell certain quantities of "bridging" crude oil, of each grade at each specified offtake point, out of the Gulf State's Basic Right to such crude oil in respect of the Concession concerned. Unless arrangements more appropriate to the Concession concerned are mutually agreed, the purchase contracts for bridging crude oil in respect of the Gulf State's initial percentage level of participation, shall be for seventy-five percent (75%) of its Basic Right to such grade of crude oil at such specified offtake point during the first year, fifty percent (50%) during the second year, and twenty-five percent (25%) during the third year. The price to be charged by the Gulf State concerned and the conditions of payment for each grade of bridging crude oil shall be specified in a collateral agreement to be executed with respect to each Concession by the Gulf State and the Company or Companies concerned prior to or concurrently with the execution of this Agreement.

(e) In response to the requirements of the Gulf State, the Company or Companies concerned or their designated subsidiaries (acting either individually or collectively as they may elect in the applicable Implementing Agreement) shall purchase from the Gulf State, for each year specified in Paragraph E of Annex 3, certain quantities of "phase-in" crude oil, of each grade at each specified offtake point, pursuant to the provisions and procedures in Paragraph E of Annex 3, in addition to the quantities of bridging crude oil to be purchased under paragraph (d) of this Article.

(f) In respect of any Concession the obligations of the Company or Companies concerned to purchase crude oil from the Gulf State pursuant to this Agreement will be satisfied by the contracts to be entered into, and duly performed, under paragraphs (d) and (e) of this Article and Paragraphs D and E of Annex 3.

(g) With effect from the Effective Date for Participation:

(1) Where any Company or Companies concerned in respect of any Concession is or are immediately prior to the Effective Date for Participation, under obligation to supply crude oil for domestic consumption requirements in the Gulf States concerned, the Gulf State participant shall supply in any year, out of the crude oil to which it has a Basic Right in respect of such Concession, that proportion of such supplies which the total quantity of its Basic Right crude oil bears to the total of both parties' Basic Right crude oil in respect of such Concession.

(2) The sum of

   (i) the quantity of crude oil, by grade and specified offtake point, taken by any Gulf State pursuant to its Basic Right in respect of any Concession in any year, and

   (ii) the quantity of such crude oil taken by such Gulf State pursuant to its right to take royalty in kind in such Concession in such year

   shall not exceed 51% of the total of such grade of crude oil available at such offtake point during such year.

(3) Existing barter oil obligations of the Company or Companies concerned shall terminate.

Article Six

(a) Each Gulf State as a participant in a Concession shall have the right to take an active part with the Company or Companies concerned in management. Major management decisions shall require the approval of an agreed number (which may be all) of the parties concerned holding, directly or indirectly, a total agreed percentage interest in the Concession, as may be provided in the Implementing Agreement. Major management decisions are those which relate to the following matters and any other matters which may be specified in the applicable Implementing Agreement:

   (1) Sale or disposition of assets above a value to be specified in the applicable Implementing Agreement.

   (2) Capital and operating expenditures and disposition of funds above a value or of a type to be specified in the applicable Implementing Agreement.

   (3) Exploration and development programs and construction of new facilities.

   (4) Selection of key personnel, and

   (5) Employee compensation and benefit plans.

   Modification or termination of any Concession and related agreements, or of the corporate or other arrangements provided in the applicable Implementing Agreement for the ownership and operation of the Concession concerned, shall require the approval of the Gulf State concerned and the Company or Companies concerned listed in Columns 2 and 3 of Annex 1; provided, however, that in any case where any Company or Companies listed in Column 2 of Annex 1 have an existing right of termination or abandonment of a Concession, such Company or Companies shall continue to have such right.

Appendix I.9

(b) If the undivided interest form of concession ownership and operation between the Company or Companies concerned and the Gulf State participant is not adopted in the applicable Implementing Agreement, such Agreement shall contain a provision to protect the interests of all shareholders, whatever their percentage holdings of the total shares may be, in respect of the declaration and payment of dividends.

(c) Matters relating to negotiation between any Gulf State, as the grantor of a Concession, and the Company or Companies concerned shall be handled as provided in the applicable Implementing Agreement, which will also provide that no party holding an interest in a Concession shall interfere with or prevent any other party from exercising any remedies under existing agreements in relation to the settlement of disputes.

(d) Decisions relating to relinquishment of Concession areas pursuant to existing agreements, including, without limitation, designation of the areas to be relinquished, shall be made by the Company or Companies concerned after consultation with the Gulf State participant.

Article Seven

(a) In accordance with provisions to be included in the applicable Implementing Agreement, each Gulf State and the Company or Companies concerned (considered as a group for administrative purposes) shall bear the costs associated with the production and delivery of crude oil in respect of each Concession as follows:

   (1) Capital requirements, including advances for working funds, in accordance with their respective percentage interests from time to time in the Concession concerned.

   (2) All other costs, including without limitation depreciation and overhead, in the proportion that their respective liftings bear to total liftings, by grade and offtake point where appropriate. A party's liftings shall include any quantities sold by it as bridging, phase-in, and forward avails crude oil and any quantities for which it is paid the overlift price as provided in Paragraphs F and G of Annex 3.

(b) The applicable Implementing Agreement shall provide that if any Company or Companies concerned have failed to pay, when due, any obligation under paragraph (a) (1) of this Article, the Gulf State participant concerned shall have the right to make such payment on behalf of such Company or Companies and to reduce by the same amount any financial obligation to such Company or Companies, individually or collectively at the Gulf State's discretion. Such Company or Companies shall have a similar right to make payment and set-off against any financial obligation to such Gulf State participant if the latter fails to pay, when due, any obligation under paragraph (a) (1) of this Article; provided that no financial obligation of such Company or Companies to the Gulf State concerned other than in its capacity as a participant hereunder shall be affected.

Article Eight

(a) Each applicable Implementing Agreement shall contain provisions pursuant to which, following ratification of such agreement, the concerned Gulf State may transfer or assign the whole or part of its participation interest in the concerned Concession. Each Gulf State undertakes that any such transfer or assignment shall be to its existing national oil company, or to any entity at least 51% owned by the concerned Gulf State or by its existing national oil company, and the balance of which is owned, directly or indirectly, by individuals who are nationals of such Gulf State. Notwithstanding any such assignment, the whole of such interest, as it may exist from time to time, and the owners of holders thereof, shall be considered as a unit and shall be represented in relation to the Company or Companies concerned by either the concerned Gulf State or its existing national oil company.

(b) Any transferee or assignee of an interest in any Concession shall assume and be subject to the concessionary and related obligations, fiscal and otherwise, in proportion to its percentage level of participation.

(c) Each Gulf State shall guarantee the performance and obligations of its transferee or assignee.

Article Nine

Each Gulf State and the Company or Companies presently holding a Concession agree that all existing agreements between them in respect of such Concession shall remain in full force and effect in accordance with their terms, the terms of this Agreement and the applicable Implementing Agreement. Each State and the Company or Companies concerned shall cause all steps to be taken to perform their respective obligations in respect of the relevant Concession.

Article Ten

In respect of each Concession, the applicable Implementing Agreement shall include appropriate provisions for the settlement of any difference or dispute which may arise concerning the interpretation or performance of such agreement between any transferee or assignee of the Gulf State's interest and the Company or Companies specified therein.

Article Eleven

All Annexes referred to in this Agreement shall be considered as fully a part hereof as though repeated herein verbatim.

Article Twelve

In a Gulf State the Effective Date for Participation shall be 1 January 1973.

Article Thirteen

The Term of this Agreement in respect of each Concession shall continue until the end of such Concession.

Done this 14th day of 1392, corresponding to the 20th day of December 1972, at Riyadh, in twenty-five originals in both Arabic and English texts.

The Agreement was signed on 20 December 1972 after preliminary meetings between HE Shaikh Ahmed Zaki Yamani, as representative of five Gulf States, and the companies which were completed on 5 October 1972, and another meeting involving OPEC countries.

Appendix I.9

ANNEX 1

Column 1 COUNTRIES	Column 2 CONCESSION HOLDERS OR PARTIES TO GOVT AGREEMENTS	Column 3 SHAREHOLDERS, DIRECT OR INDIRECT, OF CONCESSION HOLDERS OR OF PARTIES TO GOVERNMENT AGREEMENTS
ABU DHABI	(1) Abu Dhabi Marine Areas Limited	(1) The British Petroleum Company Limited Compagnie Francaise des Petroles
	(2) Abu Dhabi Petroleum Company Limited	(2) The British Petroleum Company Limited Compagnie Francaise des Petroles Standard Oil Company (New Jersey) Mobil Oil Corporation The Shell Petroleum Company Limited and Shell Petroleum N.V. Participations and Explorations Corporation
ABU DHABI/QATAR	(3) Bunduq Company Limited	(3) The British Petroleum Company Limited Compagnie Francaise des Petroles United Petroleum Development Co. (Japan) Limited
IRAQ	(4) Basrah Petroleum Company Limited	(4) Same as (2)
KUWAIT	(5) BP (Kuwait) Limited Gulf Kuwait Company	(5) The British Petroleum Company Limited Gulf Oil Corporation
QATAR	(6) Qatar Petroleum Company Limited	(6) Same as (2)
	(7) Shell Company of Qatar Limited	(7) The Shell Petroleum Company Limited and Shell Petroleum N.V.
SAUDI ARABIA	(8) Arabian American Oil Company	(8) Mobil Oil Corporation Standard Oil Company of California Standard Oil Company (New Jersey) Texaco Inc.

ANNEX 2

Increments and Percentage Levels of Participation

Increment	Percentage Increments	Percentage Levels of Participation	Earliest Dates for Acquisition of Percentage Increments
First	5%	30%	1 January 1978
Second	5%	35%	1 January 1979
Third	5%	40%	1 January 1980
Fourth	5%	45%	1 January 1981
Fifth	6%	51%	1 January 1982

In respect of each percentage increment each Gulf State will give notice to the Company or Companies concerned of its intention to exercise its right to acquire such increment, such notice to be given on or before the date on which notice of the phase-in quantities is given pursuant to Paragraph E (2) (b) of Annex 3 for the year when such increment is to become effective. Only one percentage increment may be acquired in any one year and the Effective Date for each increment shall be 1 January. If a Gulf State has not given notice as above provided, or has not satisfied all payment obligations due under paragraphs (c) (i) and (ii) and (d) of Annex 4 and paragraph (a) of Article Seven prior to 31 December of the year preceding the year when such increment would have become effective, the earliest date for acquisition of such increment and for each succeeding percentage increment shall be postponed one year.

Appendix I.9

ANNEX 3

OUTLINE OF PROCEDURES GOVERNING EXERCISE OF BASIC RIGHTS AND PRICES RELATIVE TO DISPOSITION OF CRUDE OIL

A. Each year the Gulf State and the Company or Companies concerned (considered as a group for administrative purposes) shall simultaneously table their respective offtake requirements by grade and specified offtake point for the year three years forward ("planned year"), such tablings to be made on or before an agreed date prior to 1 January, e.g., prior to 1 January 1974, for 1977. Quantities to be purchased pursuant to paragraphs (d) and (e) of Article Five shall be included in the Gulf State's tabled requirements.

B. Each party may table its requirements in any one of the following forms:

   (1) Any quantity;

   (2) A quantity with a proviso for automatic reduction if necessary to insure that its requirements shall not exceed its Basic Right percentage of the total quantity tabled by both parties; or

   (3) A quantity with a proviso for automatic increase if necessary to insure that its requirements shall not be less than its Basic Right percentage of the total quantity tabled by both parties.

C. "Planned Capacity" for each grade and specified offtake point shall be set, if feasible, at a level not less than the total quantity tabled by both parties plus a margin taking into account appropriate operational and seasonal factors, unless otherwise agreed in the applicable Implementing Agreement. If such is not feasible, Planned Capacity shall be set at the maximum reasonably feasible. In such event, tabled requirements shall be cut back to equal Planned Capacity after allowance for operational and seasonal factors, with cut-backs falling first upon the party which tabled more than its Basic Right percentage of total tabled requirements until its revised tabled requirements are equal to its Basic Right percentage of the sum of its revised requirements and the tabled requirements of the other party, and thereafter in accordance with Basic Right percentages.

D. (1) For each grade at each specified offtake point the amount of the excess, if any, of either party's tabled requirements, adjusted if necessary under Paragraph C, over its Basic Right percentage of the greater of (a) total tabled requirements for such year for that grade and specified offtake point, adjusted if necessary under Paragraph C ("Total Requirements") or (b) Planned Capacity for the preceding year, shall be known as "forward avails". (If Planned Capacity for the planned year is by reason of operational factors less than Planned Capacity for the preceding year, Planned Capacity for the planned year shall be used for this computation). Under contracts to be entered into between the Gulf State concerned on the one hand, and the Company or Companies concerned or their designated subsidiaries (acting either individually or collectively as they may elect in the applicable Implementing Agreement), on the other hand, whichever party tabled above such Basic Right percentage ("overtabler") shall purchase from the other party ("undertabler"), and such other party shall sell the forward avails established in respect of each year under this Paragraph D (1) and Paragraph D (2), subject to adjustment with respect to quantities under Paragraph D (3), at the price provided in Paragraph D (4).

(2) If in respect of any planned year the tabling procedures in Paragraphs A, B and C result in the establishment of forward avails under Paragraph D (1), then a quantity of forward avails for each grade and specified offtake point shall be calculated for each of the four succeeding years equal, respectively, to 4/5, 3/5, 2/5 and 1/5, of the quantity of forward avails calculated for such planned year. In tabling requirement for each succeeding year after the planned year, each party shall take account of the quantities calculated for each such succeeding year, with the initial undertabler including the relevant quantities (4/5, 3/5, 2/5 or 1/5 as the case may be) in its tabled requirements, and the initial overtabler excluding the relevant quantities from its tabled requirements.

(3) (a) At the conclusion of the lifting year, a quantity of forward avails shall be computed for each party by grade and specified offtake points as follows:

  (i) Calculate such party's Basic Right share of the total crude oil of the grade concerned available in the lifting year;

  (ii) Determine the cumulative sum of forward avails, calculated according to Paragraphs D (1) and D (2) in respect of such lifting year;

  (iii) Multiply the quantity determined under paragraph (ii) by a fraction of which the numerator is the quantity of the grade of crude oil concerned (excluding purchases of such grade of phase-in crude oil and any other purchases other than forward avails) taken by such party and the denominator the total of the quantities determined under paragraphs (i) and (ii);

 (b) The quantity of forward avails which such party shall purchase, and the other party shall sell to such party, shall be the quantity determined in paragraph (3) (a) (iii).

(4) The price ("contract price") for each grade of forward avails crude oil and the conditions of payment shall be specified in a collateral agreement to be executed with respect to each Concession by the Gulf State and the Company or Companies concerned prior to or concurrently with the execution of this Agreement.

(5) If a Gulf State gives notice of its intention to acquire a percentage increment of participation as provided in Annex 2 but for any reason does not acquire such increment on the date stated in such notice, any contracts for the purchase of forward avails crude oil which were entered into by the Company or Companies concerned or their designated subsidiaries in reliance on such notice may be cancelled or modified at the option of the purchasers so as to exclude therefrom a total amount of crude oil equal to the anticipated increase in the Gulf State's Basic Right which did not materialize because of the postponement provisions of Annex 2.

E. (1) (a) Pursuant to the provisions of paragraph (e) of Article Five, the quantities of "phase-in" crude oil, of each grade at each specified offtake point, in respect of the Gulf State's initial percentage level of participation, shall be as specified pursuant to Paragraph E (2) below, but shall not exceed the following stated percentage of the Gulf State's Basic Right to such crude oil in the particular participation year:

Appendix I.9

First year	15%	Sixth year	60%
Second year	30%	Seventh year	50%
Third year	50%	Eighth year	40%
Fourth year	70%	Ninth year	30%
Fifth year	65%	Tenth year	10%

In respect of the year 1976 and any subsequent year, the percentages stated above shall be applied for the purposes of this Paragraph E (1) (a) only to the quantity of crude oil available to the Gulf State as its Basic Right in the Concession concerned in the year 1975.

(b) Pursuant to the provisions of paragraph (e) of Article Five, the quantities of phase-in crude oil, of each grade at each specified offtake point, in respect of each increase in the Gulf State's Basic Right to crude oil arising out of the acquisition of percentage increments and resulting percentage levels of participation, shall be as specified pursuant to Paragraph E (2) below, subject always to the right of either party to give the notices as provided in Paragraph K. The applicable percentages, in respect of each such increment, shall for the ten-year period immediately following the Effective Date for each such increment be:

First year	90%	Sixth year	60%
Second year	80%	Seventh year	50%
Third Year	75%	Eighth year	40%
Fourth year	70%	Ninth year	30%
Fifth year	65%	Tenth year	10%

For each year in the table above, the phase-in quantity in respect of each such increment shall not exceed the quantity obtained by applying the applicable percentage increment of participation to the quantity of crude oil available in the Concession concerned in the year 1975 and multiplying the result by the percentage stated in the table above in respect of such year.

(2) (a) Prior to the Effective Date for Participation for each Concession, each Gulf State shall give notice in respect of that Concession of the annual amounts of phase-in crude oil which it requires the Company or Companies concerned to take for each of the first four calendar years beginning with such Effective Date. Amounts of phase-in crude oil may be expressed in such notices, and in any subsequent notices, either as a quantity or as a percentage of the Gulf State's Basic Right, both subject to the overall limitation of the percentages stated in Paragraph E (1) (a).

(b) In each year (including the first year of participation), at least one month before tablings of requirements are due under Paragraph A, each Gulf State shall give notice of the amount of phase-in crude oil which it requires the Company or Companies concerned to take during the fourth calendar year ("planned year") after the year in which such notice is given, e.g., in 1973 for the year 1977.

(c) Each Gulf State undertakes that, except as may be permitted by the scheduled reductions in Paragraphs E (1) (a) and E (1) (b), the amount of phase-in crude oil specified in each such notice for any planned year will not be less than three-fourths ($\frac{3}{4}$) of the amount of phase-in crude oil it requires the Company or Companies to take during the calendar year immediately preceding the planned year; provided, however, that each Gulf State shall have the right, in any notice duly given in respect of any planned year, i.e., any calendar year after the first four calendar years beginning with the Effective Date for Participation, to reduce the amounts of phase-in crude oil which it requires the Company or Companies concerned to take to zero over a four-year period, beginning with such planned year, in steps not exceeding 25% of the amount it requires the Company or Companies concerned to take during the calendar year immediately preceding such planned year (i.e., the notice would specify an amount for such planned year not less than 75% of the amount specified for the year immediately preceding the planned year, 50% in the first year following the planned year, 25% in the second year following the planned year, and zero in the third year following the planned year).

(3) In any calendar quarter the quarterly quantities of phase-in crude oil, i.e., one-quarter of the annual amount arrived at pursuant to Paragraphs E (1) and E (2), may, at the election of the Gulf State concerned and by notice given not less than one year before the beginning of such quarter, be increased or decreased by 10%, subject always to the overall limitation of the percentages stated in Paragraphs E (1) (a) and E (1) (b).

(4) The price ("contract price") to be charged by the Gulf State concerned and the conditions of payment for each grade of phase-in crude oil shall be specified in a collateral agreement to be executed with respect to each Concession by the Gulf State and the Companies concerned prior to or concurrently with the execution of this Agreement.

G. If, at the end of any calendar year, in respect of each Concession, either the Gulf State, or the Company or Companies concerned or their designated subsidiaries (regarded for this purpose as a group for administrative purposes), has lifted in total a quantity of a grade of crude oil in excess of its total Basic Right to such grade of crude oil at all specified offtake points ("overlift" crude oil), the party so overlifting shall pay to the other party the overlift price (as defined in Paragraph G) for each such overlifted barrel. Total quantities of each grade of crude oil contracted for under paragraphs (d) and (e) of Article Five and Paragraph D of this Annex shall be included in the liftings of the seller.

G. The term "overlift price" of a barrel of crude oil means:

(1) through year-end 1975, either

(a) that amount equal to the sum of the total costs (exclusive of taxes and royalties) and taxes and royalties payable by the Company or Companies concerned, and their designated subsidiaries, in respect of an identical barrel if sold by it or them for export, plus a margin equal to twenty-five percent (25%) of the difference between such sum and the posted price of such barrel determined in accordance with the Tehran Agreement of February 14, 1971, and related agreements as supplemented by the Geneva Agreement of January 20, 1972; or

(b) that lesser amount equal to the sum of the total per barrel costs (exclusive of taxes and royalties) determined under Paragraph G (1) (a), and the royalties payable in respect of such barrel, plus such amount which, when multiplied by the difference between one hundred percent (100%) and the applicable percentage tax rate, would equal the margin determined pursuant to Paragraph G (1) (a); and

(2) beginning 1 January 1976, either

    (a) an amount such as shall equal the sum of the total costs (exclusive of taxes and royalties) and taxes and royalties then payable by the Company or Companies concerned, and their designated subsidiaries, in respect of an identical barrel if then sold by it or them for export, plus a margin equal to twenty-five percent (25%) of the difference which would have been determinable for 1975, pursuant to Paragraph G (1) (a), for an identical barrel had it been overlifted on 31 December 1975, or

    (b) that lesser amount equal to the sum of the total per barrel costs (exclusive of taxes and royalties) determined under Paragraph G (2) (a) and the royalties payable in respect of such barrel, plus such amount which, when multiplied by the difference between one hundred percent (100%) and the applicable percentage tax rate, would equal the margin determined pursuant to Paragraph G (2) (a).

Which of the foregoing alternative overlift prices shall apply in respect of each Concession shall be determined according to principles to be set forth in the applicable Implementing Agreement.

H. The term "contract price" of a barrel of crude oil means either:

(1) the sum of the total per barrel costs (exclusive of taxes and royalties) and taxes and royalties payable by the Company or Companies concerned, and their designated subsidiaries in respect of an identical barrel if sold by it or them for export, plus the applicable margin agreed for the Concession concerned between the Gulf State and the Company or Companies concerned; or

(2) that lesser amount equal to the sum of the total per barrel costs (exclusive of taxes and royalties) determined under Paragraph H (1), and the royalties payable in respect of such barrel, plus such amount which, when multiplied by the difference between one hundred percent (100%) and the applicable percentage tax rate, would equal the applicable margin agreed for the Concession concerned pursuant to Paragraph H (1).

Which of the foregoing alternative contract prices shall apply in respect of each Concession shall be determined according to principles to be set forth in the applicable Implementing Agreement.

I. (1) Except as provided in paragraph (2) (a) below, the Company or Companies concerned in each Concession and their designated subsidiaries, and each of them, shall be relieved of any and all obligations to the concerned Gulf State in respect of the crude oil included within such Gulf State's Basic Right, other than the obligation to pay the applicable price for any such crude oil purchased or overlifted by any of them.

(2) If the overlift price is as defined in Paragraphs (G) (1) (b) or G (2) (b), or if the contract price is as defined in Paragraph H (2) of this Annex the Company or Companies concerned and their designated subsidiaries shall:

    (a) remain subject to the obligation under the applicable tax laws and applicable agreements to pay tax, in respect of any Gulf State's Basic Right crude oil purchased or overlifted by them or their designated subsidiaries, on the difference between the applicable posted price and such purchase price; and

(b) be relieved of any obligation to pay tax in respect of crude oil overlifted by the Gulf State pursuant to Paragraph F of this Annex on the difference between the price received from such Gulf State for such crude oil and the applicable posted price.

J. The terms "tax" and "taxes" mean the taxes imposed under applicable tax laws as well as amounts equivalent to and in lieu of such taxes payable to the Gulf State concerned under applicable agreements, and the term "tax rate" means the tax rate under applicable income tax laws as well as the rate used under applicable agreements for determining such equivalent amounts.

K. (1) Each Implementing Agreement will provide that either party to such Implementing Agreement (i.e., the Gulf State or the Company or Companies concerned) may by giving written notice to the other party not later than 1 March 1976 request the other party's agreement to a revision of the margin for phase-in, forward avails, and overlift crude oil (or any of them) to be effective 1 July 1976 and thereafter. Similar requests may be made at three-year intervals thereafter, i.e., not later than 1 March 1979 to be effective 1 July 1979 and thereafter, etc.

(2) If after any such request the parties fail to agree upon a revised margin by the 1 May before the 1 July when the revision, if any, is to be effective, the existing price arrangements shall continue unaffected (subject always to the right of either party to give notice not later than the 1 March of the third year forward, as provided in Paragraph K (1)).

(3) If, however, by 1 June either party gives to the other party written notice of dissatisfaction requesting an increase or decrease in the margin for the types of crude oil concerned (i.e., phase-in, forward avails, or overlift), then

(a) If such notice requested an increase, the party receiving such notice shall have option, exercisable by giving written notice on or before the 1 July in question, to continue arrangements for the type of crude oil concerned (subject always to the right of either party to give notice as provided in Paragraph K (1)) with an increase in the margin of four U.S. cents per barrel for phase-in and forward avails and three U.S. cents per barrel for overlift;

(b) If such notice requested a decrease, the party receiving such notice shall have the option, exercisable by giving written notice on or before the 1 July in question, to continue the arrangements for the type of crude oil concerned (subject always to the right of either party to give notice as provided in Paragraph K (1)) with a reduction in the margin of two U.S. cents per barrel. If however as a result of cumulative reductions in the margin pursuant to this Paragraph K (3) (b) a further exercise of the option under this Paragraph would have the effect of reducing the margin to a level of twenty percent (20%) or more below the margin applicable as of the Effective Date for Participation, any party receiving a notice of dissatisfaction requesting a further decrease shall have the option to continue the arrangements for the type of crude oil concerned at a margin equal to the margin applicable as of the Effective Date for Participation reduced by twenty percent (20%);

(c) If each party has given notice of dissatisfaction, and if both parties duly exercise the options available to them under (a) and (b), the arrangements for the type of crude oil concerned shall continue (subject always to the right of either party to give notice as provided in Paragraph K (1)) with the margin increased by an amount equal to half the difference between the increase and decrease specified in such options (i.e., one U.S. cent per barrel for phase-in and forward avails and one-half U.S. cent per barrel for overlift).

    (d)   If neither of the options in (a) or (b) is exercised on or before the 1 July in question, either party may then elect, by written notice given on or before 31 July, with no change in price arrangements, to

        (i)   continue any affected phase-in crude oil arrangements for three calendar years following the current calendar year, phasing out such arrangements at quantities equal to three-fourths for the first year, one-half in the second year, and one-fourth in the third year, of the quantities of phase-in crude oil committed for the current calendar year;

        (ii)   continue any affected forward avails arrangements for a period sufficient to satisfy, in the first three subsequent calendar years the existing obligations between the parties which have arisen pursuant to Paragraph D of this Annex, and to phase out such forward avails arrangements at quantities equal to two-thirds in the fourth subsequent calendar year, and one-third in the fifth subsequent calendar year, of the cumulative quantity of forward avails crude oil purchased in the third subsequent calendar year following the notice year.

    (e)   If none of the options in (a), (b) or (d) is exercised, the service of such notice of dissatisfaction shall operate effectively to terminate the existing arrangements for the type of crude oil concerned at the end of that current calendar year.

(4)   The term "margin" as used in this Paragraph K means the margin referred to in Paragraph G or H, as the case may be, of this Annex.

L.   The provisions and definitions in respect of prices contained in this Annex and in the Agreement shall be incorporated in the applicable Implementing Agreement and shall apply only to crude oil delivered at offtake points in the Gulf area.

ANNEX 4

PAYMENT OF CONSIDERATION

Each amount to be paid by each Gulf State pursuant to paragraphs (a) and (b) of Article Four shall be:

(a) Determined and expressed in United States Dollars;

(b) Paid in United States Dollars or Sterling as specified in the applicable Implementing Agreement. If any such sum or any portion thereof is to be paid in Sterling, the rate of exchange to be used in respect of such Sterling payment shall be that rate determined pursuant to the provisions of paragraph 3 (a) (i) of the Agreement executed in Geneva, Switzerland, on 20 January 1972, by representatives of certain Gulf States and certain companies, for the month prior to the month of payment;

(c) Paid in a lump sum;

   (i) in the case of the amount to be paid pursuant to paragraph (a) of Article Four, it shall be paid within thirty (30) days following the Effective Date for Participation;

   (ii) in the case of each amount to be paid pursuant to paragraph (b) of Article Four, it shall be paid on or before the December 31 preceding the date for acquisition of the percentage increment concerned;

(d) Notwithstanding (c) (i) above, the applicable Implementing Agreement may, if the Gulf State concerned so requests, provide that the amount to be paid pursuant to paragraph (a) of Article Four may be paid in three (3) instalments. These three instalments shall be paid as follows:

   (i) the first instalment shall equal thirty percent (30%) of the total amount agreed to pursuant to paragraph (a) of Article Four and shall be paid within thirty (30) days following the Effective Date for Participation;

   (ii) the second instalment shall be paid on or before the anniversary of the date of payment of the first instalment in the year following the Effective Date for Participation and shall equal thirty-five percent (35%) in the aforesaid total amount; and

   (iii) the third instalment shall be paid on or before the anniversary of the date of payment of the first instalment in the second year following the Effective Date for Participation and shall equal thirty-five percent (35%) of the aforesaid total amount.

(e) If payment is to be made in instalments as provided for in Paragraph (d) above, the amount of principal to be paid by the Gulf State concerned shall bear interest at the rate of interest specified in paragraph (h) below. Such interest shall be calculated separately for the six months' period commencing on the thirty-first (31st) day following the Effective Date for Participation and for each succeeding period of six months until the final instalment is paid. Each such calculation shall be in respect of the amount of principal outstanding at the beginning of the period concerned and the amount of interest shall be payable on the last day of the six month period concerned. If such payment due date falls on a day which is not a business day, then the interest payment shall become due on the first succeeding business day.

(f) The amounts payable by the Gulf State concerned shall be:

    (i) paid into a bank designated by the Company or Companies receiving payment of such sums;

    (ii) free of any tax or other financial imposition by such Gulf State.

(g) The amounts to be paid pursuant to paragraphs (a) and (b) of Article Four shall, if a corporate form of concession ownership and operation is adopted and if such amounts be paid to such corporation, be increased sufficiently so that the portions attributable, within such corporation, to the interests of all the shareholders other than the Gulf State will be equal to such amounts.

(h) The interest rate referred to in paragraph (e) above shall be equal to one percent (1%) per annum above the rate at which U.S. dollar deposits for six months are offered in the interbank deposit market in London, such rate to be certified by the National Westminster Bank London, for each period of six months or less that such interest is due, as the rate at which such deposits are offered to it at noon on the first day of such period, or in the event that such first day is not a business day then on the first succeeding business day.

ANNEX 5

MIDDLE EAST CONSTRUCTION PRICE FACTORS

Index Numbers and Derived Multiplier Factors

Year	Index	Multiplier	Year	Index	Multiplier
1972	140	1.00	1958	83.8	1.67
1971	126	1.11	1957	81.4	1.72
1970	117	1.20	1956	79	1.77
1969	114	1.23	1955	74.1	1.89
1968	108	1.30	1954	70.6	1.98
1967	102	1.37	1953	67.3	2.08
1966	100	1.40	1952	65.4	2.14
1965	97	1.44	1951	62	2.26
1964	94.2	1.49	1950	58.8	2.38
1963	91.5	1.53	1949	58	2.41
1962	88.9	1.57	1948	58	2.41
1961	88.1	1.59	1947	51	2.75
1960	87.5	1.60	1946	38	3.68
1959	87.5	1.60	1945	32	4.37

APPENDIX I.10

SOLEMN DECLARATION OF SOVEREIGNS AND HEADS OF STATE OF OPEC MEMBER COUNTRIES

(The following is the full text of the Solemn Declaration issued by the Sovereigns and Heads of State of the Organization of the Petroleum Exporting Countries following their Conference in Algiers on 4-6 March 1975.)

The Sovereigns and Heads of State of the Member Countries of the Organization of the Petroleum Exporting Countries met in Algiers on 4 to 6 March 1975, at the invitation of the President of the Revolutionary Council and of the Council of Ministers of the Democratic People's Republic of Algeria.

1. They reviewed the present world economic crisis, exchanged views on the causes of the crisis which has persisted for several years, and considered the measures they would take to safeguard the legitimate rights and interests of their peoples, in the context of international solidarity and cooperation.

   They stress that world peace and progress depend on the mutual respect for the sovereignty and equality of all member nations of the international community, in accordance with the UN Charter. They futher emphasize that the basic statements of this Declaration fall within the context of the decisions taken at the VIth Special Session of the General Assembly of the United Nations on problems of raw materials and development.

   The Sovereigns and Heads of State reaffirm the solidarity which unites their countries in safeguarding the legitimate rights and the interests of their peoples, reasserting the sovereign and inalienable right of their countries to the ownership, exploitation and pricing of their natural resources and rejecting any idea or attempt that challenges those fundamental rights and, thereby, the sovereignty of their countries.

   They also reaffirm that OPEC Member Countries, through the collective, steadfast and cohesive defence of the legitimate rights of their peoples, have served the larger and ultimate interest and progress of the world community and, in doing so, have acted in the direction hoped for by all developing countries, producers of raw materials, in defence of the legitimate rights of their peoples.

   They conclude that the interdependence of nations, manifested in the world economic situation, requires a new emphasis on international cooperation and declare themselves prepared to contribute with their efforts to the objectives of world economic development and stability, as stated in the Declaration and Programme of Action for the establishment of a new international economic order adopted by the General Assembly of the United Nations during its VIth Special Session.

2. The Sovereigns and Heads of State note that the cause of the present world economic crisis stems largely from the profound inequalities in the economic and social progress among peoples; such inequalities, which characterize the under-development of the developing countries, have been mainly generated and acti-vated by foreign exploitation and have become more acute over the years due to the absence of adequate international cooperation for development. This situation has fostered the drainage of natural resources of the developing countries impeding an effective transfer of capital resources and technology, and thus resulting in a basic disequilibrium in economic relations.

They note that the disequilibrium which besets the present international economic situation has been aggravated by widespread inflation, a general slowdown of economic growth and instability of the world monetary system in the absence of monetary discipline and restraint.

They reaffirm that the decisive causes of such anomalies lie in the long-standing and persistent ills which have been allowed to accumulate over the years, such as the general tendency of the developed countries to consume excessively and to waste scarce resources, as well as inappropriate and short-sighted economic policies in the industrialized world.

They, therefore, reject any allegation attributing to the price of petroleum the responsibility for the present instability of the world economy. Indeed, the oil which has contributed so significantly to the progress and prosperity of the industrialized nations for the past quarter of a century, not only is the cheapest source of energy available but the cost of imported oil constitutes an almost negligible part of the Gross National Product of the developed countries. The recent adjustment in the price of oil did not contribute but insignificantly to the high rates of inflation which have been generated within the economies of the developed countries basically by other causes. This inflation exported continuously to the developing countries has disrupted their development efforts.

3. Moreover, the Sovereigns and Heads of State condemn the threats, propaganda campaigns and other measures which have gone so far as to attribute to OPEC Member Countries the intention of undermining the economies of the developed countries; such campaigns and measures that may lead to confrontation have obstructed a clear understanding of the problems involved and have tended to create an atmosphere of tension that is not conducive to international consultation and cooperation. They also denounce any grouping of consumer nations with the aim of confrontation, and condemn any plan or strategy designed for aggression, economic or military, by such grouping or otherwise against any OPEC Member Country.

In view of such threats the Sovereigns and Heads of State reaffirm the solidarity that unites their countries in the defence of the legitimate rights of their peoples and hereby declare their readiness, within the framework of that solidarity, to take immediate and effective measures in order to counteract such threats with a united response whenever the need arises, notably in the case of aggression.

4. While anxious to satisfy the legitimate aspirations of their peoples for development and progress, the Sovereigns and Heads of State are also keenly aware of the close link which exists between the achievement of their national development and the prosperity of the world economy. Increased interdependence between nations

makes them even more mindful of the difficulties experienced by other peoples which may affect world stability. In view of this, they reaffirm their support for dialogue, cooperation and concerted action for the solution of the major problems facing the world economy.

In this spirit, the OPEC Member Countires, with increased financial resources in a relatively short period of time, have contributed through multilateral and bilateral channels, to the development efforts and balance of payments adjustments of other developing countries as well as industrialized nations. As a proportion of Gross National Product, during 1974, their financial support to other developing countries was several times greater than the average annual aid given by industrialized nations to developing countries during the last development decade. In addition, OPEC Member Countries have extended financial facilities to developed countries to help them meet their balance of payments deficits. Furthermore, the acceleration of their economic development and the trade promotion measures adopted by OPEC Member Countries have contributed to the expansion of international trade as well as balance of payments adjustments of developed countries.

5. The Sovereigns and Heads of State agree in principle to holding an international conference bringing together the developed and developing countries.

 They consider that the objective of such a conference should be to make a significant advance in action designed to alleviate the major difficulties existing in the world economy, and that consequently the conference should pay equal attention to the problems facing both the developed and developing countries.

 Therefore, the agenda of the aforementioned conference can in no case be confined to an examination of the question of energy; it evidently includes the questions of raw materials of the developing countries, the reform of the international monetary system and international cooperation in favour of development in order to achieve world stability.

 Furthermore, this conference may, for reasons of efficiency, be held in a limited framework provided that all the nations concerned by the problems dealt with are adequately and genuinely represented.

6. The Sovereigns and Heads of State stress that the exploitation of the depletable oil resources in their countries must be based, first and foremost, upon the best interests of their peoples and that oil, which is the major source of their income, constitutes a vital element in their development.

 While recognizing the vital role of oil supplies to the world economy, they believe that the conservation of petroleum resources is a fundamental requirement for the well-being of future generations and, therefore, urge the adoption of policies aimed at optimizing the use of this essential, depletable and non-renewable resource.

7. The Sovereigns and Heads of State point out that an artificially low price for petroleum in the past has prompted over-exploitation of this limited and depletable resource and that continuation of such policy would have proved to be disastrous from the point of view of conservation and world economy.

 They consider that the interest of the OPEC Member Countries as well as the rest of the world would require that the oil price, being the fundamental element in the national income of the Member Countries, should be determined taking into account the following:

 - the imperatives of the conservation of petroleum, including its depletion and increasing scarcity in the future;

- the value of oil in terms of its non-energy uses; and

- the conditions of availability, utilization and cost of alternative sources of energy.

Moreover, the price of petroleum must be maintained by linking it to certain objective criteria, including the price of manufactured goods, the rate of inflation, the terms of transfer of goods and technology for the development of OPEC Member Countries.

8. The Sovereigns and Heads of State declare that their countries are willing to continue to make positive contributions towards the solution of the major problems affecting the world economy, and to promote genuine cooperation which is the key to the establishment of a new international economic order.

In order to set in motion such international cooperation, they propose the adoption of a series of measures directed to other developing countries as well as the industrialized nations.

They, therefore, wish to stress that the series of measures proposed herein constitute an overall programme, the components of which must all be implemented if the desired objectives of equity and efficiency are to be attained.

9. The Sovereigns and Heads of State reaffirm the natural solidarity which unites their countries with the other developing countries in their struggle to overcome under-development and express their deep appreciation for the strong support given to OPEC Member Countries by all the developing nations as announced in the Conference of Developing Countries on Raw Materials, held in Dakar between 3rd and 8th February, 1975.

They recognize that the countries most affected by the world economic crisis are the developing couuntries and therefore reaffirm their decision to implement measures that will strengthen their cooperation with those countries. They are prepared to contribute within their respective possibilities to the realization of the UN Special International Programme and to extend additional special credits, loans and grants for the development of developing countries.

In this context, they have agreed to coordinate their programmes for financial co-operation in order to better assist the most affected developing countries especially in overcoming their balance of payments difficulties. They have also decided to coordinate such financial measures with long-term loans that will contribute to the development of those countries.

In the same context, and in order to contribute to a better utilization of the agricultural potential of the developing countries, the Sovereigns and Heads of State have decided to promote the production of fertilizers, with the aim of supplying such production under favourable terms and conditions, to the countries most affected by the economic crisis.

They reaffirm their willingness to cooperate with the other developing countries which are exporters of raw materials and other basic commodities in their efforts to obtain an equitable and remunerative price level for their exports.

10. To help smooth out difficulties affecting the economies of developed countries, the Sovereigns and Heads of State declare that the OPEC Member Countries will continue to make special efforts in respect of the needs of these countries.

Appendix I.10

As regards the supply of petroleum, they reaffirm their countries' readiness to ensure supplies that will meet the essential requirements of the economies of the developed countries, provided that the consuming countries do not use artificial barriers to distort the normal operation of the laws of demand and supply.

To this end, the OPEC Member Countries shall establish close cooperation and co-ordination among themselves in order to maintain balance between oil production and the needs of the world market.

With respect to the petroleum prices, they point out that in spite of the apparent magnitude of the readjustment, the high rate of inflation and currency depreciation have wiped out a major portion of the real value of price readjustment, and that the current price is markedly lower than that which would result from the development of alternative sources of energy.

Nevertheless, they are prepared to negotiate the conditions for the stabilization of oil prices which will enable the consuming countries to make necessary adjustments to their economies.

The Sovereigns and Heads of State, within the spirit of dialogue and cooperation, affirm that the OPEC Member Countries are prepared to negotiate with the most affected developed countries, bilaterally or through international organizations, the provision of financial facilities that allow the growth of the economies of those countries while ensuring both the value and security of the assets of OPEC Member Countries.

11. Recalling that a genuine international cooperation must benefit both the developing and developed countries, the Sovereigns and Heads of State declare that parallel with, and as a counterpart to, the efforts, guarantees and commitments which the OPEC Member Countries are prepared to make, the developed countries must contribute to the progress and development of the developing countries through concrete action and in particular to achieve economic and monetary stability, giving due regard to the interests of the developing countries.

In this context, they emphasize the necessity for the full implementation of the Programme of Action adopted by the United Nations General Assembly at its VIth Special Session and accordingly they emphasize the following requirements:

- Developed countries must support measures taken by developing countries which are directed towards the stabilization of the prices of their exports of raw materials and other basic commodities at equitable and remunerative levels.

- Fulfillment by the developed countries of their international commitments for the second UN Development Decade as a minimum contribution to be increased particularly by the most able of the developed countries for the benefit of the most affected developing countries.

- Formulation and implementation of an effective food programme under which the developed countries, particularly the world major producers and exporters of foodstuffs and products, extend grants and assistance to the most affected developing countries with respect to their food and agricultural requirements.

- Acceleration of the development processes of the developing countries particularly through the adequate and timely transfer of modern technology and the removal of the obstacles that slow the utilization and integration of such technology in the economies of the developing countries. Considering that in many cases obstacles to development derive from insufficient and inappropriate transfers of technology, the Sovereigns and Heads of State attach the greatest importance to the transfer of technology which, in their opinion, constitutes a major test of adherence of the developed countries to the principle of international cooperation in favour of development. The transfer of technology should not be based on a division of labour in which the developing countries would produce goods of lesser technological content. An efficient transfer of technology must enable the developing countries to overcome the considerable technological lag in their economies through the manufacture in their territories of products of a high technological content, particularly in relation to the development and transformation of their natural resources. With regard to the depletable natural resources, as OPEC's petroleum resources are, it is essential that the transfer of technology must be commensurate in speed and volume with the rate of their depletion which is being accelerated for the benefit and growth of the economies of the developed countries.

- A major portion of the planned or new petrochemical complexes, oil refineries and fertilizer plants be built in the territories of OPEC Member Countries with the co-operation of industrialized nations for export purposes to the developed countries with guaranteed access for such products to the markets of these countries.

- Adequate protection against the depreciation of the value of the external reserves of OPEC Member Countries, as well as assurance of the security of their investments in the developed countries.

Moreover, they deem it necessary that the developed countries open their markets to hydrocarbons and other primary commodities as well as manufactured goods produced by the developing countries and consider that discriminatory practices against the developing countries and among them, the OPEC Member Countries, are contrary to the spirit of cooperation and partnership.

12. The Sovereigns and Heads of State note the present disorder in the international monetary system and the absence of rules and instruments essential to safeguard the terms of trade and the value of financial assets of developing countries.

They emphasize particularly the urgent need to take the necessary steps to ensure the protection of the developing countries' legitimate interests.

They recognize that the pooling of the financial resources of both the OPEC Member Countries and the developed countries, as well as the technological ability of the latter, for the furtherance of the economy of the developing countries would substantially help in solving the international economic crisis.

They stress that fundamental and urgent measures should be taken to reform the international monetary system in such directions as to provide adequate and stable instruments for the expansion of trade, the development of productive resources and balanced growth of the world economy.

They note that the initiatives so far taken to reform the International monetary system have failed, since those initiatives have not been directed towards the removal of the inherent inequity in the structure of the system.

Decisions likely to affect the value of the reserve currencies, the Special Drawing Rights, and the price and role of gold in the international monetary system, should no longer be made on a unilateral basis or negotiated by developed countries alone; the developed countries should subscribe to a genuine reform of the international monetary and financial institution, to ensure its equitable representation and to guarantee the interests of all developing countries.

The reform of the monetary and financial system should allow a substantial increase in the share of developing countries in decision-making, management and participation, in the spirit of partnership for international development and on the basis of equality.

With this in mind, the Sovereigns and Heads of State have decided to promote amongst their countries a mechanism for consultation and coordination for full co-operation in the framework of their solidarity and with a view to achieving the goal of a genuine reform of the international monetary and financial system.

13. The Sovereigns and Heads of State attach great importance to the strengthening of OPEC and, in particular, to the coordination of the activities of their National Oil Companies within the framework of the Organization and to the role which it should play in the international economy. They consider that certain tasks of prime importance remain to be accomplished which call for concerted planning among their countries and for the coordination of their policies in the fields of production of oil, its conservation, pricing and marketing, financial matters of common interest and concerted planning and economic cooperation among Member Countries in favour of international development and stability.

14. The Sovereigns and Heads of State are deeply concerned about the present international economic crisis, which constitutes a dangerous threat to stability and peace. At the same time, they recognize that the crisis has brought about an awareness of the existence of problems whose solution will contribute to the security and well-being of humanity as a whole.

Equally aware of the hopes and aspirations of the peoples the world over for the solution of the major problems affecting their lives, the Sovereigns and Heads of State solemnly agree to commit their countries to measures aimed at opening a new era of cooperation in international relations.

It behoves the developed countries, which hold most of the instruments of progress, well-being and peace, just as they hold most of the instruments of destruction, to respond to the initiatives of the developing countries with initiatives of the same kind, by choosing to grasp the crisis situation as an historic opportunity in opening a new chapter in relations between peoples.

The anxiety generated by the uncertainty marking relations between those who hold power, coupled with the climate of uneasiness created by the confusion reigning in the world economy, would then give way to the confidence and peace resulting in an atmosphere of genuine international cooperation in which the developing countries would derive the greatest benefit and to which they would contribute their immense potentialities.

At a time when, thanks to man's genius, scientific and technological progress has endowed peoples with substantial means of surmounting natural adversity and of bringing about the most remarkable changes for the better, the future of mankind ultimately depends solely on men's capacity to mobilize their imagination and willpower in the service and interest of all.

The Sovereigns and Heads of State of the OPEC Member Countries proclaim their profound faith in the capability of all peoples to bring about a new economic order founded on justice and fraternity which will enable the world of tomorrow to enjoy progress equally shared by all in cooperation, stability and peace. They accordingly make a fervent appeal to the Governments of the other countries of the world and solemnly pledge the full support of their peoples in the pursuance of this aim.

Source: Middle East Economic Survey 7 March 1975.

APPENDIX I.11

AGREEMENT ESTABLISHING THE OPEC SPECIAL FUND

The Countries, Members of the Organization of the Petroleum Exporting Countries (OPEC), on whose behalf this Agreement is signed:

Conscious of the need for solidarity among all developing countries in the establishment of a New International Economic Order,

In keeping with the spirit of the Solemn Declaration of the Sovereigns and Heads of State of the OPEC Member Countries issued in Algiers, March 1975, of promoting the economic development of all the developing countries,

Aware of the importance of financial cooperation between OPEC Member Countries and other developing countries,

And desirous of establishing a collective financial facility to consolidate their assistance to other developing countries, in addition to the existing bilateral and multilateral channels through which they have individually extended financial cooperation to such other countries,

Have, therefore, agreed to establish a new facility for the provision of additional finance to other developing countries under the name of the OPEC SPECIAL FUND and in accordance with the following Articles:

Article One: Nature of the Fund

1.01 The OPEC SPECIAL FUND (hereinafter called the Fund) is established as an international special account collectively owned by the contributing Parties to this Agreement, each in the proportion of its contribution to the total amount of all contributions to the Fund.

1.02 Participation in the Fund shall be open to all OPEC Member Countries.

Article Two: Purpose of Establishing the Fund

2.01 The purpose of establishing the Fund is to reinforce financial cooperation between OPEC Member Countries and other developing countries by providing financial support to assist the latter countries on concessional terms.

2.02 In particular the Fund's resources shall be utilized for the following purposes:

(a) Providing interest-free long-term loans to finance balance of payments support and development programmes and projects.

(b) Making contributions by the Parties to this Agreement to international development agencies the beneficiaries of which are developing countries.

Article Three: Beneficiaries of the Fund

3.01 Eligible beneficiaries of the loans provided through the Fund shall be the Governments of developing countries other than OPEC Member Countries.

Article Four: Resources of the Fund

4.01 The Fund shall consist of Eight hundred million United States dollars contributed by OPEC Member Contries, signatory to this Agreement, in accordance with the Schedule of Contributions appended thereto.

4.02 Each contributing Party to this Agreement shall within thirty days of the entry into force of this Agreement in respect thereof, issue a letter of undertaking in the amount of its contibution to the benefit of the Fund and shall deposit such letter with the Director-General of the Fund referred to in Article 5.09 hereunder.

4.03 Contributions to the Fund shall be paid by each contributing Party in such amounts and at such dates as the Governing Committee referred to in 5.01 hereunder shall determine to allow for the timely disbursement of the loans committed through the Fund. Payments of the contribution of each Party shall be made in freely convertible currencies the total amount of which shall be equivalent to the US dollar amount of the contribution of that Party, subject to the arrangements provided for by the Governing Committee in the implementation of Article 2.02 (b) hereabove. Such payments shall be credited in each case to an account to be established under the name of Fund by the Executing National Agency designated by each contributing party pursuant to Article 5.07 for the purpose of this Agreement.

4.04 Any Party to this Agreement may at any time increase its contribution to the Fund.

Article Five: Administration of the Fund

5.01 The Fund shall be administered by a Governing Committee which shall have full powers in the administration of the Fund subject to the provisions of this Agreement.

5.02 The Governing Committee shall be composed of one representative of each contributing Party to this Agreement. Each contributing Party to this Agreement shall designate its representative to the Committee by a notice to the Secretary General of OPEC who shall notify all other Parties accordingly. Any contributing Party may thereafter replace its representative in the Committee by a notice to the Director-General of the Fund referred to in Article 5.09 hereunder.

5.03 The Governing Committee shall lay down the general policies for the utilization of the Fund's resources and shall issue guidelines and regulations according to which contributions to the Fund shall be administered and disbursed. In issuing such guidelines and regulations, the Governing Committee shall give due regard to the equitable distribution of the Fund's operations among eligible beneficiaries.

5.04 Applications for loan assistance from the Fund shall be submitted by eligible beneficiaries to the Director-General of the Fund who shall refer them to the Governing Committee for consideration in accordance with the procedure provided for in Article 5.05 hereunder. The loans approved by the Governing Committee shall be extended in each case by virtue of an agreement to be signed on behalf of the owners of the fund by the Chairman of the Governing Committee. The Governing Committee shall entrust Executing National Agencies or international development agencies of a world-wide or regional character, with the task of the administration of the loans approved by it. Such administration shall be undertaken subject to the provisions of this Agreement, the guidelines and regulations issued by the Governing Committee and the details of each loan agreement.

5.05 The Governing Committee shall, by virtue of a special agreement to be entered into for this purpose, entrust an appropriate international development agency with the task of technical, economic and financial appraisal of the projects or programmes submitted to the Governing Committee by eligible beneficiaries for financing from the Fund. However, the Governing Committee may entrust the above-mentioned task to the Executing National Agency or any other agency of a Party to this Agreement, by virtue of a special agreement to this effect.

5.06 In approving a loan for financing a development project or programme, the Governing Committee shall take into consideration the recommendations submitted to it by the appraising agency referred to in Article 5.05 hereabove.

5.07 Each contributing Party to this Agreement shall designate, by a written notice to the Governing Committee, its Executing National Agency for the purposes of this Agreement. Each Executing National Agency shall establish in its records a special account in the name of the Fund separate from its own accounts and shall keep the Fund's assets and transactions separate and apart from all its own assets and transactions. The Executing National Agency shall disburse from such an account amounts of the loans entrusted to it for administration by the Governing Committee as well as its share in the administrative expenses referred to in Article 5.10 hereunder. Repayments of the loan shall be made to the Fund's special account held by the Executing National Agency entrusted with the administration of the loan. Such repayments shall, on receipt, be remitted by the Executing National Agency to its respective State. However, in case the administration of a loan is entrusted to an international development agency, the Governing Committee shall designate the Executing National Agency which shall provide such an international development agency with the funds needed to meet the financial requirements of the loan from the Fund's special account held by it, and which shall eventually receive repayments of the loan amount.

5.08 The Governing Committee shall adopt its rules of procedures in accordance with the provisions of this Agreement. A two-thirds majority shall constitute the quorum for the meetings of the Governing Committee. Unless otherwise provided in this Agreement, decisions of the Governing Committee shall be reached by a two-thirds majority of those present at the meeting provided that they represent Parties to this Agreement contributing at least seventy percent of the total amount of the contributions to the Fund. The Secretary General of OPEC shall call the first meeting of the Governing Committee, which shall begin its work by election of the Chairman of the Governing Committee.

5.09 The Governing Committee shall appoint the Director-General of the Fund who shall be responsible for the organization of the work of the Governing Committee, the follow-up of its resolutions and the general supervision, including adequate auditing, of the adminstration of loans by the executing agencies. The Director-General shall participate in the meetings of the Governing Committee without having the right to vote and shall be the official spokesman for the Fund. He may appoint a limited number of assistants within the necessary limits authorized by the Governing Committee. The Director-General and his assistants shall, to the extent possible, operate from the Headquarters of OPEC.

5.10 Administrative expenses of the Governing Committee including cost of the appraisal envisaged in Article 5.05, of the services of the Director-General and his assistants and of the administration of loans by executing agencies, shall be covered from contributions to the Fund on a pro-rata basis according to an administrative budget to be approved by the Governing Committee. Members of the Governing Committee shall serve as such without any remuneration from the Fund.

5.11 The Governing Committee shall submit periodical reports on its activities and on the developments related thereto to the Ministers concerned of the Parties to this Agreement.

Article Six: Privileges and Immunities

6.01 The resources of the Fund shall enjoy in the territory of the Parties to this Agreement immunity from all confiscation measures, as well as from sequestration, moratoria or any form of seizure by executive or legislative action.

6.02 The Fund's resources and transactions related thereto shall also be exempted from rules and regulations applicable to national public funds as well as from exchange control regulations and all forms of taxes and duties imposed in the countries Party to this Agreement.

Article Seven: Amendments

7.01 Amendments to this Agreement may be proposed to the Ministers concerned of the Parties to this Agreement by the Governing Committee acting on the basis of a three-fourths majority of its members, on the initiative of any member or of the Director-General of the Fund. The Ministers concerned may thereafter approve the proposed amendments by a three-fourths majority of the Ministers of Parties contributing at least four-fifths of the total amount of contributions to the Fund.

7.02 Notwithstanding the provisions of Article 7.01 hereabove, no Party to this Agreement shall be obliged to increase its contribution to the Fund without its own accord.

Article Eight: Termination of the Fund

8.01 After the commitment of all the resources of the Fund to eligible beneficiaries, the Chairman of the Governing Committee shall, within three months, invite the Ministers concerned of the Parties to this Agreement to convene, in order to consider the dissolution of the Governing Committee and the adoption of appropriate measures for the legal representation regarding the Fund and the supervision of the implementation of the loans committed through it.

8.02 The Governing Committee may decide to terminate the Fund at any time by a four-fifths majority of its members representing Parties contributing at least four-fifths of the total amount of contributions to the Fund. The proposal to terminate the Fund can only be made by at least four members of the Governing Committee and shall not be put to the vote in the same session in which it is submitted. In case of the adoption of a decision to terminate the Fund, the balance of the Fund's special account established by each Executing National Agency shall accrue to its respective State subject to the losses and liabilities assumed for the Fund, being distributed among the contributing Parties to this Agreement in the proportion of the contribution of each to the total amouunt of the contributions to the Fund and in accordance with such rules and regulations as the Governing Committee shall lay down.

Article Nine: Interpretation and Settlement of Disputes

9.01 Any question of interpretation of the provisions of this Agreement, or any dispute on their application arising between the Parties to this Agreement or between an Executing National Agency and the Director-General of the Fund, shall be settled by the Governing Committee and failing this by the Ministers concerned of the Parties to this Agreement meeting for this purpose at the invitation of the Chairman of the Governing Committee.

Article Ten: Signature and Entry into Force

10.01 This Agreement shall be open for signature at the OPEC Secretariat between January 28th and February 28th, 1976. The Secretary General of OPEC shall act as the depositary and, as such, shall provide each signatory with a certified copy of this Agreement and shall notify each OPEC Member Country of each signature, acceptance or accession to this Agreement.

10.02 This Agreement shall enter into force when instruments of ratification, acceptance or accession have been deposited by at least nine Member Countries, of the Organization of the Petroleum Exporting Countries contibuting at least seventy-five percent of the total amount of contributions mentioned in the Schedule of Contributions appended to this Agreement.

Appendix II
INTERNATIONAL ENERGY AGENCY

Summary Statement

1. Foreign Ministers of Belgium, Canada, Denmark, France, the Federal Republic of Germany, Ireland, Italy, Japan, Luxembourg, The Netherlands, Norway, the United Kingdom, the United States met in Washington from February 11 to 13, 1974. The European Community was represented as such by the President of the Council and the President of the Commission. Finance Ministers, Ministers with responsibility for Energy Affairs, Economic Affairs and Science and Technology Affairs also took part in the meeting. The Secretary General of the OECD also participated in the meeting. The Ministers examined the international energy situation and its implications and charted a course of actions to meet this challenge which requires constructive and comprehensive solutions. To this end they agreed on specific steps to provide for effective international cooperation. The Ministers affirmed that solutions to the world's energy problem should be sought in consultation with producer countries and other consumers.

Analysis of the Situation

2. They noted that during the past three decades progress in improving productivity and standards of living was greatly facilitated by the ready availability of increasing supplies of energy at fairly stable prices. They recognised that the problem of meeting growing demand existed before the current situation and that the needs of the world economy for increased energy supplies require positive long-term solutions.

3. They concluded that the current energy situation results from an intensification of these underlying factors and from political developments.

4. They reviewed the problems created by the large rise in oil prices and agreed with the serious concern expressed by the International Monetary Fund's Committee of Twenty at its recent Rome meeting over the abrupt and significant changes in prospect for the world balance of payments structure.

5. They agreed that present petroleum prices presented the structure of world trade and finance with an unprecedented situation. They recognised that none of the consuming countries could hope to insulate itself from these developments, or expect to deal with the payments impact of oil prices by the adoption of monetary

Appendix II.1

or trade measures alone. In their view, the present situation, if continued, could lead to a serious deterioration in income and employment, intensify inflationary pressures, and endanger the welfare of nations. They believed that financial measures by themselves will not be able to deal with the strains of the current situation.

6. They expressed their particular concern about the consequences of the situation for the developing countries and recognised the need for efforts by the entire international community to resolve this problem. At current oil prices the additional energy costs for developing countries will cause a serious setback to the prospect for economic development of these countries.

7. General Conclusions. They affirmed, that, in the pursuit of national policies, whether in trade, monetary or energy fields, efforts should be made to harmonize the interests of each country on the one hand and the maintenance of the world economic system on the other. Concerted international cooperation between all the countries concerned including oil producing countries could help to accelerate an improvement in the supply and demand situation, ameliorate the adverse economic consequences of the existing situation and lay the groundwork for a more equitable and stable international energy relationship.

8. They felt that these considerations taken as a whole made it essential that there should be a substantial increase of international cooperation in all fields. Each participant in the Conference stated its firm intention to do its utmost to contribute to such an aim, in close cooperation both with the other consumer countries and with the producer countries.

9. They concurred in the need for a comprehensive action program to deal with all facets of the world energy situation by cooperative measures. In so doing they will build on the work of the OECD. They recognised that they may wish to invite, as appropriate, other countries to join with them in these efforts. Such an action program of international cooperation would include, as appropriate, the sharing of means and efforts, while concerting national policies, in such areas as:

    - The conservation of energy and restraint of demand.

    - A system of allocating oil supplies in times of emergency and severe shortages.

    - The acceleration of development of additional energy sources so as to diversify energy supplies.

    - The acceleration of energy research and development programs through international cooperative efforts.[1]

10. With respect to monetary and economic questions, they decided to intensify their cooperation and to give impetus to the work being undertaken in the IMF, the World Bank and the OECD on the economic and monetary consequences of the current energy situation, in particular to deal with balance of payments disequilibria. They agreed that:

    - In dealing with the balance of payments impact of oil prices they stressed the importance of avoiding competitive depreciation and the escalation of

---

1. France does not accept point 9.

restrictions on trade and payments or disruptive actions in external borrowing*[2]

- While financial cooperation can only partially alleviate the problems which have recently arisen for the international economic system, they will intensify work on short-term financial measures and possible longer-term mechanisms to reinforce existing official and market credit facilities.*

- They will pursue domestic economic policies which will reduce as much as possible the difficulties resulting from the current energy cost levels.*

- They will make strenuous efforts to maintain and enlarge the flow of development aid bilaterally and through multilateral institutions, on the basis of international solidarity embracing all countries with appropriate resources.

11. Further, they have agreed to accelerate wherever practicable their own national programs of new energy sources and technology which will help the overall worldwide supply and demand situation.

12. They agreed to examine in detail the role of international oil companies.

13. They stressed the continued importance of maintaining and improving the natural environment as part of developing energy sources and agreed to make this an important goal of their activity.

14. They further agreed that there was need to develop a cooperative multilateral relationship with producing countries, and other consuming countries that takes into account the long-term interests of all. They are ready to exchange technical information with these countries on the problem of stabilizing energy supplies with regard to quantity and prices.

15. They welcomed the initiatives in the UN to deal with the larger issues of energy and primary products at a world-wide level and in particular for a special session of the UN General Assembly.

Establishment of Follow-on Machinery

16. They agreed to establish a coordinating group headed by senior officials to direct and to coordinate the development of the actions referred to above. The coordinating group shall decide how best to organise its work. It should:

    - Monitor and give focus to the tasks that might be addressed in existing organizations;

    - Establish such ad hoc working groups as may be necessary to undertake tasks for which there are presently no suitable bodies;

    - Direct preparations of a conference of consumer and producer countries which will be held at the earliest possible opportunity and which, if necessary, will be preceded by a further meeting of consumer countries.[3]

---

2. In point 10, France does not accept paragraphs cited with asterisks.
3. France does not accept Point 16.

17. They agreed that the preparations for such meetings should involve consultations with developing countries and other consumer and producer countries.[3]

Source: US Department of State, Selected Documents No.3.

---

3. France does not accept Point 17.

APPENDIX II.2

A SELECTION OF ARTICLES OF THE IEA

Article 2

1. The Participating Countries shall establish a common emergency self-sufficiency in oil supplies. To this end, each Participating Country shall maintain emergency reserves sufficient to sustain consumption for at least 60 days with no net oil imports. Both consumption and net oil imports shall be reckoned at the average daily level of the previous calendar year.

2. The Governing Board shall, acting by special majority, not later than 1st July, 1975, decide the date from which the emergency reserve commitment of each Participating Country shall, for the purpose of calculating its supply right referred to in Article 7, be deemed to be raised to a level of 90 days. Each Participating Country shall increase its actual level or emergency reserves to 90 days and shall endeavour to do so by the date so decided.

Article 3

1. The emergency reserve commitment set out in Article 2 may be satisfied by

    - oil stocks,

    - fuel switching capacity,

    - stand-by oil production,

    in accordance with the provisions of the Annex which forms an integral part of this Agreement.

Article 5

1. Each Participating Country shall at all times have ready a program of contingent oil demand restraint measures enabling it to reduce its rate of final consumption.

Article 41

1. The Participating Countries are determined to reduce over the longer term their dependence on imported oil for meeting their total energy requirements.

Article 42

1. The Standing Group on Long Term Co-operation shall examine and report to the Management Committee on co-operative action. The following areas shall in particular be considered:

    (a) Conservation of energy, including co-operative programs on

    - exchange of national experiences and information on energy conservation;

    - ways and means for reducing the growth of energy consumption through conservation.

    (b) Development of alternative sources of energy such as domestic oil, coal, natural gas, nuclear energy and hydro-electric power, including co-operative programs on

    - exchange of information on such matters as resources, supply and demand, price and taxation;

    - ways and means for reducing the growth of consumption of imported oil through the development of alternative sources of energy;

    - concrete projects, including jointly financed projects;

    - criteria, quality objectives and standards for environmental protection.

    (c) Energy research and development, including as a matter of priority co-operative programs on

    - coal technology;

    - solar energy;

    - radioactive waste management;

    - controlled thermonuclear fusion;

    - production of hydrogen from water;

    - nuclear safety;

    - waste heat utilization;

    - conservation of energy;

    - municipal and industrial waste utilisation for energy conservation;

    - overall energy system analysis and general studies.

    (d) Uranium enrichment, including co-operative programs

    - to monitor developments in natural and enriched uranium supply;

    - to facilitate development of natural uranium resources and enrichment services;

- to encourage such consultations as may be required to deal with international issues that may arise in relation to the expansion of enriched uranium supply;

- to arrange for the requisite collection, analysis and dissemination of data related to the planning of enrichment services.

Article 44

The Participating Countries will endeavour to promote co-operative relations with oil producing countries and with other oil consuming countries, including developing countries. They will keep under review developments in the energy field with a view to identifying opportunities for and promoting a purposeful dialogue, as well as other forms of co-operation, with producer countries and with other consumer countries.

Article 45

To achieve the objectives set out in Article 44, the Participating Countries will give full consideration to the needs and interests of other oil consuming countries, particularly those of the developing countries.

Article 46

The Participating Countries will, in the context of the Program, exchange views on their relations with oil producing countries. To this end, the Participating Countries should inform each other of co-operative action on their part with producer countries which is relevant to the objectives of the Program.

Article 47

The Participating Countries will, in the context of the Program

- seek, in the light of their continuous review of developments in the international energy situation and its effect on the world economy, opportunities and means of encouraging stable international trade in oil and of promoting secure oil supplies on reasonable and equitable terms for each Participating Country;

- consider, in the light of work going on in other international organisations, other possible fields of co-operation including the prospects for co-operation in accelerated industrialisation and socio-economic development in the principal producing areas and the implications of this for international trade and investment;

- keep under review the prospects for co-operation with oil producing countries on energy questions of mutual interest, such as conservation of energy, the development of alternative sources, and research and development.

Article 49

1. The Agency shall have the following organs:

Appendix II.2

- a Governing Board
- a Management Committee
- Standing Groups on
    - Emergency Questions
    - The Oil Market
    - Long Term Co-operation
    - Relations with Producer and Other Consumer Countries.

2. The Governing Board or the Management Committee may, acting by majority, establish any other organ necessary for the implementation of the Program.

3. The Agency shall have a Secretariat to assist the organs mentioned in paragraphs 1 and 2.

Source: Decision of the OECD Council Establishing an International Energy Agency of the Organisation, 15 November 1974.

APPENDIX II.3

COMMUNIQUE

INTERNATIONAL ENERGY AGENCY

Meeting of Governing Board at Ministerial Level

22nd May, 1980

The Governing Board of the International Energy Agency (IEA) met at Ministerial Level on 22nd May 1980 in Paris under the Chairmanship of the Minister of Economics of the Federal Republic of Germany, Otto Graf Lambsdorff.

1. At the last Ministerial Meeting in December 1979 Ministers responded to the turbulent oil market conditions by establishing oil import ceilings for 1980, revising the 1985 Group Objective and establishing individual countries' contribution to this Group Objective, creating a monitoring system and emphasizing the need for further steps for restoring order in the oil market. This time Ministers met in order to review progress achieved and take additional action.

Assessment of World Energy Situation

2. Ministers expressed their concern about the level of oil prices which confronts the world economy with declinging economic activity, having serious negative results for all countries. In particular, the price increases since the end of 1979 have occurred despite falling oil demand and appear to have been made without taking into account their adverse impact on the world economy.

3. Ministers considered projections of world energy supply and demand trends through the 1980s and agreed that, in order to protect and enhance economic growth prospects, IEA countries will continue to strengthen and implement their energy policies in ways which ensure that structural changes in energy economies actually take place over the medium term, reducing the need for energy and oil in particular, and also provide protection against short-term market disruptions. They recognized that if energy problems are not resolved, the ability to manage the general economy effectively would be put seriously in question, which could damage the prospects for economic growth on a lasting basis.

Measures to Ensure Structural Change in the Medium Term

4. Ministers considered the results of the 1979 annual review of IEA countries' energy policies and programmes conducted by the IEA's Standing Group on Long-Term Co-operation and noted its ereport "Energy Policies and Programmes of IEA Countries - 1979 Review". They agreed that not all measures, as required by the IEA

Principles for Energy Policy, have been put in place or are sufficient to produce the necessary results.

5. Ministers discussed the extent to which countries have followed up on the Principles for Energy Policy. They noted the Secretariat analysis of areas where energy policies could be strengthened in individual IEA countries (as set forth in Annex I), which they regard as a useful instrument for a substantial and qualitative monitoring of IEA's countries' energy policy efforts. They recognised that the indicated areas represent a considerable potential for achieving significant results over the medium term. They therefore agreed that each Minister will give weight to this analysis within his country's process of deciding what national energy policies are required.

6. Ministers welcomed the formation of the Coal Industry Advisory Board to help governments develop programmes and policies to greatly increase coal production, trade and use. They believe that the Board should have a very important role in the formulation of policies toward coal development, welcomed its intention to develop an action programme by autumn, invited it to provide concrete recommendations on action needed to double coal production and use by 1990, and agreed to give great weight to its recommendations and consider them quickly. Ministers also noted that the expansion of nuclear power, under appropriate conditions taking into account the progress made in the International Nuclear Fuel Cycle Evaluation, is indispensable for ensuring structural change in the medium term.

7. Ministers recognised that the process of restructuring energy economies will involve major investments, which wil have to be encouraged by active policy and will best develop under conditions of economic growth. They noted the concern of some countries that under their particular circumstances this could pose difficulties, which the Governing Board at official level will consider and will report on to the next meeting of Ministers.

8. Ministers agreed that medium and long term Group Objectives are important and agreed instruments to serve as numerical indicators which point the directions for structural change, provided a framework within which to identify the measures necessary to achieve them, and form a basis for monitoring progress. They should be designed to get well ahead of the situation rather than merely keeping up with it.

9. Ministers agreed that results actually achieved by IEA countries as a group for net oil imports in 1985 should substantially undershoot the existing 1985 Group Objective (26.2 mbd of oil imports, including bunkers), to reflect both the potential for savings and oil production probabilities. Ministers agreed to quantify the reduction as part of the monitoring process, taking into account consumption and imports. Based on currently available information the Secretariat estimates this potential with all existing uncertainties at around 4 mbd.

10. Ministers agreed that efforts to reduce oil imports will be continued beyond 1985. It is expected that as a result of these efforts it will be possible to reduce the ratio between the rate of increase of energy consumption and the rate of economic growth for IEA countries as a group over the coming decade to about 0.6 and the share of oil in total energy demand from 52% at present to about 40% by 1990.

11. Ministers agreed that regular and effective monitoring is essential to ensure that existing and additional measures are being implemented and are in fact resulting in the necessary structural changes in all IEA countries.

Short Term Instruments

12. Ministers concluded that because of the time required to achieve structural change, short-term instruments must be available in the meantime to limit the damaging economic efects of short-term price or volume disruptions which could occur in the oil market. They therefore decided upon the following actions to improve the preparedness of IEA countries:

Yardsticks and Ceilings

(a) Arrangements for yardsticks and ceilings, in order to measure progress in achieving structural change and medium-term goals and to put the IEA in a position to deal at short notice with a deterioration in the oil market situation, as follows:

- Estimates of individual countries' oil requirements, derived from, consumption, stock change and indigenous production, will be developed on an annual basis.

- Under normal market conditions, these estimates will serve as yardsticks for measuring progress in implementing measures to achieve structural change. For this purpose, they will be compared with medium-term goals in order to determine whether measures and their results are tending over time towards medium and long-term objectives. They will also be compared with the short-term oil supply outlook in order to monitor oil market developments.

- If at any time tight market conditions appear imminent, Ministers will meet at short notice. If Ministers decide that tight oil market conditions exist, IEA countries will take positive, effective short-term action as necessary, in particular, measures to restrain demand in order to prevent the scramble for scarce resources which could otherwise occur. In such cases Ministers will take a decision on use of individual oil import ceilings based in part on these estimates as a means of self-imposed restraint and as a means for monitoring its effectiveness. The ceilings will represent a political commitment stating the degree of self-restraint which individual countries are willing to impose upon themselves in a tight market situation.

(b) A system for adjustment of ceilings and goals, because the need may arise to establilsh new oil import ceilings and goals and to respond quickly if changing oil market conditions require their adjustment.

Stock Policies

(c) A system for consultations between governments within the IEA and between Governments and the oil industry on stock policies, which will be used to respond to oil market conditions beginning in 1980. The Governing Board at official level will consider guidelines for the use of stocks for this purpose.

(d) Reconfirmation that the 90-day emergency reserve requirement appears to provide reasonable protection against future emergencies.

13. Ministers also considered other measures for dealing with short-term oil market disruptions, including:

- Flexible use of stocks over and above the 90-day emergency reserve requirement and normal working stocks to meet short-term market disruptions.

- Other mechanisms, including those referred to by the Ministers in December 1979.

These areas will be considered further by the Governing Board at official level.

14. Ministers reviewed the results for the first monitoring round for the first quarter of 1980, which shows that all IEA countries expect to stay within the limits of the 1980 oil import ceilings established at the meeting of the Governing Board at Ministerial level in December 1979. They discussed the present situation in the international oil market and concluded that at present the 1980 oil import ceilings established in December 1979 do not appear to require adjustment, and that ceilings for 1981 do not now appear necessary. But this could change rapidly for 1980 or 1981 if there is a deterioration in the oil supply or in the oil demand situation. Ministers will reconsider the oil market conditions in their full meeting.

Energy Research, Demonstration and Development and Commercialisation

15. Ministers will attach greater political importance to energy research, development and demonstration, as well as commercialisation of new technologies, as essential elements for ensuring that medium-term structural changes in their energy economies are carried over into the long term. They endorsed the Report of the International Energy Technology Group, and its recommendations for accelerating commercialisation of new energy technologies.

16. Ministers noted that an IEA RD & D Group Strategy has been developed. They concluded that the Governing Board at official level will pursue the strategy's accelerated scenario, which minimizes oil imports for the IEA as a whole. They agreed that IEA countries will use the IEA RD & D Group Strategy as a guide for setting national priorities and funding levels as well as for IEA collaborative project priorities. The Committee on Energy Research and Development will closely monitor and periodically consider the extent to which aggregate national RD & D efforts are consistent with the Group Strategy. The Coal Industry Advisory Board is invited to provide recommendations as to which new technologies should be pursued in order to further speed up expanded production and use of coal.

17. The political aspects of energy RD & D issues and in particular the follow-up of the Report of the International Energy Technology Group and the IEA RD & D Strategy Report, should be given high priority. Consideration should be given to a meeting of Ministers and highest level officials responsible for energy technology in IEA countries for that purpose.

International Co-operation

18. Ministers noted that a smooth medium-term transition away from an oil-based economy, accompanied by stable short-term oil market conditions, is a prerequisite for a prospering world economy in which all nations can pursue economic growth and development. They belive that action along the above lines will contribute to these results. Ministers also expressed their hope that oil-exporting countries would adopt a similar view of their contribution to better world economic conditions, and again stated the willingness of IEA countries to discuss these issues with oil-producing countries. They accepted the need for further action by the international community to help developing countries in meeting their energy requirements by the development of indigenous energy resources. This was considered to be an effort to which both industrialized countries and oil-exporting countries could contribute. Ministers recognised the importance of discussion of energy issues in the forthcoming Global Negotiations within the United Nations system, and will make every effort to contribute to their success. They will also

continue to support strongly the forthcoming United Nations Conference on New and Renewable Sources of Energy.

Annex 1

Secretariat Analysis of Areas Where Energy Policies Could be Strengthened in Individual IEA Countries

(i) Oil prices in general should reflect international oil prices. The United States should continue its progress in decontrolling oil and natural gas prices. Canada should take steps as rapidly as possible to increase domestic oil prices to a level that encourages further energy conservation, substitution of other fuels for oil, and development of alternative sources of energy. In countries with abundant gas reserves (Australia, Canada, New Zealand) and options to increase non-oil fired electricity (Australia, Canada, New Zealand and Sweden) pricing policies for those fuels should take into account the desirability of encouraging their substitution for oil in appropriate uses.

(ii) Efforts should be made in all countries (but particularly in Japan, the Netherlands and the United States) to reduce oil-fired electricity generation as rapidly as possible by substituting other fuels and restricting oil use to middle and peak loads. No new oil-fired electricity plants should be authorised except in particular circumstances where there are no practical alternatives. Existing capacity should be operated with maximum use of fuels other than oil.

(iii) Strong action is necessary in all countries to reduce the non-feedstock use of oil in industry. Careful review of the situation is warranted in Greece, Ireland, Japan, the Netherlands and the United States where forecasts suggest that oil use may grow rapidly, and in Germany, Italy and the United Kingdom where stronger action may be necessary to achieve the expected results.

(iv) Non-oil fuels, used either directly (including district heating) or converted to electricity should be substituted for oil in residential use wherever infrastructure exists or can be provided. Countries now using or considering district heating should consider greater use of coal for this purpose, including the possibility of converting oil-fired units. Australia, Canada, Germany, Italy and Japan should endeavour to replace oil by natural gas, the latter three countries through increased imports. Sweden should encourage the substitution of electricity for oil in the residential sector. The United Kingdom should ensure that its current plans to substitute natural gas for oil over the 1980s are realised.

(v) All countries should give greater emphasis to strong and comprehensive conservation programmes to encourage the rational and efficient use of energy in general and oil in particular. They must effectively inform the public about why and how to conserve energy, and they must produce results. In particular, housing insulation efficiency standards should be reviewed and increased where necessary. Countries that do not have insulation retrofitting programmes should give serious and prompt consideration to introducing them, wherever climatic conditions make them appropriate. Major conservation gains can also be made in other sectors, particularly industry and transportation, and appropriate actions should be taken to ensure that savings result.

(vi) In the transportation sector, substantial oil savings can result through continuing increases in fuel efficiency. Countries which now have fuel economy programmes (Australia, Canada, Germany, Japan, New Zealand, Sweden, the United Kingdom and the United States) should extend their programmes to ensure that efficiency gains continue through the 1980s and review the existing programmes to see if they could be strengthened. Results should be assured, by making programmes mandatory if necessary. Countries not now having programmes should consider introducing them.

Consideration should be given to setting standards for commercial and recreational vehicles, and all countries should review the level and structure of their taxes on fuel-inefficient automobiles and gasoline.

All countries not hacing reduced speed limits should consider imposing them and enforcing them, in order to impress upon motorists the need to save fuel.

(vii) Stronger actions are required to expand coal production (Australia, Canada and the United States, which should be prepared to develop further their capacity to export substantial quantities of coal); use (Germany, Italy, Japan, Spain, the United Kingdom); and trade, where greater attention to long-term contractual arrangements is necessary to provide the stability and confidence to develop new mines and transportation facilities. Positive action is required to deal with environmental considerations, including demonstration projects and other support for technologies that can reduce environmental impacts.

(viii) Efforts should be made to increase natural gas supplies through increased domestic production (Norway, the United Kingdom and the United States) and imports (Austria, Belgium, Germany, Italy, Japan, Sweden and the United States). Strategies for gas use in all countries should ensure that use is minimised in electricity generation and industry, where other alternatives to oil exist. Canada should continue and strengthen its efforts to promote increased domestic use of natural gas to replace imported oil.

(ix) Greater efforts must be made to accomplish projected nuclear programmes and to create an environment in which discussion of nuclear issues can take place in an objective and balanced way, taking account of economic and energy considerations as well as safety and proliferation aspects (Germany, Italy, Japan and the United States), and to streamline regulatory processes for the licensing of nuclear plants and for authorisations related to nuclear fuel cycle activities in other Member countries.

(x) Hydrocarbon exploration and development activities should be strengthened in order to maximize production in the longer term (Denmark, New Zealand, the Netherlands, Norway, Spain, the United Kingdom and the United States). Opportunities to increase production through enhanced recovery should be actively pursued.

(xi) Early action is required to accelerate the development and commercialisation of new energy technologies, particularly in the areas of conservation and liquid and gaseous fuels.

(xii) Countries which have not been individually mentioned above in connection with specific lines of action should actively pursue similar opportunities for new or stronger policy measures in all of the above areas which are applicable to them.

Source: IEA/OECD Press Release IEA/PRESS (80) 8 Paris, 22 May 1980.

Appendix III
REPORT OF THE CONFERENCE ON INTERNATIONAL ECONOMIC CO-OPERATION

1. The Conference on International Economic Cooperation held its final meeting in Paris, at ministerial level, from May 30 to June 2, 1977. Representatives of the following 27 members of the Conference took part: Algeria, Argentina, Australia, Brazil, Cameroon, Canada, Egypt, European Economic Community, India, Indonesia, Iran, Iraq, Jamaica, Japan, Mexico, Nigeria, Pakistan, Peru, Saudi Arabia, Spain, Sweden, Switzerland, United States, Venezuela, Yugoslavia, Zaire and Zambia. The participants welcomed the presence of the Secretary General of the United Nations. The following observers also attended the Conference: OPEC, OAPEC, IEA, UNCTAD, OECD, FAO, GATT, UNDP, UNIDO, IMF, IBRD and SELA.

2. The Honourable Allan J. MacEachen, PC, MP, President of the Privy Council of Canada, and His Excellency Dr. Manuel Perez-Guerrero, Minister of State for International Economic Affairs of Venezuela, Co-Chairman of the Conference, presided over the Ministerial Meeting. Mr. Bernard Guitton served in his capacity of Executive Secretary of the Conference.

3. The Ministerial representatives at the meeting recognized that during the course of its work, and within the framework established at the Ministerial Meeting with which the Conference was initiated in December 1975, the Conference had examined a wide variety of economic issues in the areas of Energy, Raw Materials, Development and Finance. There was recognition that the issues in each of these areas are closely interrelated and that particular attention should be given to the problems of the developing countries, especially the most seriously affected among them.

4. The Co-Chairmen of the Commissions of Energy, Mr. Stephen Bosworth and H.E. Abdulhady H. Taher; on Raw Materials, Their Excellencies Alfonso Arias Schreiber and Hiromichi Miyazaki; on Development H.E. Messaoud Ait-Chaalal and Mr. Edmund Wellenstein; and on Financial Affairs, Mr. Stanley Payton and H.E. Mohammed Yeganeh presented on May 14 the final reports of the work of the four Commissions, which were considered at a meeting of Senior Officials of the Conference on May 26-28, and subsequently submitted to the Ministerial Meeting.

5. The participants recalled their agreement that the Conference should lead to concrete proposals for an equitable and comprehensive programme for international economic co-operation including agreements, decisions, commitments and recommendations. They also recalled their agreement that action by the Con-

Appendix III.1   319

ference should constitute a significant advance in international economic cooperation and make a substantial contribution to the economic development of the developing countries.

6. The participants were able to agree on a number of issues and measures relating to:

Energy

1. Conclusion and recommendation on availability and supply in a commercial sense, except for purchasing power constraint.*

2. Recognition of depletable nature of oil and gas. Transition from oil based energy mix to more permanent and renewable sources of energy.

3. Conservation and increased efficiency of energy utilization.

4. Need to develop all forms of energy.

5. General conclusions and recommendations for national action and international cooperation in the energy field.

Raw Materials and Trade

1. Establishment of a Common Fund with purposes, objectives and other constituent elements to be further negotiated in UNCTAD.

2. Research and development and some other measures for natural products competing with synthetics.

3. Measures for international cooperation in the field of marketing and distribution of raw materials.

4. Measures to assist importing developing countries to develop and diversify their indigenous natural resources.

5. Agreement for improving generalized system of preferences schemes; identification of areas for special and more favourable treatment for developing countries in the Multilateral Trade Negotiations, and certain other trade questions.

Development

1. Volume and quality of official development assistance.

---

* Some delegations of the G. 19 consider that this item should be viewed in the context of the report of the Co-Chairmen of the Energy Commission to the Ministerial meeting and the proposal presented to the Energy Commission by the delegates of Egypt, Iran, Iraq and Venezuela.

2. Provision by developed countries of $1 billion in a Special Action Programme for individual low-income countries facing general problems of transfer of resources.

3. Food and agriculture.

4. Assistance to infrastructure development in developing countries with particular reference to Africa.

5. Several aspects of the industrialization of developing countries.

6. Industrial property, implementation of relevant UNCTAD resolutions on transfer of technology and U.N. Conference on science and technology.

Finance

1. Private foreign direct investment, except criteria for compensation, transferability of income and capital and jurisdiction and standards for settlement of disputes.

2. Developing country access to capital markets.

3. Other financial flows (monetary issues).

4. Cooperation among developing countries.

The texts agreed appear in the attached annex which is an integral part of this document.

7. The participants were not able to agree on other issues and measures relating to:

Energy

1. Price of energy and purchasing power of energy export earnings.

2. Accumulated revenues from oil exports.

3. Financial assistance to bridge external payments problems of oil importing countries or oil importing developing countries.

4. Recommendations on resources within the Law of the Sea Conference.

5. Continuing consultations on energy.

Raw Materials and Trade

1. Purchasing power of developing countries.

2. Measures related to compensatory financing.

3. Aspects of local processing and diversification.

4. Measures relating to interests of developing countries in: world shipping tonnage and trade; representation on Commodity Exchanges; a Code of Conduct for Liner Conferences, and other matters.

5. Production control and other measures concerning synthetics.

Appendix III.1   321

  6. Investment in the field of Raw Materials.

  7. Means for protecting the interest of developing countries which might be adversely affected by the implementation of the Integrated Program.

  8. Relationship of Integrated Program to New International Economic Order.

  9. Measures related to trade policies, to the institutional framework of trade, to aspects of the GSP, to the MTN, and to conditions of supply.

Development

  1. Indebtedness of developing countries.

  2. Adjustment assistance measures related to industrialisation.

  3. Access to markets for manufactured and semi-manufactured products.

  4. Transnational corporations.

Finance

  1. Criteria for compensation, transferability of income and capital and jurisdiction and standards for settlement of disputes.

  2. Measures against inflation.

  3. Financial assets of oil exporting developing countries.

The proposals made by participants or groups of participants on these matters also appear in the same annex.

8. The participants from developing countries in CIEC, while recognizing that progress has been made in CIEC to meet certain proposals of developing countries, noted with regret that most of the proposals for structural changes in the international economic system and certain of the proposals for urgent actions on pressing problems have not been agreed upon. Therefore, the Group of 19 feels that the conclusions of CIEC fall short of the objectives envisaged for a comprehensive and equitable programme of action designed to establish the New International Economic Order.

9. The participants from developed countries in CIEC welcomed the spirit of co-operation in which on the whole the Conference took place and expressed their determination to maintain that spirit as the dialogue between developing and developed countries continues in other places. They regretted that it had not proved possible to reach agreement on some important areas of the dialogue such as certain aspects of energy co-operation.

10. The participants in the Conference think that it has contributed to a broader understanding of the international economic situation and that its intensive discussions have been useful to all participants. They agreed that CIEC was only one phase in the ongoing dialogue between developed and developing countries which should continue to be pursued actively in the U.N. system and other existing, appropriate bodies.

11. The members of the Conference agreed to transmit the results of the Conference to the United Nations General Assembly at its resumed 31st Session and to all other relevant international bodies for their consideration and appropriate action.

They further agreed to recommend that intensive consideration of outstanding problems be continued within the United Nations System and other existing, appropriate bodies.

12. The participants in the Conference pledged themselves to carry out in a timely and effective manner the measures for international cooperation agreed to herein. They invite the countries which did not participate in the Conference to join in this cooperative effort.

13. Finally, the ministerial representatives at the Conference reiterated their appreciation to the President of the French Republic and to the Government of France for their hospitality and for their cooperation in facilitating the work of the Conference on International Economic Cooperation.

Source: CIEC DOCUMENT NO: CCEI-CM-5. REV. 1 Dated 2 June 1977.

APPENDIX III.2

ENERGY COMMISSION REPORT

Annex to the Report of the Conference on International Economic Cooperation

Note: Throughout this document the following uses of brackets and parentheses applies to all non agreed texts:

(( ))   Indicate G-19 proposals
( )   Indicate G-8 proposals

A.   CONCLUSIONS

The Energy Commission having assessed past and current trends in the world energy situation and having taken into account the economic interests, including energy interests, of all countries, with a view to dealing with energy-related problems on a basis of international economic cooperation recognizes that:

1. Energy availability and supply are among the important factors for the economic advancement of both industrial and developing countries.

2. Oil and gas are the most rapidly depleting among the non-renewable sources of energy owing to the concentration of worldwide demand on them. They have a number of non-energy and, at least in the short to medium term, non-substitutable uses.

3. It is in the interest of the world community that a transition should take place

    -   from the present, primarily oil-based energy mix to an energy mix primarily based on more permanent and renewable sources of energy,

    -   and from an economy in which oil and gas are predominantly used as a source of energy to an economy in which oil and gas are predominantly reserved for non-energy and non-substitutable uses, and be short enough so that these changes are brought about well before depletion of oil resources.

During this period, the world community should, as rapidly as possible and practicable, expand, develop and diversify its energy resources and implement adequate conservation policies for oil and gas, while ensuring that sufficient supplies of energy are available to meet demand, and that energy exporting countries are enabled to develop their economies sufficiently.

Failure to take actions required to achieve these objectives would lead to serious consequences for the world as a whole.

4. The availability and supply of energy resources are to be within a commercial sense and must take into account technical limitations, financial needs, replacement costs, future requirements of currently energy exporting countries and other constraints. Adequate and stable supplies of energy, both non-renewable and renewable sources, are essential to the economic well-being and progress of all countries. Within this context, all countries will need to contribute, on the basis of their individual capacities and potential - with due regard to their different forms of energy resources and technological knowhow - toward the adequate availability of energy supplies and sources for the transition period and beyond.

5. The economies of the energy importing developing countries do not possess the ability to adjust easily to the structural changes in the world economy, particularly those related to the new economic situation, which add to the constraints on their development progress. The energy importing developing countries will continue to increase their energy use as they modernize and industrialize their economies. Considering their inability to reduce energy use signficantly without hindering development, alleviation of their energy problems needs special attention.

6. External payments factors related to the present world economic situation seriously reduce, for various energy-importing developing countries, their development progress, which depends on many variables inter alia the ability to import inputs such as technology, capital and other manufactured goods, services and oil as well as world economic growth trends and other domestic and international economic factors.

7. There is scope for energy conservation in most countries, especially the developed countries. The developed countries generally possess the resources and technology needed to curtail non-essential uses of energy, to increase the efficiency of energy utilization and to concentrate oil and gas gradually in uses where possibilities of substitution are limited.

8. All countries should endeavor to develop all forms of energy, renewable as well as non renewable, to the maximum extent possible subject to technical feasibility, economic needs and efficiency, safety and security considerations, national policy and environmental constraints. This will require large volumes of capital, advanced technology and know-how. Technology and know-how are primarily available in the developed countries; capital is mostly available in the developed and some developing countries. Special attention should be given to the development of indigenous energy sources in the developing countries.

9. The developing countries, being committed to the New International Economic Order, are determined, as part of their overall economic development objectives, to develop downstream industries in order to diversify their industrial base and realize value added benefits. The oil-exporting developing countries have great potential for making substantial progress in the fields of refining, petrochemical and other hydrocarbon-based industries. Their own domestic energy requirements and financial needs will grow as they pursue their long-term economic development plans.

As global demand for oil and gas products and derivatives expands and the availability of gas and gas products grows and as existing capacities become obsolete, the oil exporting countries will contribute by increasing their share of total global supply of these products and derivatives. It is in the interest of all countries that the economic integration of the refineries, petrochemical facilities and other downstream industries of the oil exporting countries into the expanding global industrial community take place as rapidly as economically practicable. Cooperative efforts will need to be made, within the context of mutually beneficial technical and commercial arrangements and taking account of relevant national and international constraints, to bring about in an economic manner the structural and other changes in the world community, including those connected with refinery capacity and production mix, with a view to ensuring that restrictions to market access do not prevent sufficient supplies of various grades and types of oil and gas and their associated and manufactured products from being available and making their necessary contribution to world energy supply.

10. The world community requires an international energy cooperation and development program within the overall framework of an international economic cooperation program that would, recognizing relevant constraints, encourage and accelerate energy conservation and the development of additional energy supplies through, inter alia, facilitating and improving access to energy-related technology, expanding energy research and development and increasing investment flows into energy exploration and development. It is clear, on the basis of the analysis performed in the Energy Commission that without such a comprehensive program the world risks significant shortages of energy in the medium term and rapid depletion of oil and gas that will seriously jeopardize the economic progress of all countries. This comprehensive program would address financial aspects of energy development problems, energy conservation, exploration and development for non-renewable energy resources and technological research and development efforts related to both renewable and non-renewable energy sources. There is a need to initiate measures promptly and simultaneously that will produce results in the short, medium and long term. Within this comprehensive program:

a) Financial assistance will be needed to bridge external payments problems of oil importing ((developing))* countries, particularly those developing countries heavily dependent on imports of energy,** while these countries adjust to the present world economic situation. Such financial assistance is particularly important for countries where unavoidable payments problems are particularly large in relation to the size of their economies*. These problems need to be addressed within the appropriate framework of existing** institutions.

b) Efforts at energy conservation, in particular of oil and gas, will need to be intensified. In all countries where there is potential to do so, new actions will need to be taken. This is particularly true for the developed countries

---

* This bracket appeared in the draft of the Co-Chairmen of the Energy Commission.

** Wherever reference is made to developing countries heavily dependent on imports of energy, it is intended to include the non-oil exporting developing countries heavily dependent on oil.

* India and Jamaica suggested replacing "economies" with "imports".

** Zaire and Jamaica suggested replacing "existing" with "international".

who as a group have the greatest possibilities for expanded energy conservation.

- c) Additional energy supplies will need to be developed in developed and developing countries:

- i) The developing countries, in particular, will need to be encouraged and assisted in exploration, expansion, development and diversification of their indigenous energy resources, both non-renewable and renewable, as well as related infrastructure. For these purposes, the developing countries require technology primarily from the developed countries. They will also require technical assistance, particularly in the area of energy research and development and the facilitation of improved access to appropriate energy technology. Investment funds on a substantial scale will also be required from countries capable of providing them. International financial institutions will also need to play a significantly enlarged role in this regard.

- ii) Oil exporting developing countries need to be assisted in diversifying their economic base in order to fulfil their development plans and sustain their long-term economic progress. An integral part of such diversification for these developing countries is the progressive and orderly expansion of their share in downstream hydrocarbon processing. Such diversification would need appropriate action to facilitate access to technology as well as to ensure that restrictions to market acccess do not prevent access of their exports of hydrocarbon-based products to the developed countries' markets.

- d) Cooperation between developed and developing countries will need to be expanded in current and future research, development and demonstration facilities, projects and training related to both non-renewable and renewable sources of energy.

11. *There is considerable lack of knowledge on the part of all countries with respect to the world energy outlook. All countries would benefit from improved knowledge.*

    ** There is considerable lack of knowledge on the part of all countries with respect to the world energy, economic and monetary outlook. All countries would benefit from improved knowledge in these fields.**

12. The Energy Commission examined extensively the issues of energy prices and the purchasing power of energy export earnings on the basis of the following work program:

    - a) Competitive standing of various sources of energy including but not limited to:

    - i) availability;

---

* This version of the paragraph is the final draft proposed by the Co-Chairmen of the Energy Commission on which agreement was not reached.

** This version of the paragraph is alternative language proposed by one or more delegations on which agreement was not reached.

ii) depletability;
iii) prices and economic costs of existing sources of energy as well as the intrinsic value of depletable energy sources;
iv) the probable prices and economic costs of new sources of energy.

b) The Energy Commission considered proposals for the preservation of purchasing power of energy export earnings, including accumulated revenues from oil exports, within a general framework of the improvement and preservation of the purchasing power of the export earnings of developing countries vis-a-vis among other factors, inflation in industrialized countries, changes in prices of imported goods and services and other economic factors essential to the economic progress of developing countries.

c) The role of energy prices together with the prices of other major economic inputs in the world economic situation including growth, inflation and investment.

B. RECOMMENDATIONS

The Energy Commission

With the aim or providing impetus to the development and implementation of effective and comprehensive national and international efforts and measures for global energy conservation, exploration, development and transfer of related technologies and to achieve these energy objectives through an international energy cooperation and development program within a framework of an international economic cooperation program recommends the following:

1. That the availability and supply of energy are to be within a commercial sense and must take into account the technical limitations, financial needs, replacement costs, ((protection of the purchasing power of the unit value of energy export earnings,)) future energy requirements of the currently energy exporting countries and other constraints, and taking into consideration the limited availability and rapid depletion of oil and gas, the world community should earnestly endeavor to expedite in a practicable manner the process of progressive reduction of the world's dependence on oil and gas and increased reliance on other non-renewable and renewable sources of energy. In this process, effective measures such as price incentives, sufficient allocation of financial resources, direct quantitative measures, taxation, transfer of technology for development of alternative sources of energy on a non-discriminatory basis, etc. should be taken. Oil and gas throughout the world should be increasingly conserved in an efficient manner for non-energy uses and for those uses which are at least in the short to medium term non-substitutable in order to effectively and progressively reduce over-dependence of energy consumption on these rapidly depleting sources, while providing that sufficient supplies of energy are available to meet essential demand.

2. That intensified worldwide efforts should be made to increase exploration, augment reserves of conventional energy resources, increase productive capacity of conventional and non-conventional resources of energy, particularly those resources that are less rapidly depletable and more permanent and renewable, subject to technical feasibility, economic needs and efficiency, safety and security considerations, national policy and environmental constraints. The development and use of alternative conventional and non-conventional resources of energy should be expedited. In this context, the developed countries should take into account the world requirements for special and advanced technology and provide access to appropriate government-owned technology and know-how and facilitate provision of the technology and know-how available in the private sector in order to enhance

appropriate development of energy sources. The amount of capital required for energy exploration and development in the future will be substantially greater than the level of energy investment in the past. The major part of the investment funds required for such development is likely to come from the developed countries and the remainder from some developing countries. International financial institutions would also be required to play a significantly enlarged role in this regard. Particular attention should be given to the development of indigenous energy resources in the developing countries.

3. That measures be adopted worldwide to intensify exploration and to ensure efficient extraction and utilisation of oil, gas and other energy sources through advanced technology and appropriate conservation practices particularly in energy deficient developing countries.

4. That during occasional periods of inadequate energy supply, the international community should, within the limits of availability and supply of energy, give priority consideration to the particular vulnerability of developing countries most dependent on oil, especially the MSACs and those most dependent on any form of energy imports to satisfy their essential domestic, industrial and economic development requirements.

5. That the Law of the Sea Conference should continue its efforts to establish - in accordance with the principle of the common heritage of mankind - an international regime for the exploration for and exploitation of (mineral) ((natural)) resources of the (deep seabed area) ((the seabed and ocean floor and the subsoil thereof)) beyond the limits of national jurisdiction, to ensure equitable sharing by states in the benefits derived therefrom - taking into particular consideration the interests and needs of the developing countries - as provided for in Resolution 2749 (XXV) of the UN General Assembly.

6. That all countries, particularly the developed countries, should increase the efficiency of energy utilization by means of conservation and greater technical efficiency. To this end developed countries and other countries with the potential and means to do so should establish definite self-imposed objectives for conservation in energy, in particular oil and gas, and take the necessary measures to meet these objectives.

7. The Energy Commission was unable to reach agreement on a recommendation on the issue of accumulated revenues from oil exports, and recommended that the Conference take note of the situation for further consideration. The Energy Commission also recognized that this issue was on the work program of the Financial Affairs Commission where various proposals had been made.

8. That appropriate measures be undertaken for effective energy cooperation between developed and developing countries in the fields of technical assistance and technology transfer in order to assist developing countries to develop and diversify their sources of energy. In this regard, the developed countries should on a non-discriminatory basis, consistent with national and international legal, security and safety considerations:

   (i) facilitate on the most extensive basis improved access by developing countries to technology and technological support for energy exploration, expansion, development and diversification programs to developing countries;

   (ii) take action to encourage private and public holders of modern energy technology to transfer that technology to developing countries under appropriate terms and conditions.

Appendix III.2

9. That, while capital and technology required for global energy development are held in most cases primarily by the private sector, public financial institutions should have an increasingly important and effective role in facilitating the financing of energy exploration and development in developing countries, particularly energy importing developing countries. New procedures should be introduced to enhance the complementary roles of all sectors in channelling flows of capital into these countries. The International Bank for Reconstruction and Development is invited to expand its participation in the diversification and development of energy resources in developing countries, particularly energy importing developing countries, in order to:

    -   augment capital availabilities for investment in energy exploration and development in these countries;

    -   act as a catalyst to induce additional flows of capital into energy development in those developing countries which desire such capital through both participation in energy projects as well as serving as a source of lending for them;

    -   contribute to the general improvement in the investment and economic climate, thereby promoting increased efforts at exploration for energy resources, through an active expanded role in energy development projects.

10. That the IBRD/IDA be invited to evaluate on a priority basis, in consultation as appropriate with the IBRD/IMF Development Committee, how it can most effectively expand its activities in line with the preceding paragraph in order to increase capital flows, on concessional terms where appropriate, into the development of indigenous energy resources in the developing countries, particularly the energy importing developing countries.

11. That member countries through their Governors in the IBRD take account of the capital requirements associated with the expansion of its activities in the energy area when deciding on the general capital increase in the Bank's resources, while taking fully into account the need to assure that the Bank's activities in other priority areas are not prejudiced.

*   ((That the IMF should be invited to establish its proposed Supplementary Credit Facility as early as possible in 1977 and, in managing the Facility, should take account of the needs of energy importing developing countries:

    -   to avoid such restrictions on imports of petroleum and petroleum products as would prejudice the maintenance of economic growth;

    -   to have assistance available for this purpose until energy diversification measures can begin to take effect.

    The IMF should be requested to maintain an Interest Subsidy Account from which developing countries most seriously affected and most dependent on imports of energy may be assisted to make use of the Facility.))

---

*   This paragraph, had it been accepted, would have become Recommendation No. 12 with subsequent paragraphs being renumbered accordingly.

12** That other international and regional financial institutions be invited to study whether they can also play a role in contributing to greater capital flows to developing countries, particularly energy importing developing countries for energy development.**

13. The developed and developing countries should cooperate to facilitate the rapid and economic integration of the petroleum refineries, petrochemical facilities and other downstream industries that the oil exporting countries have started and intend to develop further, and the products therefrom, into the expanding global industrial community as rapidly as is economically practicable. An integral part of such diversification of the economic base of these developing countries is the progressive and orderly expansion of their share in downstream hydrocarbon processing. The developed and developing countries should also take appropriate cooperative and economically efficient action, consistent with relevant national and international constraints, to make structural and other changes as required during the transition period and beyond to ensure that restrictions to market access do not prevent sufficient supplies of various grades and types of oil and gas and their associated and manufactured products from being available to the world community.

14. That measures be undertaken by the international community to facilitate the availability and expansion of transportation, storage, harbor and marketing facilities for oil, gas, coal and their derivatives, inside and outside developing countries for the benefit of all countries, particularly the developing countries.

15. That as part of the general effort of energy technology transfer, bilateral and multilateral efforts should be made to assist developing countries interested in such transfer, particularly energy importing developing countries, to diversify and develop their energy sources through:

    - undertaking national assessments of energy resource potential and developing concrete national energy strategies, where appropriate, for interested countries by countries prepared to extend such assistance;

    - establishing an appropriate managerial and technical base through training of personnel;

    - promoting and facilitating energy resource exploration and development in the oil-importing developing countries; and

    - facilitating access to and adaptation of existing energy technology and the development of new energy technologies to meet the special needs of developing countries, particularly energy importing developing countries.

    In order to help achieve the above objectives, various proposals, such as the proposal for an International Energy Institute or expansion of energy related

---

** Some energy importing countries expressed reservations on this recommendation stating that it did not adequately address the immediate balance of payments problems of the oil importing developing countries associated with the financing of their oil imports. The delegations of Brazil, Cameroon, India, Jamaica and Zaire tabled a proposal on this subject which appears at the end of the section on energy.

Appendix III.2

activities in other existing institutions, such as the U.N. and IBRD, have been considered. Expeditious international consideration of appropriate means to achieve these objectives is recommended.

16. That international cooperation in energy research and development between developed and developing countries be intensified, particularly with regard to the technologies needed by the developing countries. This should be achieved by arrangements, in accordance with national and international policy and legal frameworks, which:

   (a) facilitate access, on a non-discriminatory basis to existing and new energy technologies, particularly for developing countries, and

   (b) provide for opportunities for active and positive participation by developing countries in energy R & D activities. In this regard, the developed countries will endeavor to make available to developing country participants in energy R & D activities - jointly identified - inter alia, the following:

   i) the results of current research;
   ii) test and demonstration facilities;
   iii) training of scientists and technicians;
   iv) joint energy R & D projects with participation by developing country scientists and technicians.

PROPOSAL SUBMITTED BY THE DELEGATIONS OF BRAZIL, CAMEROON, INDIA, JAMAICA AND ZAIRE

International, financial, scientific and technological co-operation amongst all countries for the development of energy resources, etc.

The Energy Commission recognizes:

1. that the world's limited non-renewable resources of oil and gas are being rapidly depleted owing to the concentration of worldwide energy demand on these resources;

2. that it is in the interest of the international community that the developing countries be encouraged and assisted to expand, develop and diversify their indigenous and non-conventional sources of energy and related infrastructure. For these purposes the developing countries will require technology from developed countries and investment funds on a substantial scale;

3. that the economies of the energy importing developing countries do not possess the strength to adjust easily to the new energy situation which has added to the constraints on the achievement of maximum progress in the development programmes of these countries. The use of energy in energy importing developing countries will continue to increase as these countries modernize and industrialize their economies and they have no ability to reduce oil use significantly without hindering developing.

The Energy Commission therefore recommends:

1. That an international programme of financial and technological co-operation be established including short, medium and long-term measures designed to meet the needs of developing countries.

The short-term measures should aim at enabling the energy deficient developing countries to maintain their imports of petroleum and petroleum products. The medium and long-term measures should aim at assisting and encouraging interested developing countries to explore for additional sources of energy and to expand, develop and diversify their indigenous conventional and non-conventional sources of energy with the related infrastructure.

2. Short-term Measures

   The Conference should:

   (a) Request the IMF to provide for a period of five years beginning in 1977 a Special Credit Facility to assist those developing countries which are net importers of petroleum products to finance their imports of those essential supplies;

   (b) Inform the IMF that

   i) the annual amount of the credits required for the purpose by the oil importing developing countries is of the order of 4 billion SDRs;

   ii) industrialized countries and oil exporting countries in a position to do so should be invited to contribute to the resources of the Facility;

   iii) a country's total purchases under the Facility in any year should be limited by reference to its quota in the IMF, but should be sufficient to enable it to meet the cost of its estimated imports of petroleum and petroleum products in that year;

   iv) the conditionality and terms of repurchase applicable to the drawings on this Credit Facility should be similar to those applied to the drawings on the 1975 Oil Facility.

   (c) Request the IMF to maintain, in connection with the Facility, an interest subsidy account, to be contributed voluntarily, from which the most seriously affected countries and developing countries most dependent on imports of petroleum and petroleum products may be assisted to make use of the Facility.

3. Medium and Long-Term Assistance

   (a) Priority should be given in the world programme of financial and technological co-operation to medium and long-term measures which will operate to expedite the increase and diversification of energy supply sources in developing countries especially in the present energy deficient countries.

   In particular the programme should promote:

   i) exploration for oil, natural gas, uranium, thorium, sources of geothermal energy, coal and lignite. Exploration includes the conduct of aerial, geological, geophysical surveys, the assessment of geophysical data and planning and execution of exploratory drilling programmes;

   ii) energy development projects in the developing countries, including inter alia: the development of oil, natural gas, shale oil, bituminous sands, uranium and thorium resources and the installation of the necessary transport and processing facilities; the develpment of hydropower and associated transmission lines; the development of coal, lignite and peat resources, including the installation of transport infrastructure, and the introduction of coal gasification and liquefaction technologies;

the development and introduction of nuclear technology in accordance with internationally accepted standards of safety and security safeguards;

the development and introduction of solar, wind and wave energy and non-commercial energy sources such as wood, bio-gas and organic and inorganic wastes;

the manufacturing of capital equipment for energy projects on both a national and regional basis.

For the purpose of the diversification programme industrialized countries should accept the commitment to:

i) provide technological support for energy exploration, expansion and diversification in the developing countries;

ii) take action to ensure that private companies domiciled within their jurisdiction co-operate in the diversification programmes by supplying finance and technical know-how on fair and reasonable terms.

The industrialized countries and developing countries having the capacity to do so should undertake further -

(a) to contribute not less than 300 million SDRs per annum for the next three years to an appropriate international institution to be disbursed for energy exploration in developing countries.

(b) to enable the IBRD to embark upon a long-term programme of finance for energy projects and programmes in developing countries. Before 1980 the capital of the Bank should be increased sufficiently to enable it to augment assistance for such programmes by not less than an average of 5 billion SDRs per annum over the next five years without affecting normal growth in real terms in its lending for other purposes.

Co-operation in Energy Research and Development and the Transfer of Technology

By way of co-operation with developing countries in the area of research and development and transfer of technology, the industrialized countries should undertake to adopt upon request by interested developing countries, measures to promote:

(a) research and development programmes on energy related technology, including efficient utilization and conservation and in this connection establishment of research, development and demonstration facilities and projects within the developing countries for existing and alternative energy sources,

(b) energy planning,

(c) training of personnel from the developing countries in the energy sector,

(d) exchange of and access to information by the developing countries on scientific and technological research related to production, use and conservation of energy,

(e) establishment of appropriate manpower and technical bases in the developing countries sufficiently early to facilitate introduction in these countries of industrialization and other related energy technologies of the future.

17. The Energy Commission considered various proposals submitted by delegations for recommendations on the issues of energy prices and the purchasing power of energy export earnings and was unable to reach agreement on them. Two of these proposals, appearing below, were given for consideration by the Conference.

PROPOSAL SUBMITTED IN THE ENERGY COMMISSION BY THE DELEGATIONS OF EGYPT, IRAN, IRAQ AND VENEZUELA

Energy Prices and the Purchasing Power of Energy Export Earnings

The Energy Commission recognizes:

That the principle of improving and preserving the purchasing power of the unit value of the export earnings of raw materials of the developing countries has already been agreed upon in several international fora;

That the development needs for all developing countries are vitally linked with their export earnings and that erosion of the purchasing power of the unit value of such earnings seriously hinders their development plans and projects and that increased volumes of exports are not an acceptable solution to compensate for this erosion;

That so far the price structure for energy, mainly oil, has been neither conducive to the development of alternative energy sources nor to a balanced mix of its components nor to the adoption of sufficient or effective conservation measures, mainly in industrialized countries.

That the oil price, whose competitive standing with the price of alternative sources was partially re-aligned in 1973 and 1974, has since been affected by serious erosion in its purchasing power.

The Energy Commission, acknowledging that pricing of raw materials is the sovereign right of the respective producing exporting countries, and accordingly, it is the sovereign right of the oil exporting countries to determine the prices of their oil,

Further recognizes, without prejudice to the aforesaid sovereign right:

1) That the prices of oil be established, taking into consideration, inter alia, the competitive standing of oil vis-a-vis other sources of energy; and

2) That the purchasing power of the unit value of energy export earnings, including accumulated revenues from oil exports be protected within a general framework of the improvement and protection of the purchasing power of the unit value of export earnings of developing countries - through indexation or any other appropriate methods, taking into consideration, inter alia, the rising cost of imported goods and services, the erosion of purchasing power due to inflation and currency depreciation;

3) That measures for such adjustment are entirely different from changes in prices of oil due to changes in the competitive standing of oil vis-a-vis other sources of energy mentioned in (1) above.

Appendix III.2

PROPOSAL SUBMITTED IN THE ENERGY COMMISSION BY THE JAPANESE DELEGATION

Energy Prices and the Purchasing Power of Energy Export Earnings

1. Energy prices in general and - given the important share of oil in total energy consumption - oil prices in particular exercise and will continue to exercise great influence on the world economic situation, especially concerning growth, employment, inflation, the allocation of investment capital and the evolution of payments balances.

2. The price of energy is of particular importance to the economic progress of both developing and developed countries which are either dependent on imports for a major part of their energy requirements or dependent on energy export earnings for the financing of their development programmes.

3. Participants in CIEC recognize the common interest of all countries in energy prices which are fair both to consumers and producers. In particular they recognize the need for reducing uncertainty about the future and avoiding large and sudden changes in energy price levels.

4. The concern of oil producers with the purchasing power of oil revenues, both current and accumulated, is recognized; as is the necessity to reduce world price-inflation as well as cost increases resulting from other elements.

5. Taking into account these general considerations, the participating countries recognize that the following interrelated elements are relevant to the formation and trend of energy prices:

    - energy supply and demand trends,

    - the costs of existing alternative sources of energy, including such aspects of each form of energy as investment requirements, transportability, pollution characteristics and versatility of use and degree of substitutability,

    - the range and costs of developing and utilizing new energy sources, with emphasis on lead times, investment requirements, likely technological progress and economies of scale,

    - impact on the world economy as a whole as well as on the economies of individual countries, bearing in mind the problems of energy-importing developing countries and of industrial countries heavily dependent on energy imports,

    - the concern of oil exporting countries with the purchasing power of their energy earnings.

* The measures suggested above may involve various forms of contacts. This question remains to be considered further at a later stage in the Conference.*

---

* Proposal made by the Co-Chairmen of the Energy Commission on which agreement was not reached.

**   ((The measures suggested above may involve various forms of contact amongst energy exporting and energy importing countries both developed and developing. The possibility of establishing institutional arrangements to facilitate this contact and continue consultation on energy issues remains to be considered further at a later stage in the Conference.))**

***   ((The implementation of above conclusions and recommendations are to be carried out by the governments of the respective countries, with due regard to the linkage between energy problems and those of raw materials, development and finance.))**

SOURCE: CIEC DOCUMENT - Final Conference Report 2nd June 1977.

**   Proposal made by a delegation of the Group of 19.

***   Proposal made by a number of delegations of the Group of 19.

APPENDIX III.3

LIST OF STUDIES AND PROPOSALS SUBMITTED TO THE ENERGY COMMISSION OF THE CIEC DURING ITS WORKING SESSIONS - 1976 to 1977.

LIST OF STUDIES AND PROPOSALS SUBMITTED BY THE GROUP OF 19

Energy Commission Framework, CCEI-EN-1, February 1976.

General Analysis of the Energy Situation Supplement No. 1 - Saudi Arabia, March 1976.

Historical Analysis of the Energy Situation, presented by the Group of 19, 19 March 1976.

Energy Resources and Reserves - Group of 19, 22 March 1976.

Competitive Standing of the Price of Each Source of Energy - Group of 19, April 1976.

Protection of Purchasing Power, April 1976.

Availability and Supply of Energy to the Developed Countries - Group of 19, June 1976.

Availability and Supply of Energy to the Developing Countries - Group of 19, June 1976.

Enumeration of Areas of Concentration for work in the second phase of the four commissions set up by CIEC (CCEI-CP-6), presented at the High Level Officials Meeting, July 1976.

Proposal by the Co-Chairmen of the Energy Commission, 17 July 1976.

Statement of the Group of 19, CCEI-CP-13, 13 September 1976.

Statement of the Delegation of Mexico, CCEI-CP-14, 14 September 1976.

Statement of Brazil, CCEI-CP-15, 15 September 1976.

Resource Availability, Supply and Development, CCEI-EN-2, 21 October 1976.

LIST OF STUDIES AND PROPOSALS SUBMITTED BY THE GROUP OF 8

Analysis of Energy Demand and Supply - EEC, 11 February 1976.

Role and objectives of an analysis of the problems arising from the energy situation and their inter-relationship with the world economy - Switzerland, 12 February 1976.

Supply and Energy - Japan, 16 February 1976.

Areas of Co-operation - USA, 16 February 1976.

The Price of Energy - Canada, 14 February 1976.

Energy Commission Work Program Areas of Concentration - Group of 8, July 1976.

Modifications to Areas of Concentration for the Second Phase - Group of 8, July 1976.

Energy Supply and Demand in the Past - USA, March 1976.

Japan's Energy Supply and Demand Situation in the Past - Japan, 19 March 1976.

Review of Trends in the Supply and Demand of Energy in the Community 1963-1975 - EEC, March 1976.

US Energy Outlook - USA, March 1976.

Review of Trends in the Supply and Demand for Energy in the Community, Paper II - EEC, 18 March 1976.

Japan's Energy Demand - Supply Plan - Japan, March 1976.

An Examination of the Competitive Standing and Prices of Various Energy Sources - USA, 21 April 1976.

The Impact of the 1973-74 Oil Price Increase on the United States Economy to 1980 - USA, 21 April 1976.

Oil Prices and the Japanese Economy - Japan, 23 April 1976.

The Importance of the Price of Energy for the Economy of the EEC - EEC, 24 April 1976.

Implications of Oil Price Increases of 1973-74 for the Energy-deficient Developing Countries in 1975 - USA, 24 April 1976.

Methodology utilized in preparation of foreign exchange loss to energy-deficient developing countries in 1975 due to oil price increase in 1973/74 - USA, 26 April 1976.

Statement by the US Delegation on Protection of Purchasing Power - USA, 9 June 1976.

Availabilty and Adequacy of Energy Supply - USA, 10 June 1976.

Supply of Energy in the European Community - EEC, 11 June 1976.

Energy Supply in Japan - Japan, 11 June 1976.

Statement of the United States, CCEI-CP-12 - USA, 13 September 1976.

Statement of Japan, CCEI-CP-10 - Japan, 11 September 1976.

Statement of the EEC, CCEI-CP-11, 13 September 1976.

Proposed Conclusions for Part I of the Work Program, CCEI-EN-3 - Switzerland, 20 October 1976.

Proposed Conclusions for Part II of the Work Program, CCEI-EN-5 - Japan, 22 October 1976.

Outline of US Proposal for the Establishment of an International Resources Bank (IRB) - USA, 15 September 1976, (first submitted in Raw Materials Commission on 14 June 1976).

US Proposal for an International Energy Institute - USA, 12 July (circulated September 1976).

The Framework for Establishing an International Energy Institute - USA, 23 October 1976.

Proposal for an International Energy Institute, CCEI-EN-4 - USA, 23 October.

Energy Research and Development, CCEI-EN-6 - Canada, 26 October 1976.

Proposal for Cooperation in Energy Investment, CCEI-EN-7 - United States, 27 October.

Energy Cooperation - EEC, 17 September 1976.

Proposed Conclusions on Continuing Cooperation and Consultation, CCEI-EN-8, 18 November 1976.

LIST OF STUDIES AND PROPOSALS SUBMITTED BY OBSERVERS

IEA — Statistical Data and Projections - World Energy Consumption and Supply, 18 March 1976.

IEA — Supplements I, II and III to Statistical Data, 24 March 1976.

UN Secretariat - World Energy Supplies 1970-1973, March 1976.

UN Secretariat - An Extract of Global Energy Statistics, March 1976.

OAPEC - Perspective on the Pricing of Oil, 27 April 1976.

OECD Secretariat - The Impact of Developments in Energy Prices on the World Economy, 22 April 1976.

OPEC — Cost composition of a barrel of imported oil in Western Europe in 1973-75 and inflationary effect of oil price adjustment, 26 April 1976.

UNDP — The Role of UNDP and the UN Revolving Fund for Natural Resources Exploration in the Exploration for Hydrocarbons, 15 July 1976.

World Bank - Energy Development and the World Bank Statement by the World Bank Observer, 12 July 1976.

UN Secretariat - World Energy Supplies 1950-1974, April 1976.

UN Secretariat - Commentary on Energy Production and Consumption Trends, April 1976.

OPEC - Past Trend of Prices in International Trade, 26 April 1976.

IEA - Current IEA Activities in Energy Research and Development, 26 April 1976.

World Bank - Energy and Petroleum in non-OPEC Developing Countries 1974-80, July 1976.

World Bank - Financing Energy in Developing Countries, July 1976.

World Bank - International Energy and Petroleum Prospects, July 1976.

UNDP - UNDP Activities in the Field of Energy, 15 September 1976.

SOURCE: OFFICE OF THE CO-CHAIRMAN OF ENERGY COMMISSION (G-19) JUNE 1977

Appendix IV

ENERGY: DEFINITION, GLOSSARY, EXPLANATORY NOTES, COSTS, SUPPLY LEAD TIMES, UNITS AND CONVERSION FACTORS

DEFINITION OF ENERGY

Energy is defined as the ability of matter or radiation to do work.

GLOSSARY

This glossary is intended as a quick source of reference to a wide range of topics.

API:	American Petroleum Institute
ARAMCO:	Arabian American Oil Company
CIF:	Cost, insurance and freight
CMEA:	Council for Mutual Economic Assistance (also COMECON)
CPE:	Centrally Planned Economies
ECU:	European Currency Unit
EEC:	European Economic Community
FOB:	Free on board
GDP:	Gross Domestic Product
GNP:	Gross National Product
IEA:	International Energy Agency
LNG:	Liquefied Natural Gas
LPG:	Liquefied Petroleum Gas
LWR:	Light-Water Reactor
MBD:	Million Barrels per Day
MHD:	Magnetohydrodynamics
OAPEC:	Organization of Arab Petroleum Exporting Countries
OECD:	Organisation for Economic Co-operation and Development
OPEC:	Organization of the Petroleum Exporting Countries
SDR:	Special Drawing Right
TOE:	Metric Tons Oil Equivalent
UAE:	United Arab Emirates

USGS: United States Geological Survey
VLCC: Very Large Crude Carrier

API gravity: An arbitrary scale expressing the gravity or density of liquid petroleum. The scale is expressed in degrees API (American Petroleum Institute) and is related to specific gravity at standard conditions by the following formula: Deg. API = (141.5/sp gr)-131.5.

Baseload: A subdivision of the total demand profile for electricity. This load category refers to those plants which operate continuously, except for maintenance requirements, to satisfy demand.

Biomass resource: Includes wood, wood byproducts, crop residues, animal manure, and urban solid waste.

British thermal unit (Btu): The amount of heat required to raise the temperature of 1 pound of water $1°F$.

Coal slurry: A pulverised coal-liquid mixture transported by pipeline.

Cogeneration: The generation of both steam and electric energy in the same facility.

Combined cycle plant: A two-stage electricity generating plant. The first stage is composed of combustion turbines and the second stage is a waste heat-steam generator system that operates with the exhaust heat of the first stage.

Crude oil: A mixture of hydrocarbons that exists in the liquid phase in natural underground reservoirs and remains liquid at atmospheric pressure after passing through surface separating facilities. Statistically, crude oil reported at refineries, in pipelines, at pipeline terminals, and on leases may include lease condensates.

Crude runs: Quantity of crude oil and petroleum liquids processed through a refinery's crude oil distillation units.

Developmental well: A well drilled within the known or proved productive area of a reservoir, as indicated by reasonable interpretation of data, with the objective of obtaining oil or gas from that reservoir.

Distillate fuel oil: A light fuel oil distilled off during the refining process. Included are products known as No. 1 and No. 2 heating oils, diesel fuels, and No. 4 fuel oil. These products are used primarily for space heating, on- and off-highway diesel engine fuel (including railway engine fuel), and electric power generation.

Dry hole: A well that does not yield oil or gas in commercially marketable quantities.

Economic rent: The difference between the marginal cost and the price of a depletable resource. Rent provides the incentive for the resource owner to produce today rather than postponing investment and production in anticipation of higher prices.

Elasticity: The rate of change in the quantity demanded of a good divided by the rate of change in an economic variable, such as price or income. An elasticity can be used to estimate the impact of a change in an economic variable on the quantity demanded. For example, a price elasticity of -0.2 indicates a 10 per cent increase in prices will result in a 2 per cent decrease in demand.

o   <u>Price elasticity</u> The economic variable is price.

o   <u>Income elasticity</u> The economic variable is income.

Appendix IV

- o **Short term elasticity** An elasticity, usually a price elasticity, reflecting the change in demand for a good that occurs over a time span so short as not to allow changes in capital stock.

- o **Long term elasticity** An elasticity, usually a price elasticity, reflecting the changes in demand for a good occurring over a time span long enough to allow for adjustments in capital stocks. Usually, long term elasticities are larger in absolute value than short term elasticities.

- o **Feedback elasticity** A long term elasticity that has been estimated by a process reflecting the impacts on economic growth.

- o **System elasticity** A price elasticity derived from an alternative equilibrium energy model that allows all energy prices to change, with all other exogenous input variables, such as income, remaining unchanged.

End-use demand: Energy consumption measured at the final consuming sectors -- residential, commercial, industrial, agricultural and transportation -- consisting of marketed fuels.

Energy balance: An account of the quantities of energy supplied and consumed during a specified time period.

Energy converters: Industries that convert fuels from one form into another more usable form, such as refineries and electric utilities.

Energy losses: The difference between primary energy supply and final end-use demand. Losses result from conversion processes (such as electricity generation).

Enhanced gas recovery (EGR): Increased recovery of natural gas from a reservoir through the external application of physical or chemical processes. An example of an EGR process is hydraulic fractioning.

Enhanced oil recovery (EOR): The recovery of oil from a petroleum reservoir resulting from application of a recovery process beyond secondary oil recovery. Examples of enhanced oil recovery processes are steam injection, chemical flooding, miscible flooding and thermal recovery.

Enrichment: A process whereby the percentage of a given uranium isotope (^{235}U) present in a material is artificially increased to a higher percentage of that isotope naturally found in the material.

Exploratory well: A well drilled to find oil or gas in an unproved area; to find a new reservoir in a field known to contain productive oil or gas reservoirs; or to extend the limit of a known oil or gas reservoir.

Extraction loss: The loss of energy when natural gas is processed to remove some of its constituents. These constituents include the natural gas plant liquids such as ethane, propane, butane, natural gasoline, and undesirable gases such as hydrogen sulphide.

Feedstock: A raw material in production. For example, petroleum distillates used for producing petrochemicals are referred to as petrochemical feedstocks.

Fissile material: Capable of being fissioned (split into several parts) by neutrons, resulting in the release of energy. The only naturally occurring fissile material is ^{235}U, an isotope of uranium with an atomic mass of 235.

Fluidised-bed combustion boiler: A furnace design in which the fuel is buoyed up by air. It offers advantages in the removal of sulphur during combustion.

Fossil fuel: Any naturally occurring fuel such as coal, oil, and natural gas, derived from the remains of ancient plants and animals. These sometimes are called conventional fuels or conventional energy sources (as compared with nuclear power, solar, and wind energy) because they provide the bulk of today's energy for most of the world's industrial economies.

Front end of the fuel cycle: Those activities involving the preparation of nuclear fuel, encompassing the range from exploration for natural uranium to the fabrication of nuclear fuel assemblies.

Fuel cell: A device that produces electrical energy directly from the controlled electro-chemical oxidation of the fuel. It does not contain an intermediate heat cycle, as do most other electrical generation techniques.

Fusion: The combining of atomic nuclei of very light elements by high-speed collision to form new and heavier elements, the result being the release of energy.

Gas flaring: The burning of excess or undeliverable gases. Natural gas produced in conjunction with crude oil is sometimes flared at the well-head, though under the influence of rising real prices the practice should become less common.

Gasohol: A mixture of gasoline and alcohol. Ratios may vary but typically it is 90 per cent gasoline and 10 per cent alcohol.

Geothermal energy: Energy from the internal heat of the earth, which may be residual heat, friction heat, or a result of radioactive decay. The heat is found in rocks and fluids at various depths and can be extracted by drilling and/or pumping.

Heat pump: A mechanically driven device that uses a refrigeration cycle to raise a low-grade heat source to a higher temperature. (The heat pump may also provide air cooling, dehumidifying, circulating and air cleaning).

Heavy crude oil: Crude oil containing a weighted average gravity of 20.0 degrees API or less, corrected to $60°F$.

Heavy fuel oil: A liquid product produced in refining crude oil that is used as fuel, rather than as asphalt for road building or tar for roofing. (See Residual fuel oil).

High-Btu gas: High-Btu gas is predominantly methane and has a heat content greater than 800 Btu per cubic foot. High-Btu gas can be produced from coal through chemical reactions (coal gasification). Natural gas, a high-Btu gas, has a heat content in the range of 900-1100 Btu per cubic foot.

Hubbert Factor: A method used to estimate the growth of ultimate recovery over time from known oil and gas fields. Estimation method is based on average growth curves for those fields, derived from historical changes in estimates of ultimate oil and gas recovery with time since the known fields were discovered.

Hydropower: Electricity generation using water flow to drive a turbine.

Implicit GNP deflator: A measure of the change in price levels, which is the ratio of the current value of goods and services to the base-year value for the same goods and services for a specific country or area.

Appendix IV

International Energy Agency (IEA): A 21-member body composed of Austria, Australia, Belgium, Canada, Denmark, West Germany, Greece, Ireland, Italy, Japan, Luxembourg, the Netherlands, New Zealand, Norway, Portugal, Spain, Sweden, Switzerland, Turkey, the United Kingdom, and the United States. The Agency's purpose is to carry out a comprehensive programme of energy co-operation among its members, and to promote co-operative relations with oil producing nations and other oil consuming countries.

Lease condensate: Natural gas liquids recovered from wells (including those associated with crude oil reservoirs) in lease separators or field facilities. Lease condensates consist primarily of pentanes and heavier hydrocarbons and are comingled with crude oil in shipment to refining facilities. In this analysis, production of crude oil is defined to include lease condensate with crude oil for refining.

Light oil: Natural gas liquids and all light oil products, including gasoline, distillates, and jet fuel.

Light-water reactor (LWR): A nuclear reactor in which water is the primary coolant-moderator, with slightly enriched uranium fuel.

Lignite: A brownish-black coal in which the alternation of vegetal materials has proceeded further than peat, but not as far as sub-bituminous coal. The heat value of lignite is below 8300 Btu per pound.

Liquefied natural gas (LNG): Natural gas that has been cooled to about $-160°C$ for storage or shipment as a liquid in high pressure cryogenic containers.

Liquefied petroleum gas (LPG): A gas containing certain specific hydrocarbons that are gaseous under normal atmospheric conditions, but can be liquefied under moderate pressure at normal temperatures. The principal examples of LPG are propane and butane.

Load factor: The ratio of average electricity demand to the highest, or "peak", demand.

Low-Btu gas: A fuel gas with a heat content in the range of 100-250 Btu per cubic foot. A gaseous fuel produced from coal or other material.

Magnetohydrodynamics (MHD): An advanced power generation system that operates by forcing a hot ionised gas through a magnetic field to induce an electric voltage. This system is analogous to a typical generator that passes a conductor through a magnetic field. Direct current is produced and therefore must be passed through an inverter. The exhaust gas is used in a conventional steam-turbine cycle.

Market-clearing price: The estimated price of a commodity at which its demand equals its supply.

Medium-Btu gas: Gas with a heat content of 300-750 Btu per cubic foot. A gaseous fuel produced from coal or biomass that can be used in boilers or direct heat applications.

Metallurgical coal: Coal used to produce metallurgical coke, a primary input in steel production.

Metallurgical coke: A porous, carbonaceous material produced from coal and used in the steel industry.

National Energy Act of 1978: A package of five bills affecting the US energy markets. The five acts are:

o   The National Energy Conservation Policy Act
o   The Powerplant and Industrial Fuel Use Act
o   The Public Utilities Regulatory Policy Act
o   The Natural Gas Policy Act
o   The Energy Tax Act

Natural gas liquids: Those portions of reservoir gas that are liquefied at the surface in lease separators, field facilities, or gas-processing plants - natural gas plant liquids (NGPL). Includes ethanes, propanes, butanes, pentanes and natural gasoline.

Natural gas production, dry: The natural gas remaining after the natural gas liquids have been removed. It represents the amount of domestic natural gas production that is available to be marketed and consumed.

Nominal prices: Those prices actually observed in the market-place at any point in time. Nominal prices are sometimes referred to as market prices.

Non-associated natural gas: Natural gas not in contact with crude oil in the reservoir.

Nuclear fuel cycle: The term for all stages of nuclear fuel processing from uranium exploration through disposition of radioactive waste disposal.

Nuclear fuel reprocessing: The chemical separation of spent (used) nuclear fuel into salvageable fuel material and radioactive waste.

Oil shale: A range of shale materials containing organic matter (kerogen) that can be converted into crude shale oil, gas, and carbonaceous residue by destructive distillation.

Organisation for Economic Co-operation and Development: A 24-member body composed of the United States, Canada, Japan, Australia, New Zealand and the Western European countries comprising: West Germany, France, United Kingdom, Italy, Austria, Belgium, Denmark, Finland, Greece, Iceland, Eire, Luxembourg, Netherlands, Norway, Portugal, Spain, Sweden, Switzerland and Turkey. The Organisation's purpose is to promote mutual economic development, and contribute to the development of the world economy.

Organization of Arab Petroleum Exporting Countries: A group of Arab exporting countries seven members of which are also members of OPEC. The list of members of OAPEC is: Algerian Democratic People's Republic, State of Bahrain, Republic of Iraq, State of Kuwait, Socialist People's Libyan Arab Jamahiriya, State of Qatar, Kingdom of Saudi Arabia, Syrian Arab Republic and United Arab Emirates.

Organization of the Petroleum Exporting Countries: A group of oil exporting countries consisting of Algeria, Ecuador, Gabon, Indonesia, Iran, Iraq, Kuwait, SP Libya AJ, Nigeria, Qatar, Saudi Arabia, United Arab Emirates and Venezuela. The member countries meet from time to time to discuss the level of prices for oil exports. They do not act in concert to control the volume of oil production and/or exports. Thus OPEC is not a cartel in the normal sense of that word.

Passive solar heating: Systems that use heat flows, evaporation, or other natural processes to collect and transfer heat. (South-facing windows and greenhouses are two examples).

Peakload: A subdivision of the total demand profile for electricity. This load category requires intermittent operation of plants designed to respond to the highest levels of demand.

Petrochemical: Any chemical derived from petroleum or natural gas, such as polyethylene.

Photovoltaics: Devices that directly generate electrical current when exposed to sunlight. They are constructed of semiconductor materials that react to light or heat energy by allowing electrons to be accelerated across a junction.

Present value: A measure of today's worth of a future income stream, discounted at a given interest rate.

Pressurised-water reactor: A light-water reactor design in which water in the nuclear fuel core is pressurised to prevent boiling. Heat is transferred from the core by circulating the pressurised water to a steam generator, in order to fuel a turbine for electricity generation.

Primary oil production: Crude oil production from a reservoir where the flow of oil into the well is due to natural pressure in the reservoir.

Product mix (refined): Combination of products resulting from the refinery process. Also used in relation to the pattern of final demand for products.

Proved reserves: (See Reserves).

Pyrolysis: A process for conversion of coal and other materials that applies heat in the absence of oxygen. The products are coke, liquids and gases.

Rankine cycle system: The theoretical cycle that describes the conversion of heat energy to work and uses vapour as the working medium. It is the cycle employed in the typical steam-turbine generating plant and may be considered an external combustion engine cycle.

Real disposable income: The figure, which is expressed in money such as dollars of constant value within the National Income Accounting framework, is obtained by subtracting corporate earnings not paid out as dividends, depreciation, or taxes from gross national product. Transfer payments and Government interest payments are then added. This figure is intended to represent the income that the public has available for making purchases.

Real prices: Nominal prices adjusted for the effects of inflation on the purchasing power of a currency, often the dollar. These prices are always referenced by a particular year, such as real 1979 dollars, and sometimes are referred to as constant prices.

Real oil price: The nominal price of oil deflated by an index of prices such as US GNP, OECD consumer or export prices, or OPEC import costs. These different indices may vary considerably.

Refinery utilization rate: The percent of total crude oil throughput capacity at which a refinery is operated.

Renewable resources: Sources of energy not subject to exhaustion, such as solar, hydro and wind.

Replacement cost price: The cost of energy material to replace the last unit used.

Reserves: Identified deposits of minerals known to be recoverable using current technology and under present economic conditions. Categories of reserves are:

- <u>Extensions</u> Reserves credited to a reservoir because of enlargement of its proved area, generally due to additional drilling activity.

- <u>Indicated reserves</u> Reserves that include additional recoveries in known reservoirs (in excess of the measured reserves), which engineering knowledge and judgement indicate will be economically available by application of fluid injection, whether or not such a programme is currently installed (API, 1974).

- <u>Inferred reserves</u> Reserves based on broad geological research for which quantitative measurements are not available. Such reserves are estimated to be recoverable in future years as a result of extensions, revisions and additional drilling in known fields.

- <u>Measured reserves</u> (or <u>proved reserves</u>) Identified sources from which an energy commodity can be economically extracted with existing technology, and whose location, quality, and quantity are known on the basis of geological evidence supported by engineering evidence.

- <u>Revisions</u> Changes in earlier proved reserve estimates, either upward or downward, resulting from new information, not necessarily from additional drilling.

Residual fuel oil: Topped crude oil obtained in refinery operations, including oils used for generation of heat and/or power.

Resources: Concentration of economically valuable materials occurring in or on the Earth's crust in forms that make economic extraction potentially possible. Categories of resources are:

- <u>Identified resources</u> Specific bodies of materials whose location, quality, and quantity are known from geological evidence supported by exploratory probes into the deposits. This category of resources is frequently subdivided based on the estimated cost of recovery or the certainty of supporting evidence of existence of the deposits. (See Reserves).

- <u>Undiscovered resources</u> Unspecified bodies of materials surmised to exist on the basis of broad geological knowledge and theory, but which have not been identified by drilling. (Through exploration of such resources, they are moved into reserves subject to satisfying appropriate technological and economic criteria). In classifying uranium resources, this category is further subdivided into the following categories:

    - <u>Probable resources</u> Uranium estimated to occur in known productive areas, which are either extensions of known deposits or in undiscovered deposits or in undiscovered deposits within known geological trends or areas of mineralization.

    - <u>Possible resources</u> Uranium estimated to occur in undiscovered or partly defined deposits in formations or geological settings that are productive elsewhere within the same geological province or sub-province.

    - <u>Speculative resources</u> Uranium estimated to occur in undiscovered or partly defined deposits in formations or geological settings not previously productive.

Royalty: Payment to the owner of mineral rights by the producer, in compensation for the extraction of the mineral.

Scenario: Specification of assumptions in forecasting pertaining to states of nature (e.g. size of reserve base), economics (e.g. gross national product) and Government policy (e.g. price controls), used in making projections.

Scrubber: Equipment used to remove sulphur from flue gas emissions.

Secondary oil production: A method of recovery in which part of the energy employed to move hydrocarbons through the reservoir into the production wells is obtained by injecting liquids or gases into the reservoir.

Shale oil: A liquid similar to conventional crude oil but obtained by processing an organic mineral (kerogen) in oil shale.

Softness (in markets): Refers to an economic market situation characterized by excess supply.

Spot market: Sales available for immediate delivery, not generally recurring under fixed-term contracts.

Spot prices: The price of a commodity (such as crude oil and products) applying to immediate delivery, distinguished from future delivery under a long term contract.

Standard deviation: A measure of dispersion in a frequency distribution. It equals the square root of the mean of the squared deviations from the arithmetic mean of the distribution.

Sub-bituminous coal: Coal with a heat content of 7500 to 10,000 Btu per pound.

Syncrude: The liquid hydrocarbons produced from organic deposits, such as shale, tar sands, and coal.

Syngas: A High-Btu gas resulting from the manufacture, conversion, or reforming of petroleum hydrocarbons or coal. Syngas may be easily substituted for, or interchanged with, pipeline-quality natural gas. (See High-Btu gas).

Synthetic Natural Gas (SNG): Gas manufactured from coal, petroleum, or biomass. SNG from naphtha is the most common today. (See High-Btu gas).

Tar sands: Consolidated or unconsolidated rocks with interstices containing bitumen that ranges from very viscous to solid. In its natural state, tar sands cannot be recovered through primary methods of petroleum production.

Tertiary recovery: Enhanced recovery of crude oil from a reservoir, through the external application of heat or chemical processes that supplement naturally occurring or simple-fluid injection processes. (See Enhanced oil recovery).

Thermal recovery process: An enhanced oil recovery technique using injection of steam into a petroleum reservoir (steam drive), or propagation of a combustion zone (in-situ combustion) through a reservoir by air injection into the reservoir.

Tight formations: Sandstone deposits containing natural gas. (Prospective reservoirs generally have low porosities and permeabilities not amenable to conventional completion techniques).

$U_3 O_8$: Uranium oxide, or yellowcake, is the international standard for the form in which uranium concentrate is marketed. Conversion and enrichment of $U_3 O_8$ results in fuel for the light-water reactor.

Uranium milling: The process of crushing, grinding and chemically treating uranium ore to remove the uranium oxide.

Waterflooding: Pressured water injected into reservoirs to provide energy to drive the oil and gas into producing wells, a secondary recovery method.

Well-head: The point at which oil or natural gas is transferred from the well to pipeline or other non-well facility. This term is used in the USA to refer to "well-head price", which is the price producers of oil and natural gas receive.

Yellowcake: An oxide of uranium used to make nuclear fuel rods after further refinement. Approximately 520 tons of ore must be milled to obtain 1 ton of yellowcake. (See U_3O_8).

Source: Adapted from 1979 and 1980 Annual Reports to US Congress, US Department of Energy, Energy Information Administration. A variety of other authoritative sources for a few specific items.

EXPLANATORY NOTES

These explanatory notes are intended as a source of reference to energy in its various forms.

Importance of Energy for Global Welfare

Energy is essential for global welfare. Because all human life depends upon energy, the world could not do without it. Denied a sufficient supply of energy, any country would progressively revert to a primitive state.

Anyone who doubts the importance of energy should try to imagine what conditions would be like without it, particularly in an industrialized country. With no energy, an industrialized nation's economy would come to a standstill. There would be no production of raw materials, no industrial activity, no manufacturing, and no commercial enterprise. If none of the primary sources of energy were available, it would be impossible to generate electricity. And the manifold needs for electricity could not be accommodated.

Because there would be virtually no agricultural activity without energy, very little food could be produced. And the food could not be cooked even if it were available. Lacking energy, homes and all other buildings could not be heated - or cooled.

Without energy any nation would be virtually defenceless against aggression on the part of another country mobilizing abundant energy resources. Any defence system would be rendered largely ineffective if its operations were limited by a lack of fuel, or if necessary support activities were interrupted for the same reason.

Although a total lack of energy is not a realistic prospect, there is an actual and growing potential for an inadequate supply. And a lasting shortage or even a temporarily interrupted supply can have a devastating impact upon a nation's economy, its standard of living, and its defence posture.

Source: The report Outlook for Energy in the United States to 1985, The Chase Manhattan Bank, Energy Economics Division, June 1972.

Appendix IV 351

Historical Evolution of Energy Sources

In the beginning, man had to depend solely upon his own muscles to provide the energy to satisfy his limited wants and needs. Later, he discovered new sources of energy to do his work for him. He learned how to use the energy of the wind and falling water. And he also learned how to harness the energy of animals. But man's most rewarding discovery was the use of fire. For it has been the many uses of the energy released by burning fuel that have done the most to enrich man's life and raise his standard of living.

Man first used fire to keep himself warm. And today the use of fire for heating continues to rank among the most important of all uses. At an early date, man also began to use fire for cooking food. And that too continues to be among the most essential uses. Fire was also man's earliest source of light.

By applying the heat of fire to water man learned how to produce steam, thereby setting the stage for the invention of the piston steam engine and later the turbine steam engine. With these inventions a vast new world opened up for man. His capability for production was increased enormously. For the first time it became possible to manufacture at lower cost a wide variety of merchandise in large factories with machinery powered by steam engines instead of producing a limited quantity of expensive goods by hand in small shops or in the home. And the products of the factories could be transported much farther and faster with steam-powered locomotives and ships. Later, man learned how to produce electricity. And steam engines enabled him to generate electricity on a large scale, thereby further expanding his productivity and well-being.

Still another vitally important use of fire began with the invention and development of the internal combustion engine. Once more man's productive capacity was greatly enlarged. And his ability to transport himself and the goods he produced was revolutionized. Today, automobiles, trucks, buses, aircraft, and railroad locomotives are virtually all powered with internal combustion engines.

Most agricultural machinery and mobile construction equipment is also powered with such engines.

At first, primitive man used wood for burning and for a long time thereafter it continued to be an important source of primary energy. But, with the growing availability of other sources, the use of wood declined.

The ability of coal to support combustion was discovered early in history and eventually coal became the world's principal source of energy. As recently as 1949, coal was still the single most important source of primary energy utilized in the USA.

Oil was discovered in Pennsylvania in 1859 and for many years thereafter it was used primarily as a fuel for lamps. But with the invention and development of the automobile, the need for oil began to expand rapidly. Because it is a liquid, oil is more versatile than other forms of primary energy. And, because of that characteristic, many uses for oil rapidly developed in industry, commerce, agriculture, and the home. By 1950, oil had displaced coal as the leading source of energy.

Originally, natural gas was a by-product of the search for oil. It was found associated with oil and also in reservoirs by itself. At first it was considered of little value and for a long time only the gas associated with oil was produced. Part of that production was utilized locally and the rest was simply flared to the atmosphere. But, after World War II, the true value of natural gas as a primary source of energy and as a feedstock for fertilizers and petrochemicals was recognized and there started vigorous efforts to develop widespread markets for it.

Falling water was another of man's earliest sources of energy and it continues to be used extensively. At one time the energy created by the movement of water was applied directly to operate machinery of various kinds. But today, it is used almost exclusively for generating electricity (hydroelectricity).

Nuclear power is a comparatively new source of energy. The practical application of nuclear power to peaceful purposes became a reality only within the past twenty years. With minor exceptions, the direct use of nuclear energy is limited to the generation of electricity.

New sources of energy which are currently under various stages of research, development and use and have the potential of making a growing contribution to the global energy supply in the long term future include oil from shale, tar sands, heavy oils, synthetic oil and gas, geothermal and solar, biomass, nuclear fast (breeder) reactors and nuclear fusion, wave power and wind power and hydrogen.

Source: Adapted from the report, Outlook for Energy in the United States to 1985, Chase Manhattan Bank, June 1972.

Sources of Energy

The conventional primary sources of energy include the fossil fuels - crude oil and natural gas liquids (NGL), natural gas and coal (including peat) and non-fossils - water power and nuclear energy. Electricity is a secondary source because a primary source necessarily must be utilized to produce it. Non-conventional sources of energy include other fossil fuels like oil from shale, tar sands and heavy oils and non-fossil fuel forms like biomass, geothermal, solar, fast-breeder nuclear reactors, nuclear fusion, wave and wind power and hydrogen. All fossil fuel energy resources, being finite and exhaustible, are called non-renewable. Hydroelectricity, solar, geothermal and hydrogen being inexhaustible resources are examples of renewable energy resources.

Thus sources of energy may be classified according to whether they are primary or secondary; conventional or non-conventional; and renewable or non-renewable. Time, place or even policy or lack of it can influence the appropriate categorization of a particular energy form. For example, wood was formerly classified as a conventional solid fuel form of energy, along with hard coal, lignite and peat. But in recent years there has been a growing tendency to classify wood as a renewable source of energy. Intrinsically this is justified. Nevertheless, active government policy and/or commercial enterprise initiatives have to be taken to ensure the replacement of trees cut for energy uses or other purposes, otherwise soil erosion and other disastrous consequences can follow. In this sense wood is not a naturally renewable energy resource of a similar qualitative, quantitative or immediately replaced character as some other renewable energy forms described below. Although tree plantations for commercial energy are mentioned under biomass in a following section, until these plantations start to be developed on a commercially significant scale worldwide I have included some provision for wood in Table 37, solid fuels, rather than in Table 39 which shows the actual and projected balance for renewable energy forms.

Source: Author

Resource Base and Reserves of Fossil Fuels

The total quantity of a fossil fuel originally contained in the earth's crust is commonly referred to as the resource base. At any point in time that portion of the resource base which has been positively identified through exploration, or "discovered", can be calculated, and that portion of the discovered resources which can be recovered economically can be estimated. After accounting for that portion of the resources which have already been extracted -- cumulative production -- the result is an estimate of remaining recoverable reserves of the fossil fuel. Reserves are depleted through production and may be increased through new discoveries or improvements in recovery efficiencies. Since the nature of geological occurrence and therefore of the exploration process differs substantially among the fossil fuels, and since the technology of recovery differs also, it is quite difficult to make consistent comparisons of reserves or resources for the different fossil fuels.

Source: Energy working paper - Group of Seven Revision No. 1 Nov. 1975, prepared by Saudi Arabia (co-ordinator), Iran, Algeria and Zaire.

Petroleum

The word petroleum is derived from the Latin petra (rock) and oleum (oil), and by modern definition includes compounds of hydrogen and carbon (hydrocarbons) found in the ground in various forms ranging from solid bitumen to normal liquids and to gases.

The origin of petroleum has been the subject of many theories in the past and it cannot be said that the problem has yet been solved to the complete satisfaction of all the theorists. However, it is generally believed that it is derived from organic material, such as marine animal organisms and plants which have been buried in the earth by the deposition of sediments.

Crude oils vary considerably in viscosity, sulphur content, waxiness and other characteristics from country to country and from field to field. In colour they range from brownish-yellow to black, some are viscous, others are limpid, while a few carry paraffin wax in suspension. For example, Mexican crude is black and viscous, whereas the Saharan and Pennsylvanian crudes are usually yellowish-brown in colour and are of lower viscosity. Middle East crudes are intermediate between these two types. However, whatever their appearance and origin, crude oils consist almost entirely of compounds of carbon and hydrogen, with varying small amounts of organic sulphur, nitrogen and oxygen compounds, and ash.

The hydrocarbons present in crude oil are of three types, paraffinic, naphthenic, and aromatic. Others known as 'unsaturated' or 'olefinic' are absent but are manufactured in the refinery by processes known as cracking or reforming.

Source: Adapted from The Petroleum Industry by George Sell, F. Inst. Pet., Oxford University Press.

It is a popular misconception that oil occurs in vast underground pools or lakes that, once discovered, can be pumped dry with relative ease. Nothing could be further from the truth. Oil is found trapped in the small spaces or pores between individual rock grains, rather like water in a sponge. Over time oil and water seep through porous rocks until impervious rock is reached. The porous rock is capped by an impervious layer of rock or cap rock. If this cap rock is in the form of a dome, then fluids, mainly water but often including oil, gradually accumulate under it. Oil gradually separates from the water and finds its way to the top of the structure, accumulating in the porous rock under the cap rock and above the layer of water in the porous rock. This is the oil field or reservoir.

Drilling a well into an oil-bearing structure releases the natural pressure in the reservoir, forcing oil into the well. Oil produced by this natural pressure is known as primary production. The portion of oil in the reservoir that can be produced in this way varies from field to field. It depends on such factors as the porosity of the rock and viscosity of the oil. Faults in the rock structure may also affect primary production. It is difficult to state a global recovery average for primary production, but in the USA for example, it yields an average of about 25% of the oil in place.

The amount of oil recovered can be enhanced by pumping water or gas into the reservoir to increase or maintain the reservoir pressure. Such techniques are known as secondary recovery, and their success varies. Some fields respond well because of their physical characteristics. Others do not. In the USA secondary recovery is mainly responsible for increasing recovery from 25% of the oil in place in the 1940s to about 32% by 1975. In some Middle East countries, where large reserves of natural gas are available, injection of gas into oil fields to improve recovery has already begun.

A further method of improving recovery rates from oil fields is to lower the viscosity of the oil so that it flows more easily through the pores of the rock. This can be done by heating the oil for example, by injecting steam, or by injecting chemicals to dilute the oil. This is called tertiary recovery. It is not widely used yet since the technology is costly and not well developed. In tertiary recovery, energy is injected into the oil field in the form of heat or chemicals. This results in a reduction in the net incremental energy recovered in the form of oil by the amount used in the heat or chemicals.

Proven reserves of oil are usually defined as oil that is recoverable from known reserves with today's technology and prices. Therefore, in addition to primary recovery, proven reserves include potential production based on the use of secondary and tertiary techniques -- where such techniques have been evaluated and are expected to be used in the fields.

Ultimately recoverable reserves are an estimate of how much oil will eventually be produced. They usually include new discoveries, plus an allowance for enhanced recovery as secondary recovery becomes more widely used, and as tertiary recovery techniques are developed.

Source: Adapted from Energy: Global Prospects 1985-2000, WAES Report, 1977.

Natural gas liquids (NGL) are by-products of natural gas production. They generally include hydrocarbons that can be extracted in liquid form from natural gas when it is produced. They are generally blended with crude oil and its products to produce butane, propane and natural gasoline and contribute to oil supply rather than to natural gas supply.

Source: Energy: Global Prospects 1985-2000, WAES Report, 1977.

Natural Gas

There are many types of gas below ground. Some come to the surface of their own accord as sulphurous vapours during volcanic eruptions or, in a less violent manner, bubble up from swampy areas as marsh gas or emanate from "mineral" springs and are easily detected by their odour reminiscent of rotten eggs. These are all "natural gases". But the natural gas of commerce is a gas which contains valuable quantities of methane, a colourless, odourless hydrocarbon compound (a combination of hydrogen and carbon). Methane is not the only gas found in nature, but it is the chief component of what we familiarly call "natural gas". Other gases often found associated with methane are ethane, propane, butane, hydrogen sulphide, carbon dioxide, and nitrogen. Their presence (or absence) can affect the value or the applications of the particular natural gas mixture.

Natural gas does not occur in great open spaces deep down in the earth. It is contained in porous rocks which contain millions of tiny spaces, such as between the grains of sand in a sandstone. If the layer of sandstone is thick, and if it is covered by a layer of rock through which the gas cannot escape, it is possible for billions of cubic feet of natural gas to accumulate over a long period.

It is believed that the origins of petroleum and those of natural gas may be closely related because these two resources frequently occur together in rock reservoirs. Actually, petroleum deposits always contain some natural gas, but the reverse does not follow. There are accumulations of natural gas without petroleum.

Source: Adapted from Natural Gas: Energy Perspectives, Number 2. Sept. 1973, Battelle Memorial Inst.

Non-Conventional Oil & Gas

These include synthetic oil and gas from coal and oil production from oil shale, tar sands and heavy oil. For the last three, recovery methods include both mining and in-situ methods.

Oil shale is a sedimentary deposit rich in organic matter, which yields oil when heated in the absence of air. Practically all sediments form oil upon distillation but a sediment must produce 15 gallons of oil per ton to be classified as oil shale. Commercial grades of oil shale give 25 - 50 gallons per ton and an extreme high yield of 150 gallons has been achieved. The largest known reserves are located in the United States.

There are porous rocks (e.g. sandstones, sands, fractured limestones) at the surface that contain asphalt filling the pore spaces between the grains. The grains in some of these rocks may be cemented together, (sandstones) whereas in others (sands) the asphalt may be the only binding agent. The rocks are stained black from the entrained asphalt and frequently have the distinct odour of asphalt. They are sometimes crushed for use in road surfacing. The largest known deposits are the Athabasca tar sands in Northern Alberta in Canada.

Source: Author and adapted from Encyclopaedia Britannica.

Coal

Coal is a family name and as such is a broad generalization. The term alone simply designates a family of rocks. What modern technology can do with coal depends, to a large extent, on which "rank" of coal we are talking about since all coals are not alike.

To understand the differences between various types of coal is to appreciate the history of this unique material. The beginnings of coal were plants - trees, bushes, grasses, and the accumulation of this material in a swamp or a bog, under certain weather conditions, gradually becomes a soggy mass of plant debris which we call peat. The first step in a long chain of events had now taken place - plants become peat.

The great peat swamps of North America 300 million years ago (Carboniferous system in geological time) extended over enormous areas along a wide coastal plain. From time to time, changes in sea level took place, and on occasion, the sea invaded the swamps. Salt water killed the plants and the peat accumulations were buried beneath clay and sand. The weight of the overlying sediment compressed the peat, compacted it, and changed its colour and appearance. The second step had been taken: peat became lignite (the lowest rank of the coal family).

Successive invasions of the sea and the piling on of layer upon layer of sedimentary material resulted in the deep burial of the lignite deposits. With the passage of time, new changes occurred. Deep burial often results in a rise in temperature, and some of the original "swamp" gas contained in the peat, and still retained through the change to lignite, was expelled. Much of the trapped moisture was also squeezed out, and the third stage in the metamorphosis of plant debris was completed. The lignite became bituminous coal.

In some areas and under special circumstances, still another step occurred. The layers of coal, together with the underlying and overlying strata were subjected to further compressive forces, as great slabs of the earth's crust moved and pushed against each other. The layers wrinkled into great folds. This wrinkling of the rocks produced high temperatures, and the coal, thus heated and compressed, changed once again; this time the resulting product was anthracite.

The material we call coal, therefore, is sometimes a lignite, sometimes a bituminous coal, and sometimes an anthracite, or an intermediate stage, as for example, partway between lignite and bituminous coal, a material called "sub-bituminous" coal. These classifications of coal by rank are based upon such properties as the percentage of fixed carbon calculated on a dry, mineral-matter-free basis, or upon calorific value on a moist mineral-free basis.

The chemical composition of coal varies widely because coal is not a mineral of fixed composition. It is not homogeneous; its chemical composition depends largely on its origin, and on changes caused by temperature and pressure as the formation of coal proceeded from the original accumulated plant debris.

Carbon and hydrogen are the main constituents, with small but important quantities of oxygen, nitrogen, and sulphur. Many investigators have demonstrated the presence of various chemical groupings, such as methoxyl and carboxyl groups, carbonyl groups, hydroxyl and non-reactive oxygen, and others. But the apparently most direct measurement of all, the molecular weight of coal, has eluded all investigators. Molecular weights varying from 400 to 2500 have been determined, but these are evidently minimum values. Since coal is a polymer, the true size of the coal molecule may be very large indeed.

Source: Adapted from Coal. What is That? Energy Perspectives, No.1 Battelle Memorial Institute, Aug. 1973.

Hydroelectricity

Producing electricity by utilizing the motive power of water is an important source of energy. In 1972 hydroelectricity capacity represented about 24 per cent of the world's electrical capacity.

There are many problems associated with the development of hydroelectric power. Frequently, the potential sources of hydroelectric power are very remote from areas of high demand. The size of projects is usually quite large and not necessarily compatible with the growth in electricity demand in the region. Because most hydroelectric developments are such massive undertakings, their capital costs are high and difficult for developing countries to finance.

Source: General Analysis of the Energy Situation, Working Paper submitted to the CIEC Energy Commission by the Saudi Delegation, 1976.

Nuclear Energy

Nuclear plants are similar to fossil fuel plants in that both systems generate steam to turn turbines connected to generators that produce electricity. The difference is that nuclear plants use uranium and plutonium. Natural uranium fuel is made up of two types of atoms. The majority (known as U-238) do not readily undergo fission, but a small proportion (known as U-235) are capable of fission when hit by a neutron in a reactor.

A nuclear energy plant -- a reactor -- is based on controlling a chain reaction in fissile material. "U-235" is "diluted" with other material (U-238), but the all-important fissile nuclei (U-235) are the nuclear fuel. When a neutron is absorbed into a U-235 nucleus, the nucleus splits, with the release of a tremendous amount of energy. This is ultimately converted into heat and also ejects 2 - 3 neutrons. Three things may happen to these neutrons: (1) some may go out through the surface of the uranium mass and get lost; (2) some are absorbed by a nucleus of U-238, which then becomes plutonium (Pu-239) which is a fissile material like U-235; or (3) some will be absorbed into other U-235 nuclei. At least one neutron must enter a U-235 nucleus to keep the chain reaction going. The U-235 nucleus splits, releases heat and ejects 2 - 3 neutrons.

The "desired reproduction factor of 1" means that from each U-235 nucleus which splits at least one neutron is absorbed into another U-235 nucleus to produce fission. When a reactor has a reproduction factor of 1, the system is "critical" and the chain reaction, which produces the heat, continues.

The system depends on a large enough quanitity of uranium, the critical mass, to contain enough nuclei to maintain a chain reaction. Naturally-occurring uranium (an element found in natural deposits) contains only 0.7% fissile U-235 nuclei. Enrichment increases this concentration to about 3%. This low concentration is adequate to sustain a chain reaction in a reactor provided the neutrons from fission are slowed down by a "moderator" - a material with light nuclei, such as ordinary water, heavy water (D_2O) or carbon in the form of graphite.

A nuclear reactor, then, consists of fissile material (U-235), usually in the form of rods, in association with moderators. Adjustment of control rods which are high neutron absorbers in the reactor "core" where the reaction takes place keeps the system "critical" and allows shutdown when the control rods are pushed into the core to absorb neutrons, thus stopping the reaction.

Some understanding of this fission phenomenon is needed to appreciate the reasons for the differences in the fuel cycles in the Light Water Reactor and the Heavy Water Reactor. Natural uranium must be enriched from 0.7% U-235 to about 3.0% U-235 to make fuel for LWRs. Present enrichment plants are very expensive and use a lot of energy. New enrichment plants based on the certrifuge process require less energy. The HWR uses natural uranium as fuel but requires a large initial stock of heavy water (D_2O) and small amounts during operation to make up for losses. Plants to make heavy water are also very expensive and use a lot of energy.

The Light Water Reactor - LWR. There are two types of LWRs - the Pressurized Water Reactor (PWR) and the Boiling Water Reactor (BWR). They use the same fuel cycle. The fuel cycle for the Light Water Reactor (LWR) begins with uranium ore, a natural resource that must be discovered, mined and processed to form a concentrate U_3O_8 which is called yellowcake.

The yellowcake is then converted to a gas (uranium hexafluoride) which goes to an enrichment plant where the concentration of U-235 is increased from 0.7% in natural uranium to about 3.0% for LWR fuel. Uranium tails, which are uranium depleted into 0.2 or 0.3% U-235, are a by-product of the enrichment processes and are saved for use in Fast (Breeder) Reactors. Fabrication follows enrichment and includes converting UF_6 to uranium dioxide (UO_2) which is the final fuel, pelletising, encapsulating in tubes, and assembling the uranium dioxide tubes into fuel elements. Finally, the fuel elements are loaded into reactors, which use the heat from nuclear fission to produce steam to drive turbines and generators to produce electricity.

The steps described above are known as the "front end" of the nuclear fuel cycle. The "back end" of the cycle begins when the used fuel elements are removed from the reactor. According to a predetermined schedule, 20 - 30% of the fuel is removed each year. Used fuel is highly radioactive and produces some heat because of the slow decay of the radioactive fission products. The used fuel must be stored under water at the reactor site for many months to ensure removal of the heat, and to shield against the radiation emitted by the fission products.

The used fuel elements can then be transported in specially shielded containers to a reprocessing plant. Here they are mechanically chopped up, dissolved in acid, and go through a chemical process to separate out three components. (1) The remaining uranium; (2) plutonium; and (3) the radioactive fission products. Such reprocessing plants are complex, because the materials to be recovered are mixed with highly radioactive fission products. Although these eventually reach the point where they can be handled without shielding, many operations must be performed by remote control by operators protected by thick materials against hazards of radioactivity. Such plants have been operating for up to 30 years to recover Pu for weapons, so the technology for remote handling has been well developed.

At this stage, there are a number of possible and important choices. The recovered uranium can be converted to uranium hexafluoride as feed for an enrichment plant or it can be stored. The uranium in reprocessed fuel from LWRs is normally richer in U-235 (0.8%) than in natural uranium. The plutonium can be recycled into the LWR for use with uranium as reactor fuel; it can be stored; or it can be used as fuel for the Fast Breeder Reactor. The radioactive waste must be stored in permanent and 'leakproof' facilities. Another alternative is not to reprocess the used fuel elements but to put them directly into permanent storage repositories. This is a single-use system.

Plutonium does not occur in nature. It is a transuranic element which is a by-product of all nuclear reactor operations where U-238 is present and absorbs neutrons with the result that it is transmuted into plutonium. If recovered plutonium and uranium are recycled into reactors as fuel, they could replace about 20% of the fresh uranium required for fueling the LWR. Requirements for natural uranium could be reduced proportionately.

The Heavy Water Reactor - HWR (CANDU-type). The Heavy Water Reactor fuel cycle differs substantially from that of the LWR. HWR nuclear power plants using natural uranium as fuel and heavy water as the moderator have been developed and have been operating in Canada in the four CANDU units at Toronto (2.0 GW capacity) since 1972.

In the HWR fuel cycle, the uranium fuel is used only once. The used fuel is now stored without plans for reprocessing (single use).

The HWR cycle uses natural uranium as the fuel, thereby eliminating the conversion to uranium hexafluoride the enrichment step, and the creation of a stockpile of depleted uranium tails (0.2 - 0.3% U-235). In Canada today, used fuel elements are put into storage where they await future decisions as to whether they will; (1) go to reprocessing operations to recover uranium and plutonium and disposal of radioactive wastes; or (2) be placed in permanent storage repositories without reprocessing.

The Fast (Breeder) Reactor - FBR. The LWR and HWR are thermal reactors because they operate with thermal neutron energies. Heat is generated by energy released during nuclear fission. This heat is carried away by a coolant and used to raise steam. What then is a breeder reactor and how does it differ from a thermal reactor?

The Fast (Breeder) Reactor is so named because it operates with fast neutrons (energies 10^7 x thermal) and can be used to "breed" more fuel than it consumes. It also operates as a power plant to generate steam. It provides a way to use most of the 99% of natural uranium that is the non-fissionable U-238. The breeder converts uranium (U-238) by absorption of neutrons to produce plutonium - which is fissionable. Thus, by extracting energy from most of the uranium (U-238) that is unused by present reactors, fast breeders offer the possibility of multiplying (perhaps 50 times) the usability of natural uranium resources. The initial fuel loading of an FBR requires large amounts of plutonium which must be obtained by reprocessing fuel elements from thermal reactors.

The High Temperature Reactor - HTR. Successful development of the High Temperature Reactor could create new uses for nuclear energy because such reactor systems are expected to deliver heat at high temperatures for such industrial purposes as coal gasification and hydrogen production. Large-scale HTR prototypes have been built in Germany and in the USA. However, problems encountered in commercial scale units and the accompanying fuel cycle have not yet been resolved.

Some Special Characteristics of the Nuclear Safety Debate. Development of nuclear power is being carried forward in an atmosphere of concern about the radioactivity which is the main difference between nuclear and other forms of energy. In spite of the positive safety record of reactor operation up to this time in terms of human life, the nuclear industry has been under intense pressure to prove the safety of its reactors and the safety of every step in the fuel cycle.

The question of safety has been debated publicly and often in terms similar to those which have dominated discussions of atomic weapons and the threats that they represent to human survival.

There are several serious problems around which debate has turned. Probably the most serious is that of containment of radioactivity. The safe transport, storage and treatment of spent fuel elements and the resulting highly radioactive wastes which remain active for hundreds and thousands of years is a matter of great public concern. Further research and demonstration is required before the most acceptable processing methods and the most suitable sites for waste disposal can be selected.

Still another concern is the safety of nuclear power plants themselves, particularly the consequences of a failure which might lead to a release of radioactivity. Although it may become widely accepted that the probability of such a failure is very low, there is a fear of the unpredictable and possibly widespread effect of such an accident. The proposal to make prototype fast breeder reactors which produce heat in a small space has heightened this concern over reactor safety, despite the required introduction of special safety features in the design to provide for the possibilities of melting down of the core. The accident at Three Mile Island in the USA in March 1979 served to exacerbate public concern about nuclear reactors.

Many people are concerned about which materials are usable for nuclear weapons. The explanation given here may help the reader. Uranium for power reactor fuel at 2% to 4% enrichment in U-235 is useless for weapons. Similarly, U-238 mixed with natural uranium to the level needed for reactor fuel (below 15% U-238) is useless for weapons. Such fuel grade uranium can only be made into weapon grade materials in an enrichment plant. Uranium enriched significantly above a U-235 content of 20%, or U-238 made from thorium, or plutonium all these are all usable in nuclear weapons.

Regarding plutonium, a conventional chemical process can be used for obtaining materials from spent fuel elements for use in weapons, although it requires relatively complicated equipment. No means for rendering plutonium useless for weapons is known.

Nations that made nuclear weapons have first used nuclear reactors aiming only at producing plutonium. Other countries could, without unsurmountable difficulties, obtain the same result in the near future, without resorting to nuclear power generating technology. Most nuclear power reactors can also partially use plutonium in the form of oxide as fuel, and large fast breeder reactors (FBRs) will require several tons of plutonium (or highly enriched U-235). There is public concern about this part of the nuclear fuel cycle.

Another area of concern is the "front end" of the nuclear fuel cycle where the fissionable fraction (Uranium 235) in natural uranium can be concentrated; nuclear power reactors require 2% to 4% U-235 concentration as fuel in the form of uranium oxide. Enrichment plants using the same technology can be used for producing the highly enriched U-235 for use in weapons and have been doing so for the past twenty-five years. It should be noted that this latter method for obtaining weapon grade material is more difficult than the plutonium route mentioned above.

The public debate of such issues is widespread and goes on in many countries at different levels of intensity. There are those who argue that there should be a cessation of such nuclear activities as additions to fuel reprocessing facilities and the development of the fast breeder reactor. There are others who point out that, in spite of the extensive experience and small-scale experiment to date, these issues can only be resolved by continuing development and demonstration. The attitudes of the general public on these questions are beginning to be reflected in the political process. In some countries, political action has prevented or delayed the siting of nuclear plants. On the other hand, referenda in seven states in the USA in 1976 all gave a large majority of support for a qualified continuation of nuclear development. More recently, in Sweden, a similar outcome narrowly favoured completion of existing nuclear plants. However, in Austria a completed nuclear reactor has been idle for several years since completion because of a narrow referendum decision against its operation. The first announcements, by the new French government in 1981 seemed to promise a reduced role for nuclear in French energy policy in the longer term.

The increasing desire of many countries, including some in the developing world, for nuclear capacity to meet their energy needs has been accompanied by a rising international concern about the spread of the capability for making weapons based on plutonium and a move towards more effective controls, including the possibility of multilateral operation and control of fuel reprocessing plants and stricter international control of waste disposal.

Nuclear Fusion. In theory controlled nuclear fusion of deuterium and tritium holds a promise of furnishing an almost limitless amount of energy. The engineering problems are formidable in using a heat source at 50 million degrees C to make steam to drive turbines. The intense neutron flux in the fusion reaction makes everything nearby intensely radioactive. Energy from fusion promises solutions to the safety, environmental and fuel problems associated with fission nuclear power. However, this technology is still very much in the scientific research stage and commercial development is not expected before the year 2000. If a large and growing R & D effort is continued for several decades, fusion power might be added to the global energy supply in the 21st Century.

Source: Adapted from Energy: Global Prospects 1985-2000, WAES Report, 1977.

Geothermal Energy

High temperatures, high enough to melt rocks, can be found close to the surface of the earth. This tremendous generation of heat, once thought to be left over from an ancient molten earth, is now believed to be produced by the slow decay of radioactive materials deep within the core of our planet. It has been going on for millions of years, and will continue for millions more. The result of this heat generation inside the earth is to make the temperature higher the greater the depth.

The distance from the surface to zones hot enough to be of value to us as sources of energy varies widely. In some areas, such as the vicinity of active volcanoes, the distance from the surface to hot rock can be a matter of only a few hundred feet, while in "quieter" portions of the earth, the rate at which the temperature increases with depth (called the geothermal gradient) is very gradual, and depths of several thousand feet or more can be reached with no more than a small increase in rock temperature. ("Normal" heat gradients range from $8°$ to $50°C$ per kilometer.)

Since harnessing the heat of a volcano is at best a risky business, the greatest advances in the use of geothermal energy to date have been made in areas where hot rock, although not exposed, is close to the surface. Here the presence of hot springs and geysers is a tell-tale sign that hot rock lies below. Yellowstone National Park in USA is one of the most famous examples of such a situation, but here the "industrialization" of the heat source is out of the question. Other hot spring regions have encouraged man to explore for and, in a few areas, actually capture natural steam.

The oldest commercial sized installation to generate electricity from natural steam is the Larderello development in Italy. Here natural steam was first used to operate a small generator (250 kilowatts) 60 years ago. The district now produces 390,000 kilowatts, and is currently the world's largest electric generating installation based on natural steam.

There are at least five other places in the world where geothermal electric power stations are now in operation: Iceland, Japan, New Zealand, the USSR, and the United States. Several more are under construction, and a complete list of nations in which geothermal potentials have been recognized and are under investigation would embody a significant portion of the UN membership.

In the United States, the use of natural steam to generate electricity is limited to one area in north-central California. The Geysers, located some 90 miles north-east of San Francisco. Steam from a large number of wells here is piped to a 302 MW generating station operated by the Pacific Gas & Electric Company. A number of buildings in Klamath Falls, Oregon, are heated by hot well water, and there are examples of hot spring utilization for space heating elsewhere in Oregon, in California and in Idaho.

The absence of naturally occurring hot water or steam does not mean that a given area lacks geothermal resources. It does indicate that insufficient quantities of ground water are available for circulation through the hot zone and from there to the surface. In some regions of known geothermal activity, prolonged periods of drought may result in diminished flow from hot springs. It is apparent that rain water, seeping through the ground, ultimately reaches the zone of hot rock, is heated, and then returns to the surface as a flow of hot water.

Source: Adapted from Heat From the Earth, Energy Perspectives No.3, Battelle Memorial Institute Oct. 1973.

Solar Energy

The earth receives most of its energy from the sun in the form of an electro-magnetic radiation which consists of 3% ultra-violet rays, 42% visible light and 55% infra-red rays. Everything derives its energy ultimately from the sun: either indirectly (e.g. oil, gas and coal) or directly (e.g. solar collector panels).

As opposed to fossil energies, the sun is inexhaustible on a human scale, although its mass decreases at the speed of about five million tons a second. The earth collects about a ten thousand millionth of the energy it radiates, which represents annually from five to ten times the entire known fossil power reserve (uranium included). Above the clouds, a one square meter flat horizontal surface collects an average of 1350 watts; at ground level, this power ranges from 0 to 1100 watts. We thus have at our disposal a clean, practically non-polluting energy which is free of charge once the investments for collecting and storing equipments have been made. However, efficient extraction of this abundant resource is difficult. Solar energy is a very dilute source, and it is available only at intermittent intervals. In addition, solar climates vary greatly. In tropical latitudes sunshine is abundant and demand for space heating is small, while in more temperate latitudes winter days are short and overcast skies reduce the solar radiation reaching the collector. The need is to overcome these obstacles by finding ways to reduce the collector area required to capture this dispersed energy and by developing efficient systems to store collected solar heat for use at night or during overcast periods.

The best known application of solar energy is for hot water heating. Improved technology, lower manufacturing costs, and strong marketing efforts -- enhanced by public policy -- may lead to substantial entry of solar hot water systems into markets in areas where sunshine is strong during much of the year.

Systems for the solar heating and cooling of buildings will probably develop more slowly than solar hot water systems. For efficient use of solar heating, buildings should be properly oriented, well insulated, and equipped with supplementary heating and energy storage. Currently, small production levels for solar units usually result in high costs. When produced in greater volume and with more experience, capital costs should be substantially reduced, and maintenance and reliability requirements will be better understood.

Programmes for applying modern technology to the collection of solar energy and for storage systems which allow economical recovery and use are now being started. Such collectors may be passive, such as heat-absorbing walls or roofs of buildings, or active, as when a moving fluid collects the solar energy and transfers the heat to warm the building and stores the heat to meet heating demands at night or during heavily overcast days. An active system involves collectors which should be efficient, have long life with low maintenance, and be aesthetically acceptable because very large collector areas will be needed. Much progress is being made in developing collectors and other parts of a solar heating system including pumps, controls and efficient means for storage.

There are many possible technologies available for the conversion of solar energy to electricity. Schemes are under consideration for producing electricity both on a decentralized scale and on a centralized scale (central station power plants).

Solar energy can be converted directly into electricity in solar cells made from special materials such as silicon crystals. Such cells have provided electricity in spacecraft.

Two basic kinds of technology for central station conversion of solar energy to electricity are under consideration; thermal and photovoltaic (ground and space).

One method involves concentrating sunlight on a boiler to superheat a fluid which then drives a turbine generator. Because of cost considerations, these designs may have to be relatively large to reduce the cost of power; 100 megawatts of electricity (MWe) and above.

Engineering feasbility studies indicated that "power tower" designs may be preferable. These use a field of individually guided mirrors (heliostats) to focus sunlight on a boiler at the top of a large tower. When generating electricity, these towers use only direct sunlight and therefore are restricted to areas of low cloud cover. High capital costs, intermittent operation dependent on daylight, coupled with uncertainties in maintenance and useful life, are major obstacles.

Studies are now being made of large central station photovoltaic systems. Implementation will have to await further breakthroughs in the cost of manufacturing solar cells.

Another proposal calls for satellites, placed in synchronous earth orbit, to convert solar energy into microwave energy to be transmitted to earth for conversion to electricity. This requires no new technologies, and avoids the problems of the earth's atmosphere and cloud cover. However, the satellites would have to be enormous - with solar panels in space more than 6 miles square - to produce 5,000 megawatts at the ground receiving station. The costs for transporting the required materials into orbit, and for building the system are very large.

Sources: Adapted from Solar Energy from France, Delegation Aux Energies Nouvelles, Ministre De L'Industrie et de la Recherche, Paris and from WAES Report, 1977.

Solar Energy from the Seas. Three schemes proposed for the extraction of some form of "renewable" energy from the sea are briefly mentioned below. The capital requirements are large. The possibility of such sources making any significant contribution to meeting energy needs in this century appears to be very small.

Ocean thermal systems would use the small temperature differences (about $40°F$) between the sun-heated upper ocean layers and the colder, deeper waters to generate electricity. The temperature difference is largest in the tropics. While there appear to be economic possibilities for such schemes, the required systems are large and expensive, and must operate in a very unfavourable and corrosive environment.

Other systems harness the energy in ocean waves. Although the total amount of energy stored in waves is much smaller than that in some of the other forms of solar energy, it might provide modest amounts of energy in certain countries (like Great Britain) where much of the research in this field is currently being done.

Although the tides result from earth-moon, rather than earth-sun interactions, they are possible "renewable" sources for power generation. Two existing modern tidal power systems - one 240 MW(e) scheme at La Rance in France and a smaller one at Kislaya Guba in the USSR - have demonstrated that tidal power can be economically harnessed. However, since there are only a few major estuaries in the world which offer prospects for moderately large tidal projects, power from tides is expected to be small and localized.

Source: WAES Report, 1977

Biomass (or Bioconversion). This is based on the concept of extracting solar energy from organic matter. According to one estimate, 17 times more energy is stored in plant matter (by photosynthesis) than the world now consumes.

Biomass involves the use of agricultural, municipal and forestry wastes both as fuel for direct combustion and as raw materials for synthesis into chemicals or energy products such as methanol and ammonia. It also involves growing plants for fuel.

Bagasse from sugar cane has been used for many years as a fuel. Burning municipal wastes for energy or conversion into methane is practised in many parts of Europe. The same is beginning in the USA.

Growing plants for fuel is potentially a promising approach, especially in developing countries. Tree plantations and floating kelp farms are variants being studied. Land based energy plantations may compete with agriculture for limited arable land. Agricultural wastes in some cases may be sufficient in quantity and density to be used as a fuel, as bagasse from sugar cane has been used for many years.

Source: Adapted from WAES Report, 1977.

Wind power once provided a significant portion of the energy consumed in rural areas in some countries. Yet there are many possible ways of extracting useful energy from the wind. Research and development is now being devoted to many modern windmill technologies (in blade dynamics and engine control) - with some emphasis on devices with capacities of at least 1 MW(e). New concepts are focusing on the use of wind concentrators, diffusers, and vortex generators which increase ambient wind velocity and so decrease the size of the turbines, the most expensive element in the system. Even with presently visualized cost reductions, the materials required suggest that electrical generators must either be large-scale or the price of electricity must increase substantially, if wind power is to become a significant source of energy.

Even with large-scale wind driven generators, electricity output is small. It has been calculated that about 50,000 windmills with propeller diameters of 56 meters, and with average power of 500W, would be needed to produce energy equivalent to 1 thousand barrels a day oil equivalent.

Although the prospects for large-scale windmills are uncertain, local, small-scale uses could be important in many countries in the years ahead.

Source: Adapted from Energy: Global Prospects 1985-2000, WAES Report, 1977.

Hydrogen

According to certain expert estimates hydrogen has the highest energy content, on a pound for pound basis, of any of the fuels. It is responsible for a large fraction of the energy produced by the combustion of nearly all liquid and gaseous fuels. It is a non-polluting fuel that can be produced from water or coal, both amply abundant. Hydrogen is also renewable insofar as the combustion process (oxidation) produces water once more. As such, many researchers feel that hydrogen constitutes the most promising long term successor to petroleum as a heating, motor and especially as a jet fuel.

One of hydrogen's most valuable roles in the medium term could be as an energy vector rather than a fuel in itself. Over long distances in particular, energy could be transported in the form of hydrogen produced with the use of off-peak electricity; gas-carrying pipelines are less costly to construct than electrical transmission lines, and often simpler to install. The hydrogen could then be burnt at point of consumption to produce peak-time electricity.

Hydrogen is essential for the production of certain chemicals such as ammonia and methanol. Natural gas used at present will soon become too expensive, and by the end of the century hydrogen could well be the prime feedstock for chemical production, especially for fertilizers. The anticipated shortfall in natural gas supply by the year 2000 could be met by hydrogen or SNG.

In the direct reduction of iron ore, hydrogen could replace coke. But like many potential uses, while this is techically feasible, it could not be economically viable in the foreseeable future, according to industry sources. Another such use would be the blending, in liquid form, of hydrogen with synthetic liquid fuels (synthetic gasoline or methanol). However, hydrogen is not easy to liberate from water, at least with present technology.

The two predominant commercial processes for producing hydrogen are electrolysis and the thermal chemical splitting of water. According to a recent IEA analysis of production methods, the first could be available for large scale operation by 1985, and the second by the end of the century. In the long term, plants using one or other method will be necessary additions to energy production, if only to provide hydrogen for the manufacture of synthetic fuels such as methane and methanol from coal.

Electrolysis is the method of hydrogen production most commonly used in industry. But as the net energy input for electrolysing water is, by definition, higher than for producing the electricity needed in the process, there is little point at present in using hydrogen as a source of primary energy where it is possible to use electricity. Research continues, meanwhile, into the best materials for electrodes and electrolytes as well as the optimum operating conditions. Ultimately, however, production of hydrogen by electrolysis of water will be most likely where electricity can be generated cheaply. If the French nuclear programme continues to expand, for example, and electricity capacity exceeds base-load demand, it would be economic to use nuclear-generated power to produce hydrogen used in the production of synthetic fuels from coal. This could happen as early as 1995.

Thermochemical processes split water with a combination of heat and catalytic reactions. The majority of such processes suffer from being costly, energy-inefficient and very complex, however, and the problems of using heat from a nuclear reactor to achieve the high temperatures for the reactions (direct splitting by heat alone would require an operating temperature of $2000^{\circ}C$) add to these drawbacks. A more practical use of nuclear power would be to generate electricity to electrolyse water. Coal, on the other hand, may be used to split water directly by a process of reduction (the water gas reaction) in countries with abundant coal resources.

Under serious consideration is the use of concentrated solar power for the high temperatures necessary to disassociate water or to generate electricity for electrolysis. Although the technology involved in solar photovoltaics is highly expensive at present, production of hydrogen in sufficiently large quantities to substitute for hydrocarbons may be most practical using this form, given the objections to the use of nuclear fission power. Indeed, supporters of the solar option claim that solar generated electricity will be cheaper than nuclear power by the end of the century, provided investment in commercial development of the necessary technology is made immediately.

Various agencies which are currently doing active research on hydrogen are as follows:

The US Department of Energy: Over $20 million per year R & D programme for production of hydrogen by electrolytic, thermochemical and solar power and possible uses in ammonia and alcohol production, replacement of natural gas, fuelling of cars and small-scale power plants.

DFVLR -. The West German research organization concerned with aerospace in conjunction with Stuttgart University and Los Alamos Scientific Laboratories in New Mexico: Research on cars that can run on gasoline or hydrogen.

The IEA: Study of various techniques for producing hydrogen.

The Commission of the European Communities.

A Project initiated by five countries: At a conference on liquid hydrogen as a kerosene replacement in Stuttgart in the autumn of 1979. Lockheed mooted the idea of a new cargo airliner run entirely on liquid hydrogen.

The most dramatic and technically appealing substitution is undoubtedly as a jet fuel. Once combustion techniques and the safety aspects (danger of explosion in particular) have been resolved, the major problem lies in the necessity of creating a vast infrastructure for the storage and handling of liquid hydrogen. Massive investment would be required from airport authorities to adapt and build new facilities. The authorities are not prepared to spend that money at present, and so any prospects for widespread use of hydrogen as a jet fuel have been pushed back to at least forty years hence.

Source: Adapted from Hydrogen's Long Term Potential by Francine Stock. Published in Petroleum Eocnomist, January 1981.

Gasohol

Gasohol is a mixture usually 10 per cent ethyl alcohol and 90 per cent conventional, lead-free gasoline. It has emerged in 1980 as a new energy option for the United States, having already gained acceptance in Brazil.

In January 1980, following shortly upon his embargo of US grain sales to the Soviets, President Carter announced a major commitment to corn-derived alcohol fuels. He targeted 500 million gallons/year by 1981, anticipating a rapid increase thereafter to two billion gallons, an amount ultimately comparable to the lost grain exports to the USSR. The target is regarded as feasible because there is surplus capacity in the distilling industry which can be easily adapted to produce fuel-grade alcohol. Moreover, construction lead times for grass-roots distilleries are very short - 15 to 18 months - compared with the 4-7 years so typical of other synfuels technologies.

Alcohol does provide offsetting energy savings. Its high blending octane value, adding two pump octane numbers to the 10/90 mix, permits less severe reforming which adds almost one third to the energy value of the alcohol in the refinery-vehicle system. Congressional advocates, corn-state representatives, and some gasohol marketers, claim significant improvements in auto fuel economy - from 7 to 15% - citing one of two sets of tests from MIT or the corn-growing state of Nebraska. Engines set too rich do exhibit a modest fuel economy improvement with any oxygenated fuel; but, otherwise, for properly tuned engines, the overwhelming body of laboratory and fleet test data indicates instead a fuel economy loss of some 2-4%, and the claims of improved mileage, while widely repeated in the media and Congress, remain quite unsubstantiated.

A recent study* which included the gamut of energy costs and gains concluded that under the best circumstances it costs about US$100-$125 to "save" one barrel of imported oil with gasohol. In many cases each gallon of corn-derived ethanol actually increases US oil imports, but possibly only by 0.1-0.25 gallons per gallon.

The impact upon the US energy balance will be negligible. Even doubling the goal to two billion gallons, equivalent on paper to 2% of US gasoline consumption, yields a saving, net of the oil and gas consumed in the manufacture of fertilizer, and in processing, of at most 50,000 barrels a day (against oil imports of around 7 million barrels a day) forecast for the early 1980s.

* Gasohol: The Costly Road to Autarky, by Thomas R. Stauffer, 9 August 1980.

Source: Adapted from Gasohol Option for Motor Transport by Dr. Thomas R. Stauffer, published in Petroleum Economist, March 1981.

ENERGY COSTS

The cost of energy from some current and prospective alternatives to oil are based on Shell estimates. Oil varies considerably in price and cost, according to quality and location; but, as an indication, Arabian Light crude oil in the Middle East had a Government sales price of $32 a barrel in June 1981 and $11.51 a barrel in 1976. The element of production cost in this price is around the low end of the $1 to $3 per barrel range shown in the table below. In most other locations, because of geographical, physical and other factors, the production cost element in the price of crude oil is considerably higher; for example, the production costs of most of the existing North Sea fields are above $5 a barrel.

Appendix IV

Comparative Energy Costs

1980 dollars per barrel of oil equivalent on a thermal basis*	Technical production cost
Middle East oil (existing fields)	1 - 3
North Sea oil (existing fields)	5 - 20
Liquids from oil sands/shale (N America)	15 - 35
Indigenous coal (US)	4 - 8
Imported coal (NW Europe)	10 - 14
Indigenous coal (NW Europe)	9 - 20
Nuclear input break-even value**	7 - 20
Liquefied natural gas imports, high Btu (Europe, Japan, US)	25 - 35
Synthetic natural gas (high Btu) from indigenous coal (US)	35 - 50
Liquids from imported coal (NW Europe)	45 - 65
Biomass (crops grown for fuel)	45 - 80+
Solar heat	120+
Electricity (based on conventional fossil fuel and nuclear generation)	60 - 130
Electricity (based on solar/wind/tidal)	120+***

* These estimates do not include refining, storage, transmission and distribution costs to final consumers.

** The fuel input cost required for fossil-fuelled plants to produce electricity at the same cost as nuclear stations.

***Possibly on-site 1990s.

Source: Energy in Profile, Shell Briefing Service, Shell International Petroleum Company Limited, London, December 1980. Prepared as an information brief for staff of the companies of the Royal Dutch/Shell Group.

The costs (not prices) shown above are all expressed in terms of a standard energy unit (one barrel of oil equivalent, equal to 5.8 million British thermal units), but their outputs vary, not only in form (heat or electricity), but also in terms of quality, location and availability. The figures above exclude refining, storage, transmission or distribution costs and various other factors that enter into the decision process on energy choice. The costs shown can also change considerably over time as a reflection of, for example, the differential inflation often affecting new complex technological processes or the potential learning curve effect involved with processes that are at present still in the research or pilot stage.

ENERGY SUPPLY LEAD TIMES

There has been a tendency for lead times to lengthen during recent years, especially for nuclear energy.

The long lead times for energy supply development limit the rate at which changes in the pattern of the energy balance can be brought about. Lead times are long in part because of the large scale and increasingly complex physical nature of major projects in more and more difficult operating environments. But they also reflect the difficulty of resolving public policy issues, especially in demonstrating that a balance between energy and environmental objectives can be achieved.

The paragraphs below give some indication of the lead times required to develop new supplies, though these estimates may now be rather optimistic. The measure here is the time required between leasing or other access to the resource-bearing acreage -- or between the decision to build as in the case of a nuclear plant -- and the time of major commercial-scale production. The variability in the lead time required is indicated by the ranges mentioned, and reflects physical, economic and policy variables. In the case of oil and gas, the lead times reflect successful exploration and subsequent development, but it should be recognized that the majority of exploration ventures do not result in discoveries.

Oil and Gas

As a benchmark, the minimum time required to find and develop new oil and gas reserves in the US Lower 48 states onshore, or to develop discovered reserves in the Middle East that have already undergone some development might be six months to two years. Deep gas production can require 6 - 10 years.

For areas such as the offshore US Gulf Coast where exploration takes place in new sectors, but where developments are relatively easily integrated with existing operations, new production may be commercially available in 3 - 4 years. In similar situations onshore, such as the southern portion of Canada's Alberta province, lead times range from 1 - 4 years.

Other oil and gas areas which are remote from existing systems will require 5 - 10 years for exploration and development. In this category are new areas in the Middle East - or others which are close to existing production but which involve technically difficult operating conditions - for example, Alaska and the North Sea.

For frontier areas, such as the outer continental shelf off the US East Coast or the harsher Arctic, where conditions are difficult and the total infrastructure has to be developed, production may take 6 - 12 years. The wide range reflects a particularly wide variability in the length of time for first discovery, for delineation of sufficient reserves to justify the very large capital outlays for field development, for obtaining environmental permits, and for installation of logistics support facilities including major pipelines.

Other Supplies

The lead time to develop a coal mine varies from 4 - 8 years. This depends on whether it is a deep mine or a surface mine, on the time required to resolve environmental and safety issues, and on whether the area is near established transportation facilities. To construct an associated power plant or other consuming equipment as well as the mining capability, the upper end of the range would apply.

Appendix IV

For synthetics, the lead times assumed are for construction of commercial plants once technology is in hand. This is at present the case only for very heavy oils, coal gasification and possibly shale oil. The lead time estimated for various synthetic processes ranges from 6 - 10 years, reflecting the time to resolve environmental issues and to build infrastructure and the necessary mining and processing facilities. If standardization were to be developed as experience is gained, the time might be reduced.

For nuclear power plants, lead times now range from 6 - 12 years, the longer time period is required in the US and several other countries where regulatory and environmental delays add several years.

Thus, since most major energy supply projects require from 5 - 12 years from inception to commercial production, it is clear that the maximum supply through to 1990 is already largely determined. Delays in commencing operations of new projects would reduce projections of supplies. The only substantial sources which can be developed in a relatively short time are already discovered oil reserves, such as some of the oil fields in the Middle East, or coal which can be surface-mined.

Source: Adapted from World Energy Outlook. Exxon Background Series, April 1978.

ENERGY UNITS AND CONVERSION FACTORS

Basic Unit of Energy (System International)

The international oil industry has tended to use barrels a day of oil, or metric tons per year of oil, or oil equivalent. These are the measures which have been used in this book. However, the basic SI unit for energy is called the Joule. It is likely to come into increasing use in the future.

When a force capable of giving a mass of one kilogram an acceleration of one metre per second moves a mass of one kilogram a distance of one meter, then the work done is one Joule. Another way of describing a Joule is as a power output of 1 watt during one second. In other words, an electric light bulb of 100 watts capacity burning for one hour is using an amount of energy equal to 360,000 Joules.

Since the Joule is a very small energy unit, a very large multiple of Joule, such as Mega (10^6) or Tera (10^{12}) or Exa (10^{18}), must be used when dealing with energy consumption in very large quantities.

All other energy units can be converted to a multiple of the Joule.

Conversion Factors and Energy Units

```
1 Mega Joule =    238.8458    kilo calories
             =    947.8169    British thermal units (Btu)
             =      0.27778   kilowatt hour (kwh)
```

Like many other measurable commodities, energy and fuels are measured by a seemingly innumerable variety of units. Attempts to achieve wide acceptance and use of a common convention - for example, the International System of Units (SI) - have met with some degree of success. But different people are accustomed to different units of measurement; changes in training, habit and usage in these matters come slowly.

Not only are different systems of units in use, but there is also little consensus on precisely how to convert from one system to another. Given the nearly infinite variety in the qualities and characteristics of energy and fuels in the world, perhaps this is to be expected.

In this book I have generally expressed the rate of energy use and supply and demand of primary energy in terms of millions of barrels per day of oil equivalent. This measure is based upon the conventional unit of a "barrel of oil equivalent" with a gross calorific value of about 5.8 million British thermal units (Btu).

The calorific value of the various types and sources of coal, crude oil and natural gas varies widely. The standard equivalents used in converting from one unit of measurement to another and shown in the table below are subject to this proviso.

To produce an electrical output of 620×10^9 kWh per year (1 million barrels a day oil equivalent) would need power stations of 100 GWe (= 100,000 megawatts) installed capacity, given an average load factor of 70%. If the average efficiency of the power stations was 35%, one would need a fuel input of 1/0.35 (= approximately 2.8) million barrels a day oil equivalent. The difference between the input of energy as fuel and the output of energy as electricity (in this example, 1.8 million barrels a day oil equivalent) is the transformation loss in electricity generation, which we include under the heading of 'processing losses', along with the losses which occur in all other energy conversion processes.

In our discussion of primary energy supply and demand, we have adopted the convention of expressing electricity from primary sources (nuclear, hydro, geothermal, etc.) in terms of the fuel input (generally expressed in million barrels a day oil equivalent) that would be required to produce the equivalent amount of electricity output in fossil-fueled power stations.

Standard Energy Conversions

Energy: A Global Outlook

Standard Energy
(Equivalent Values Lie in

Unit									
Barrels per Day of Oil Equivalent[1]	–	–	–	–	–	–	–	–	–
Tons of Oil Equivalent[2]	–	–	–	–	–	–	–	0.022	0.023
Metric Tons of Coal Equivalent[3]	–	–	–	–	–	–	–	0.034	0.036
Barrels of Oil Equivalent[4]	–	–	–	–	–	0.0064	0.02	0.16	0.17
Cubic Meters of "Average" Natural Gas[5]	–	–	–	0.027	0.09	1	2.7	25	27
Kilowatt hours	–	–	–	0.3	1	11	29	280	293
Cubic Feet of "Average" Natural Gas[5]	–	–	–	1	3.4	37.3	100	950	1000
Kilocalories	0.24	0.25	1	252	860	9400	25200	0.24 mill	0.25 mill
British Thermal Units	0.95	1	4.0	1000	3400	37300	1 therm	0.95 mill	1 mill
Kilojoules	1	1.06	4.2	1055	3600	39400	105500	1 mill	1.06 mill

Calorific values are measured gross. Rounded equivalents only are given.
[1] Equivalents in other units are shown on a per annum basis
[2] of 43 million Btu (approx = 10,000 kcal/kg net cal.val).
[3] of 12,000 Btu/lb = 7000 kcal/kg.
[4] of 5.8 million Btu.
[5] of 1000 Btu/ft^3 or 9400 kcal/m^3

Source: Adapted from: Energy: Global Prospects 1985 – 2000 WAES Report, 1977. Shell International Petroleum Company Limited.

Appendix IV

Conversions
Vertical Columns)

								Abbreviation
−	0.003	0.013	0.02	1	2.7	18000	470 mill	BDOE
0.09	0.13	0.65	1	50	135	0.9 mill	23 x 10^9	TOE
0.14	0.21	1	1.5	76	209	1.3 mill	36 x 10^9	MTCE
0.68	1	4.8	7.4	365	1 TBOE	6.4 mill	170 x 10^9	BOE
106	155	745	1150	57000	0.155 mill	1 x 10^9	27 x 10^{12}	Nm^3NG
1160	1700	8140	12600	0.62 mill	1.7 mill	11 x 10^9	290 x 10^{12}	kWh
4000	5800	27800	43000	2.1 mill	5.8 mill	37.3 x 10^9	1 x 10^{15}	ft^3NG
1 mill	1.5 mill	7 mill	10.8 mill	530 mill	1.5 x 10^9	9.4 x 10^{12}	250 x 10^{15}	kcal
4 mill	5.8 mill	27.3 mill	43 mill	2.1 x 10^9	5.8 x 10^9	37.3 x 10^{12}	1 Q	Btu
4.2 mill	6.1 mill	29.3 mill	45.4 mill	2.2 x 10^9	6.1 x 10^9	39.4 x 10^{12}	1.06 x 10^{18}	kJ

− = insignificant
1 therm = 100,000 Btu
1 TBOE = 1000 BOE
1 Q = 10^{18} Btu
1 Quad = 10^{15} Btu

Bibliography

ABOLFATHI Farid, and others, 'The OPEC Market to 1985', Lexington Books, 1977.
AKINS James E., 'World Energy Supply Cooperation with OPEC or a New War For Resources', a Speech at Green Seminar, Laval University, Quebec, Canada, 1979.
AL-CHALABI Dr. Fadhil J., 'OPEC and the International Oil Industry: A Changing Structure', Oxford University Press, 1980.
AL-CHALABI Dr. Fadhil J., 'Problems of World Energy Transition: A Producer's Point of View', at Seminar on 'Development Through Co-operation' in Rome, April 1981, under the joint sponsorship of OAPEC, Italy and South European countries.
ALGERIA, Memorandum submitted to The Conference of Sovereigns and Heads of State of OPEC Member Countries, Algiers, March 1975.
ALLEN Loring, 'OPEC Oil', University of Missouri, Oelgeschlager, Gunn and Hain, Publishing, Cambridge, Massachusetts, USA, 1979.
ALTENPOHL Dr. Dieter, 'The Hard Truth About Europe's Energy Future', Swiss Aluminium Ltd., Zurich, a paper presented at European Management Forum, Davos, January 1978.
AMERICAN PETROLEUM INSTITUTE
AOG RESEARCH DEPARTMENT, 'OAPEC Objectives, Organization and Joint Companies', studies and documents.
ARAB FUND FOR ECONOMIC AND SOCIAL DEVELOPMENT and OAPEC, 'Energy in the Arab World', Volumes 1-3. Proceedings of the First Arab Energy Conference, March 1979. OAPEC, 1980.
ATTIGA Dr. Ali Ahmed, (the Secretary-General of OAPEC) writing in 'The Voice of the Arab World', London, 11 August 1976.
BACHMAN W.A., 'Carter Plan Places Emphasis on Synthetics', Oil and Gas Journal, 23 July 1979.
BATTELLE MEMORIAL INSTITUTE, Energy Perspectives, 'Coal. What is That?', No.1, August 1973.
BATTELLE MEMORIAL INSTITUTE, Energy Perspectives, 'Natural Gas', No.2, September 1973.
BATTELLE MEMORIAL INSTITUTE, Energy Perspectives, 'Heat From The Earth', No.3, October 1973.
BAUM Vladimir, 'Introduction', a speech at UN Symposium, Vienna, 7-15 March 1978.
BENARD Andre, 'World Oil and Cold Reality', Harvard Business Review, November-December 1980.
BP, 'Our Industry', British Petroleum Company Limited.

Bibliography

BP, 'Statistical Review of the World Oil Industry' - Statistics covering years 1967 to 1979.
BP, 'Oil crisis ..... again?', a brief by the Policy Review Unit, BP, September 1979.
BRANDT Willy, Chairman, 'North-South: A Programme for Survival', The Report of the Independent Commission on International Development Issues, Pan Books, 1980.
BRITANNICA, Encyclopaedia.
BRUYNE D. de, President of the Royal Dutch Petroleum Company, 'Financing Problems in the Oil Industry' Opening Address to the 10th World Petroleum Congress, Bucharest, 9 September 1979.
BULLETIN MENSUEL, various issues, Comite Professionnel du Petrole, Paris.
CANADIAN PETROLEUM INSTITUTE
CARSON Rachel, 'Silent Spring', Harvey Miller, 1963.
CARTER President Jimmy, 'A National Energy Plan for the USA', Address, 20 April 1977.
CHASE MANHATTAN BANK, 'Capital Investments of the World Petroleum Industry 1978', Energy Economics Division.
CHASE MANHATTAN BANK, 'Financial Analysis of a Group of Petroleum Companies 1979', Energy Economics Division, December 1980.
CHASE MANHATTAN BANK, 'Outlook for Energy in the United States to 1985', Report by Energy Economics Division, June 1972.
CHINA, Official Statement of natural gas production in 1979, July 1980.
CHINA, PEOPLES REPUBLIC OF, Communique from State Statistical Bureau.
CIEC, The Conference on International Economic Co-operation, Report No: CCEI-CM-5, Rev.1, 2 June 1977.
CIEC, Report of The Energy Commission: Conclusions and Recommendations, 2 June 1977, Annex to CIEC Report No: CCEI-CM-5.
CLERON Jean Paul, 'Saudi Arabia 2000: A Strategy For Growth', Croom Helm, 1978.
DARMSTADTER Joel, with TEITELBAUM Perry D. and POLACH Jaroslav G., 'Energy in the World Economy', a statistical review of trends in output, trade and consumption since 1925. Published for Resources for the Future, Inc., The John Hopkins Press.
DEAGLE Edwin A., MOSSAVAR-RAHMANI Bijan, HUFF Richard, 'Energy in the 1980's: An Analysis of Recent Studies', Occasional Papers No.4, Group of Thirty, New York, 1981.
DEGOLYER AND MACNAUGHTON, 'Twentieth Century Petroleum Statistics 1980'.
DELL Sidney, LAWRENCE Roger, 'The Balance of Payments Adjustment Process in Developing Countries', Pergamon Press, 1980, published in co-operation with the United Nations.
DIEL R., RADTKE G., STOSSEL R., 'Investment Requirements in the Energy Sector and their Financing', Dresdner Bank, Federal Republic of Germany, 11th World Energy Conference, 8-12 September 1980, Munich.
EDEN Richard, and others, 'Energy Economics: Growth, Resources and Policies', Cambridge University Press, 1981.
EL MALLAKH Ragaei and Dorothea H., (editors) 'New Policy Imperatives for Energy Producers', a publication of The International Research Centre for Energy and Economic Development, University of Colorado, Boulder, Colorado, 1980.
EL MALLAKH Ragaei, 'The Organization of The Arab Petroleum Exporting Countries: Objectives and Potential', a publication of The International Research Centre For Energy and Economic Development, University of Colorado, Boulder, Colorado.
ENER DATA, 'Energy in Countries with Planned Economies', Berne, Switzerland.
EUROPEAN COMMUNITY, 'The European Community and the Energy Problem', January 1980.
EXXON, Background Series, 'World Energy Outlook', April 1978, December 1979, December 1980.
EXXON, Background Series, 'Middle East Oil' second edition, 1980.

FAHD IBN 'ABD AL-AZIZ, HRH Crown Prince, interview with the Beirut daily al-Bayrak published in Saudi Press Agency of 28 June 1981 and reported in English in Middle East Economic Survey Vol.XXIV No.38 dated 6 July 1981.

FRANKEL Paul H., 'The Essentials of Petroleum: A Key to Oil Economics', Frank Cass, London, 1969.

FRANKEL Paul H., 'The Rationale of National Oil Companies', Keynote Address at UN Symposium, Vienna, 7-15 March 1978.

GATT, 'International Trade 1979/80' and earlier issues, General Agreement on Tariffs and Trade, Geneva.

GEROLDE Steven (compiled and edited by), 'A Handbook of Universal Conversion Factors', The Petroleum Publishing Co., 1971.

GOLDMAN Marshall I., 'The Enigma of Soviet Petroleum', Allen and Unwin, 1980.

GROUP OF SEVEN, Energy Working Paper, Rev. 1, Prepared by Saudi Arabia (Co-ordinator), Iran, Algeria and Zaire.

GROUPE DE RECHERCHE EN ECONOMIE DE L'ENERGIE (GREEN), Energy: International Co-operation or Crisis, Laval University, Quebec, Canada, 1979.

HALLWOOD Paul and SINCLAIR Stuart, 'Oil, Debt and Development, OPEC in the Third World', George Allen and Unwin, 1981.

HARDESTY C. Howard Jr., 'The Role of Independents in the International Oil Industry', Speech at OPEC Seminar, Vienna, 10-12 October 1977.

HARTSHORN J.E., 'Oil Companies and Governments', An Account of the International Oil Industry and its Political Environment, Faber and Faber, London, 1962.

HELLER C.A., 'The Birth and Growth of the Public Sector and State Enterprises in The Petroleum Industry', a Speech at UN Symposium, Vienna, 7-15 March 1978.

HENDERSON P.D., 'India: The Energy Sector', A World Bank Publication, Oxford University Press, 1975.

HUBBERT M. King, 'Energy Resources', A Report to the Committee on Natural Resources, Publication No. 1000-D, National Academy of Sciences - National Research Council, 1962.

INDIA Government of, 'Report of the Fuel Policy Committee', 1975.

IEA, Communique of the International Energy Agency, Meeting of Governing Board at Ministerial Level, 22 May 1980.

IEA, 'Outlook for the Eighties', Summary of 1979 Review of Energy Policies and Programmes of IEA Countries.

IEA, 'Principles for Energy Policy', adopted by the Governing Board of the IEA, October 1977, Annex 1.

IEA/OECD Press Release, IEA/Press (80)8, Paris, 22 May 1980.

IEA/OECD, 'Workshop on Energy Data of Developing Countries', 2 Vols.

INSTITUTE FRANCAIS DE PETROLE, International Petroleum Seminar, March 1981.

INTERNATIONAL CURRENCY REVIEW, Vol. 12, No. 3, July 1980.

INTERNATIONAL HERALD TRIBUNE AND THE OIL DAILY, Conference on 'The Energy Emergency: Oil and Money', London, 1980.

INTERNATIONAL INSTITUTE FOR APPLIED SYSTEMS ANALYSIS, 'Energy in a Finite World':
    Executive Summary written by Alan McDonald
    Paths to a Sustainable Future, Program Leader, Wolf Hafele
    A Global Systems Analysis, Program Leader, Wolf Hafele.

INTERNATIONAL MONETARY FUND, 'Annual Report', 1980.

INTERNATIONAL MONETARY FUND, 'International Financial Statistics Yearbook', 1980.

INTERNATIONAL MONETARY FUND, 'International Financial Statistics', Monthly, various issues.

JAIDAH Ali, 'Problems and Prospects of State Petroleum Enterprises in OPEC Countries', a Keynote Speech at UN Sympsium, Vienna, 7-15 March, 1978.

JAPAN PETROLEUM AND ENERGY WEEKLY, 'In Search of Japan's Oil Strategy for the 1980s', Interim Report of Study Group on Basic Oil Industry Issues, Natural Resources and Energy Agency, Ministry of International Trade and Industry (MITI). Translated into English in Japan Petroleum and Energy Weekly, 13 May 1981 and subsequent issues.

JAPAN PETROLEUM AND ENERGY WEEKLY, 'Annual Five Year Oil Supply and Demand Plan for Fiscal Years 1981-1985 inclusive,' Ministry of International Trade and Industry (MITI). Translated into English in Japan Petroleum and Energy Weekly, 1 and 8 June 1981.

JENSEN James T., President, Jenson Associates Inc., Boston, Mass., Consulting Petroleum Economist, 'Why US Oil Policy is Such a Riddle', Petroleum Intelligence Weekly, 29 September 1980.

JOHANY Ali D., 'The Myth of The OPEC Cartel - The Role of Saudi Arabia', University of Petroleum and Minerals, Dhahran, Saudi Arabia and John Wiley and Sons, Chichester, USA, 1980.

JONES Aubrey, 'Oil, The Missed Opportunity', Andre Deutsch, 1981.

KELLY J.B., 'Arabia The Gulf and The West. A Critical View of The Arabs and Their Oil Policy', Weidenfeld and Nicolson, 1980.

KNAPELS Edward N., 'Oil Crisis Management Strategic Stockpiling for International Security', John Hopkins University Press, 1980.

KUBBAH Abdul, Amir, 'OPEC: Past and Present', Petro-Economic Research Centre, Vienna, 1974.

LANGENKAMP Robert D., (ed) 'The Illustrated Petroleum Reference Dictionary', Penn-Well Publishing Co., 1980.

LEBLOND Doris, 'European Energy Report', Petroleum Economist, January and March 1981.

LEEMAN Wayne A., 'The Price of Middle East Oil', Ithaca N.Y.: Cornell University Press, 1962.

LEVITT Theodore H., 'Marketing Myopia', Harvard Business Review, July-August 1960.

MABRO Robert, 'World Energy Issues and Policies', Proceedings of The First Oxford Energy Seminar, St. Catherine's College, Oxford, 1980.

MARUBENI Petroleum Reports, Marubeni Corporation, Japan, 1 October and 16 December 1979.

MIDDLE EAST ECONOMIC SURVEY, various issues.

MINISTRE DE L'INDUSTRIE ET DE LA RECHERCHE, 'Solar Energy from France', Delegation aux Energies Nouvelles, Paris.

MORGAN GUARANTY TRUST COMPANY OF NEW YORK, 'World Financial Markets', various issues.

NATIONAL ACADEMY OF SCIENCES, 'Energy in Transition 1985-2010', Final Report of the Committee on Nuclear and Alternative Energy Systems, 1980.

NAWWAB Ismail I., SPEERS Peter C., HOYE Paul F., (editors), 'Aramco And Its World, Arabia and the Middle East', Aramco, Dhahran, Saudi Arabia, 1980.

OAPEC, An Address by George Tomeh (Consultant for International Relations to OAPEC) to a Meeting of CAABU at The House of Commons, Westminster, 22 February 1977.

OAPEC, 'Basic Facts about the Organization of Arab Petroleum Exporting Countries', OAPEC Department of Information, Kuwait, 1976.

OAPEC, 'OAPEC Bulletin', various issues.

OAPEC, 'Reservoir Engineering, Its Role in Hydrocarbon Resources Development', OAPEC, Kuwait, 1979.

OAPEC, 'Seventh Annual Statistical Report 1978-1979', OAPEC, 1980.

OAPEC, 'The Ideal Utilization of Natural Gas in the Arab World', OAPEC, Kuwait, 1980.

OECD, Decision of The OECD Council Establishing An International Energy Agency of The Organisation, 15 November 1974.

OECD, 'OECD Economic Outlook', July 1981 and earlier issues.

OECD Report, 'Energy Prospects to 1985', February 1975.

OECD, 'Main Economic Indicators', monthly, various issues,

OECD/IEA, 'Energy Statistics 1974/1978', OECD, Paris, 1980, and earlier issues.

OECD/IEA, 'Quarterly Oil Statistics', OECD, 1981/No.1, and earlier issues.
OIL AND ENERGY TRENDS, Statistics.
OIL AND GAS JOURNAL, 'Worldwide Oil and Gas at a Glance', 31 December 1979 and 29 December 1980, and earlier issues.
OIL AND GAS JOURNAL, 'Synfuel from tar, oil sands, shale seen promising', 23 June 1980.
OIL AND GAS JOURNAL, 'Soviet Oil, Gas Exports Value Hit High', 30 July 1979.
OIL AND GAS JOURNAL, 'Senate Clears Massive Crude Excise Tax', 31 March 1980.
OIL AND GAS JOURNAL, 'Carter Signs $227 Billion Excise Tax Measures', 7 April 1980.
OIL AND GAS JOURNAL, 'Reagan: End Controls on Crude Products', 2 February 1981.
OPEC, 'Annual Statistical Bulletin 1979', OPEC, 1980, and earlier issues.
OPEC, 'Future Energy Markets', Seminar proceedings, October 1979, published by Macmillan Press Ltd., 1980.
OPEC, 'OPEC Official Resolutions and Press Releases 1960-1980', Pergamom Press, 1980, for OPEC.
OPEC, 'OPEC Papers Volume 1, No.1, Domestic Energy Requirements in OPEC Member Countries', August 1980.
OPEC, 'OPEC Papers, Volume 1, No.2, Energy in Developing Countries - Present and Future', 1980.
OPEC, 'Selected Documents of the International Petroleum Industry, Saudi Arabia --Pre-1966', OPEC, Vienna, 1976.
OPEC, Various Seminar Proceedings, Bulletins, Energy and Economic Reviews and Forums, Published by Public Relations Department, OPEC, Vienna.
O'SULLIVAN Edmund, 'Saudi Third Plan', Middle East Economic Digest, 7 November 1980.
OXFORD Dictionary.
PECHMAN Joseph A., Editor of The Brookings Institute, 'Setting National Priorities Agenda for 1980's'.
PENROSE Dr. Edith, 'Government Partnership in the Major Concessions of the Middle East: the Nature of the Problem', Middle East Economic Survey, 30 August 1968.
PETROLEUM ECONOMIST, 'EEC Commission urges Action on Energy', November 1978.
PETROLEUM ECONOMIST, 'EEC Commission Urges Action on Energy', November 1978.
PETROLEUM ECONOMIST, 'China, Oil Production Targets Lowered', October 1980.
PETROLEUM INTELLIGENCE WEEKLY, various issues.
PETROLEUM INTELLIGENCE WEEKLY, Statistics, 19 May, 16 June, 1980.
PETROLEUM INTELLIGENCE WEEKLY, 'Reagan Landslide Signals Shift in US Energy Focus', 10 November 1980.
PETROLEUM INTELLIGENCE WEEKLY, 'Slide in US Imports May Signal Start of a Lasting Decline', 19 January 1981.
PETROLEUM INTELLIGENCE WEEKLY, 'United States Explains Its New Energy Act', Special Supplement, 30 October 1978.
PETROMIN, 'Petromin Handbook 1382H-1397H, 1962 AD-1977 AD', The General Petroleum and Mineral Organization, 1977.
PLATT'S 'European Marketscan', McGraw-Hill, various issues.
PLATT'S 'Oilgram News', McGraw-Hill, various issues.
PLATT'S 'Oilgram Price Report', McGraw-Hill, various issues.
PLATT'S 'Oil Price Handbook and Oilmanac', 57th Edition, McGraw-Hill, 1981.
POCOCK C.C., 'The Role of The International Oil Company', a Speech at OPEC Seminar, Vienna, 10-12 October 1977.
RUSTOW Dankwart A. and MUGNO John F., 'OPEC Success and Prospects', Council on Foreign Relations Inc., New York University Press in USA and Martin Robertson, London in UK, 1976.
SAUDI ARABIA, KINGDOM OF, 'Petroleum Statistical Bulletin', Economics Department, The Ministry of Petroleum and Mineral Resources, 1979, and earlier issues.
SAUDI ARABIAN MONETARY AGENCY, 'Saudi Arabian Monetary Agency Annual Report 1400 (1980)', Saudi Arabian Printing Company Ltd., 1980.

SCOTT Professor Bruce, 'OPEC, The American Scapegoat', Harvard Business Review, January-February 1981.
SEGAL Jeffrey, 'Long-Term Oil Strategy Succeeding For Japan', Petroleum Economist, September 1979.
SELL George, F. Inst.PET., 'The Petroleum Industry', Oxford University Press, London, 1963.
SERVAN-SCHREIBER Jean-Jacques, 'The World Challenge', Simon and Schuster, 1981.
SEYMOUR Ian, 'OPEC: Instrument of Change', an OPEC publication, Macmillan Press, 1980.
SHELL INTERNATIONAL PETROLEUM COMPANY LIMITED, 'Energy in Profile', Shell Briefing Service, December 1980 and other Shell Briefing Service papers.
SHELL INTERNATIONAL PETROLEUM COMPANY LIMITED, 'Energy Conversion Equivalents'.
SIVARD Ruth Leger, 'World Energy Survey' second edition, 1981, published under the auspices of the Rockefeller Foundation.
STAUFFER Dr. Thomas R., 'Gasohol Option for Motor Transport', Petroleum Economist, March 1981.
STOBAUGH Robert and YERGIN Daniel (editors), 'Energy Future', Report of the Energy Project at The Harvard Business School, Random House, 1979, and paperback edition (updated) Ballantine Books, New York, 1980.
STOCK Francine, 'Hydrogen's Long Term Potential', Petroleum Economist, January 1981.
TAHER Abdulhady Hassan, 'Income Determination in the International Petroleum Industry', Pergamon, 1964, and Speeches at various International Seminars.
TASS Radio Broadcast, Estimates of Soviet Production of Oil and Gas in 1980, 23 January 1981.
UNITED NATIONS CENTRE FOR NATURAL RESOURCES, ENERGY AND TRANSPORT, 'State Petroleum Enterprises in Developing Countries', Pergamon Policy Studies, 1980.
UNITED NATIONS, Monthly Bulletin of Statistics, various issues.
UNITED NATIONS, 'World Economic Survey 1979-1980', UN, 1980.
UNITED NATIONS, 'Population Statistics and Forecasts, 1950-2000', Selected Demographic Indicators, 1980.
UNITED NATIONS, World Energy Supplies 1973-78 Series J. No.22. World Energy Supplies 1950-74 Series J. No.19.
US CENTRAL INTELLIGENCE AGENCY, 'The World Oil Market in the Years Ahead', CIA, August 1979.
US DEPARTMENT OF COMMERCE, 'Survey of Current Business', December 1980.
US DEPARTMENT OF ENERGY, '1980 Annual Report to Congress', Volume Three Forecasts, US Government Printing Office, March 1981.
US DEPARTMENT OF ENERGY, Monthly Energy Review, DOE/EIA, Various Issues.
US DEPARTMENT OF ENERGY, 'Securing America's Energy Future. The National Energy Policy Plan', July 1981.
US DEPARTMENT OF STATE, Selected Documents No.3, Bureau of Public Affairs, Office of Media Services.
US DEPARTMENT OF STATE, 'US International Energy Policy October 1973 - November 1975', Selected Documents No.3, Released December 1975.
US FEDERAL ENERGY ADMINISTRATION, 'National Energy Outlook', 1976.
US INFORMATION SERVICE, US Official Text of President Carter's Energy Program --A Factual Survey, 22 April 1977.
VAN MEURS and Associates Limited, 'World Energy Supply and Demand Analysis 1973-2025', Ottawa, Ontario, Canada.
VITTACHI Varindra Tarzie, Chief of the Information Division of the United Nations Fund for Population Activities, 'Is Altruism in Retreat?', Newsweek, 25 August 1980.
WHELAN John, 'Third Development Plan', Middle East Economic Digest, Special Report, July 1980.
WILSON Carroll L., 'Coal - Bridge to the Future', Ballinger Publishing Company 1980.

WILSON Carroll L. (Project Director), 'Energy: Global Prospects 1985 - 2000', Report on the Workshop on Alternative Energy Strategies (WAES), a project sponsored by the Massachusetts Institute of Technology, McGraw-Hill, May 1977.

WORLD BANK, Report on 'Energy and Petroleum in Non-OPEC Developing Countries, 1974-1980', Staff Working Paper No: 229, Washington DC., February 1976.

WORLD BANK, Report on 'Energy In The Developing Countries', August 1980.

WORLD BANK, 'World Development Report, 1980', August 1980.

WORLD BANK, 'World Bank Atlas', 1980 and earlier issues.

WORLD ENERGY CONFERENCE, Survey of Energy Resources, 1974 and 1980, Papers, 11th Conference, Munich 1980.

WORLD OIL, 'Resource Estimates and World Politics Don't Mix', by William Stannage, October 1979.

YAMANI HE Shaikh Ahmed Zaki, Minister of Petroleum and Mineral Resources, Kingdom of Saudi Arabia, 'Petroleum: A Look Into the Future'. Unofficial Translation of a lecture by HE Shaikh Yamani from Petroleum Intelligence Weekly, Special Supplement, 9 March 1981.

ZAKARIYA Hassan S., 'State Petroleum Enterprises: Some Aspects of Their Rationale, Legal Structure, Management and Jurisdiction', a Speech at UN Symposium, Vienna, 7-15 March 1978.

Index

Abu Dhabi 34, 150, 152
Afghanistan 163
Air pollution 124
Alaskan North Slope 48, 92, 96, 125, 135
Algeria 17, 34, 39, 152, 154
Alternative Energy Corporation 149
Alternative energy sources 21, 43, 51, 52, 60, 75, 79, 81, 82, 84, 88
  see also under specific sources
Anglo-Iranian Oil Company 11
Anglo-Persian Oil Company 4
Angola-Cabinda 162
Arab-Israeli war 41
  see also Israel
Arab League 11, 12, 33
Arab light marker crude 73, 107
Arab Maritime Petroleum Transport Company (AMPTC) 37, 60
Arab Oil Congress (Cairo 1959) 33
Arab oil policies 79
Arab Petroleum Congress (Cairo) 12
Arab Petroleum Investments Corporation (APICORP) 38
Arab Petroleum Services Company (APSC) 37, 38
Arab Shipbuilding and Repair Yard Company (ASRY) 37
Aramco (Arabian American Oil Company) 16, 22, 56-7, 89
Argentina 5, 160-3
Australia 149
Austria 5, 100, 139

Bahrain 34, 39, 161
Balance of payments 31, 107
Bangladesh 159, 161

Basrah Petroleum Company 16
Belgium 139, 142, 143
Bolivia 163
Botswana 161
BP (British Petroleum) 4, 11, 96
Brandt Commission 71, 87, 106, 109, 163
Brazil 160, 162, 163
Bretton Woods Agreement 19
British National Oil Corporation 18
Brunei 161, 162, 163
Bulgaria 176

Canada 6, 35, 97, 141, 149
Capital account flows 105, 107
Carter, President 123, 127, 128, 130-2
Chemical industry 52, 98
China 7, 98, 160, 180-3
  coal 180, 181
  confrontation scenario 183
  co-operation scenario 182
  energy consumption 183
  energy dilemma 180
  energy problem summarized 180, 181
  energy resources 180-1
  energy supply and demand 181
  neutral scenario 182
  oil consumption 182-3
  oil production 182-3
  policy options 181-2
Clean Air Act 1977 125
Coal 7-9, 42, 43, 48
  China 180-1
  Japan 147, 150, 151
  non-OPEC developing countries 163
  United States 124, 126, 130, 136

USSR 171, 172
Western Europe 139, 140, 142, 143
world 117, 118
Colombia 162
Comecon countries 112
Committe of Experts 33
Communist bloc 8, 101
   confrontation scenario 113, 122
   co-operation scenario 112-3, 121
   neutral scenario 113, 121
Concession agreement 19, 54
Conference on International Economic Co-operation (CIEC) 16, 46, 52, 68-72, 77, 154
   commissions of 69
   objectives 69
   overall assessment 71
   points of agreement and disagreement 70
   representation 68
   results of other commissions' work 70
   working sessions 69
Confrontation scenario 106
   China 183
   Communist bloc 113, 122
   Eastern Europe 178-9
   industrialized countries 112, 120-1
   Japan 151
   non-Communist countries 111, 119
   non-OPEC developing countries 168
   OPEC 157
   United States 137
   USSR 174
   Western Europe 146
   world 109-10, 118
Co-operation scenario 105
   China 182
   Communist bloc 112-3, 121
   Eastern Europe 178
   industrialized countries 111, 119-20
   Japan 150
   non-Communist countries 110, 118-9
   non-OPEC developing countries 165-7
   OPEC 155-6
   United States 135-6
   USSR 173-4
   Western Europe 144-5
   world 108, 116-7
Council for Mutual Economic Assistance (CMEA) 170-2
Crude Oil Equalization Tax (COET) 130
Crude oil reserves 8, 39
Czechoslovakia 177

Denmark 139, 140, 142, 143
Developed Countries. See Industrialized Countries
Developing countries 12, 16, 19-20, 24, 41, 46, 49, 51, 56, 67-72, 82-7, 102, 105, 106
   classification 165-6
   primary energy consumption 101
   see also Non-OPEC developing countries; OPEC
Development criteria 51
Development projects 67
Diversification 52, 68
Drake, 'Colonel' Edwin I 4
Dubai 34, 152

East Mediterranean Agreements 14
Eastern Europe 7, 172, 176-9
   basic problem 177-8
   confrontation scenario 178-9
   co-operation scenario 178
   energy consumption 176, 178
   energy problem summarized 176-7
   energy resources 177
   energy supply and demand 177
   neutral scenario 178
   oil consumption 178
   oil production 177
Economic conditions 105, 120
Economic development 25, 86-90
Economic factors 102, 120
Economic growth 87, 100, 111, 116
Economic recession 124
Economic system 77
Ecuador 152
EEC (European Economic Community) 6, 141-4
Egypt 162, 163
Electricity generation 43
Energy conservation 43, 51, 105
   USA 124, 130
   Western Europe 140, 142, 143
Energy consumption 108, 116
   China 183
   Eastern Europe 176, 178
   historical pattern 7
   non-OPEC developing countries 165-7
   OPEC developing countries 157
   USSR 174
   Western Europe 140, 141, 146
Energy demand. See Energy supply and demand
Energy efficiency 102, 126, 143, 177
'Energy Emergency: Oil and Money 1980' (conference) 47
Energy forecast 74, 77-80

Energy in the Developing Countries 163
Energy Mobilization Board (EMB) 132
Energy policies and programmes 82-3
Energy problems 82-4, 88, 89
Energy projects 83, 84
Energy requirements 55
Energy resources 8-9, 19, 77, 82
  China 180, 181
  Eastern Europe 177
  Japan 147
  non-OPEC developing countries 161
  OPEC developing countries 152-3
  USA 123
  USSR 170-1
  Western Europe 139-40
Energy savings 130
Energy scenarios. See Confrontation; Co-operation; and Neutral scenarios
Energy Security Corporation (ESC) 132
Energy supply and demand 7-10, 53, 79, 94, 99, 108, 109, 116
  China 181
  Eastern Europe 177
  future prospects 100-2
  industrialized countries 92
  Japan 148
  non-Communist world 92
  non-OPEC developing countries 163
  OPEC developing countries 153
  review 91-103
  USA 124-6
  USSR 171-2
  Western Europe 140-1
  world 91-2, 94, 117
Energy Tax Act 1978 131
Energy transportation 52
Energy transportation industry 60
Environmental Regulations 127
Europe. See EEC; Eastern Europe; Western Europe
Exports 98, 100, 110, 157, 163, 171, 172, 181-2
Exxon 4

Fahd ibn 'Abd al-Aziz Al Sa'ud, HRH Crown Prince 165
Faisal ibn 'Abd al-Aziz Al Sa'ud, HM King 62
Fertilizers 8, 52
Financial assets 84
Financial climate 105, 107
Financial institutions 54
Financial problems 84
Financial resources 85
Finland 140
Flaring 52, 60
Foreign oil companies 53

co-operation with oil exporting countries 54
position of 54
France 5, 46, 139-43
Free trade 19
'Front-end loading' practices 96

Gabon 152, 153
Gas. See Natural gas, Liquefied natural gas, Liquefied petroleum gas
GATT (General Agreement on Tariffs and Trade) 49
General Participation Agreement 16
Geothermal energy 117-9, 148
Giant fields 93, 173
GNP growth 106, 107, 111, 116, 120, 136, 145, 147, 150, 168
GNP per capita 18, 34, 67, 102, 107, 117, 159, 166
Greece 139, 142, 143
'Group of 8' 69, 70
'Group of 19' 68, 154
'Group of 77' 68, 87
Growth rates 19, 20, 74, 76
Gulf 4

Health and safety controls 127
Heat pumps 102
Historical evolution of international oil industry 4-6
Historical pattern of energy consumption 7
Historical trends 91-4
Hydroelectricity 7, 8, 99, 117-9, 124, 140, 147, 161, 163, 168, 171, 174

Independent Commission of International Development Issues 71, See also Brandt Commission
India 159-64
Indonesia 4, 16, 150, 152, 154
Industrialized countries 18-9, 21, 24, 35, 45, 46, 67-73, 75, 78, 79, 83, 86, 87, 100, 101, 116
  confrontation scenario 112, 120-1
  co-operation scenario 111, 119-20
  neutral scenario 111-2, 120
  primary energy demand 92
Inflation rates 81, 84
Information system 42, 45
Integration 5, 22-3
International co-operation 83-5
International economic co-operation, conference 16
International Energy Agency (IEA) 6, 41-50, 96
  emergence 41
  formation 42

membership 45
nature of 45-9
performance 45-9
Principles of energy Policy (1977) 42-5
International energy co-operation 86
International energy co-operation and development programme 77, 81-5
International energy industries 51-4
International Energy Programme 42
International Monetary Fund 49
International monetary system 20, 68
International oil industry 11, 22, 73
  historical evolution 4-6
International petroleum industry 16, 22-3, 51
International petroleum marketing 58
International relations 84, 86, 104-6, 109
Investment 53, 54, 79, 107, 143
Iran 11, 14, 16, 17, 21, 47, 86, 96, 100, 152, 154
Iraq 11, 14, 16, 17, 39, 47, 81, 100, 152
Ireland 142
Islamic oil policies 79
Israel 41, 79, 88, 104, 109
Italy 5, 35, 46, 139-142

Japan 5, 20, 46, 71, 88, 96, 98, 99, 102, 104, 109, 116, 141, 147-51, 163, 171, 172
  coal 147, 150, 151
  confrontation scenario 151
  co-operation scenario 150
  energy consumption 150, 151
  energy policy 148-9
  energy problem summarized 147
  energy resources 147
  energy supply and demand 148
  Five Year Supply and Demand Plan 150
  neutral scenario 151
  non-conventional energy resources 148
  oil imports 149
  oil strategy 149
Japanese National Oil Corporation (JNOC) 148
Jerusalem 79, 84, 88
Jordan 161

Kelly J B 89
Khalid ibn 'Abd al-Aziz Al Sa'ud, HM King 62
Kissinger, Dr Henry 46-8
Kuwait 11, 16, 17, 34, 39, 152, 154

Lagos Agreement 14
Libya 13, 14, 16, 34, 152
Liquefied natural gas (LNG) 149, 150
Liquefied petroleum gas (LPG) 52, 149, 150
Living standards 87, 102, 109, 154, 157, 165
Los Angeles 127
Lubricants 8
Luxembourg 143

Malaysia 161, 162
Market premiums 96
Market strategies 75, 76
Marshall Plan 19
Master Gas Gathering and Processing System 60, 62
Mexico 5, 97, 160-3, 178
Middle East 4, 12, 87, 93, 98, 152
Military options 87
Mobil 4
Morocco 161

National Energy Act 130, 131
National Energy Conservation Policy Act 1978 131
National oil companies 5-6, 23, 47, 73-5
National sovereignty 61, 89
Natural gas 7, 8, 9, 44, 55
  Communist countries 121
  Eastern Europe 178, 179
  exploration 53, 54, 57
  industrialized countries 112, 119-20
  liquids 48, 52, 60, 76
  non-Communist countries 118-9
  non-OPEC developing countries 163, 168
  OPEC 153, 156
  prices 126, 130
  production 25, 60
  reserves 25, 39, 60, 106
  statistics 99
  USA 125, 126, 128, 130, 133, 136, 137
  USSR 170-4
  Western Europe 139, 140, 143, 145, 146
  world 110
Natural Gas Act 1938 126
Natural Gas Policy Act 1978 131
Natural resources 61, 68, 89
Netherlands 143
Neutral scenario 100, 106
  China 182
  Communist bloc 113, 121
  Eastern Europe 178
  Industrialized countries 111-2, 120

Japan 151
non-Communist countries 110, 119
non-OPEC developing countries 168
OPEC 156
United States 136-7
USSR 174
Western Europe 145
world 108, 117-8
New International Economic Order 68
New strategies 74-7
Nigeria 152, 154
Nixon, President 127
Non-commercial energy sources 161
Non-Communist countries 7-8, 73, 95, 101
confrontation scenario 111, 119
co-operation scenario 110, 118-9
neutral scenario 110, 119
primary energy demand 92
Non-OPEC developing countries 159-169
basic problems 163-5
coal 163
confrontation scenario 168
co-operation scenario 165-7
energy consumption 165-8
energy problem summarized 159-61
energy production 166
energy resources 161
energy supply and demand 163
fundamental problem 167
neutral scenario 168
non-commercial energy 161
oil consumption 162-3
oil production 162-3
oil reserves 162-3
policy options 163-5
special case 166
North Africa 4
North Korea 171
North Sea 5, 48, 139, 142, 144, 146
North-South: A Programme for Survival 52
North-South dialogue 67-72, 84, 87, 102, 104-7, 163
Norway 6, 18
Nuclear energy 7, 8, 44, 53, 60, 116-21, 123-4
growth deceleration 100
Japan 148, 149
non-Communist world 110
non-OPEC developing countries 168
statistics 99
USA 134, 136
USSR 171, 173, 174
Western Europe 140-6
Nuclear fuel 44
Nuclear proliferation 127

Nuclear war 109
Nuclear weapons 44

OAPEC (Organization of Arab Petroleum Exporting Countries) 15, 33-9, 41, 45, 85, 100
Council of Ministers 35
Energy Department 37
Executive Bureau 36
Exploration and Production Unit 36-7
formation of 33
Founding Agreement 34, 35
Judicial Board 37
membership 34
new projects 38
objectives 34
oil and gas proved reserves 39
organization 35
relationship to OPEC 35
Secretariat 36
specialized joint companies 37-8
training 37
Occidental 14
Oceanic energy 148
OECD (Organisation for Economic Co-operation and Development) 6, 18, 34, 35, 42, 45, 46, 49, 96, 101, 112, 119, 120, 140, 141
OEEC (Organisation for European Economic Co-operation) 46
Oil consumption 7, 8, 9, 13, 78
China 182, 183
Eastern Europe 178
non-OPEC developing countries 162-3
OPEC 153
USA 124, 126, 128, 129, 135-7
USSR 173, 174
Western Europe 146
Oil demand 21, 94-8
Oil discoveries 93, 116
Oil exploration 53, 54, 57
Oil exporting countries 53, 82
Oil prices 12-5, 20, 21, 24-30, 35, 41, 45-9, 56, 57, 73-8, 81, 83, 89, 95-7, 104-8, 115, 116, 126-30, 133, 134, 149, 150
Oil production 15, 34, 51, 57, 73, 81 82, 88
China 182, 183
Eastern Europe 178
NODC 162-3
OPEC 156
Oil reserves 9, 24, 34, 39, 51, 60, 82, 93, 100, 106, 110
Oil spills 127
Oil stocks 48-9, 95, 96

Oil supply and demand 73-5, 78-82, 95-7
Oman 162
OPEC (Organization of the Petroleum Exporting Countries) 5-6, 9, 11-32, 35, 41, 45-9, 56, 68, 71, 73, 75, 79, 83-9, 96, 97, 100-6, 112, 116, 121, 149, 151-8, 178
  achievements during the seventies 13-6
  basic problems of members 23-5, 154
  confrontation scenario 157
  co-operation scenario 155-6
  creation of 11-2, 20
  energy consumption 155-7
  energy problems summarized 152
  energy resources 152-3
  energy scenario projections 10
  energy supply and demand 153
  evolution of petroleum policy in member countries 13
  first decade 12-3
  formation 12
  governmental participation 16
  in world environment 29-31
  integration 22-3
  integration policy 23
  Long Term Strategy Committee 17, 29, 30, 48, 76, 78, 165, 166
  member developing countries 152
  membership 12
  neutral scenario 156
  non-Arab members 34
  oil consumption 153
  oil price actions 1973 20, 21
  oil price actions 1979-1980 21
  oil production 156
  participation 21-2
  policy options 154
  present and future options in price and production policies 25-9
  price actions 1978-1980 21
  relationship to OAPEC 36
  sole oil price administrator 15
  Solemn Declaration 16
  Special Fund 17, 166
  Summit Conference (Algiers) 68
  Tehran and related agreements 13-4
  Tehran conference 1973 15
  21st meeting (Caracas) 14
  see also Oil prices
Operation Dropshot 87
Organic waste 148

Pakistan 159, 161, 163
Palestinian question 79, 88
Participation 21-2

Penrose, Dr Edith 22
Peru 162
Petrochemicals 8, 52, 61
Petroleum derivatives 76
Petroleum marketing 58
Petroleum products 8, 52, 76
Petromin (General Petroleum and Mineral Organization) 55-60, 62
  future role in world oil industry 55
  international significance 55
  national objectives 55-6
  policies and their implementation 56-7
  relationship with Aramco 56-7
  significant projects 58-60
Pharmaceuticals 8
Philippines 160
Political factors 102, 109, 110, 120, 170
Portugal 140
Power Plant & Industrial Fuel Use Act 1978 131
Private oil companies 23
Public Utilities Regulatory Policy Act 1978 131

Qatar 16, 34, 39, 152

Railroad industries 78
Raw materials 68
Reagan, President 133-5
Red Sea Coast 60
Refinery projects 59
Research and development 44, 130, 132, 148
Residual fuel oil 43, 94, 145, 153
Risk analysis 115
Romania 177
Royal Dutch/Shell 4, 79
  see also Shell
Royalties 12, 56
Russian offensive in Asia and Africa 84, 88
Russian offensive in Middle East 87

Sabic 62
Santa Barbara 127
Saudi Arabia 11, 14, 16, 17, 22, 34, 39, 52, 55-66, 74, 78, 81, 89, 96, 97, 152-4, 157, 165
  domestic oil and gas policies 57
  financing of joint ventures 63-4
  future outlook 78-9
  gas exploration 57
  industrialization through joint ventures 61-6
  international petroleum marketing 58

management of joint ventures 65
manpower 66
marketing provisions 64-5
national objectives 55-6
oil exploration 57
oil policies 56
oil price policies 57
oil production level 57
philosophy underlying joint ventures 61-2
refinery projects 59
relationship with Aramco 56-7
resource economy 62-3
significant projects 58
supply assurance 62-3
see also Petromin
Saudi crude 74
Saudi master gas system 60, 62
Shale oil reserves 124, 125, 140, 161, 171, 177, 181
Sharia law 61
Sharjah 152
Shell 96
see also Royal Dutch
Sherman Anti-Trust Law 14
Sidon 13, 14
'Smog' problem 127
Socony - Vacuum 4
Solar energy 53, 60, 117-9, 130, 148, 153
Solemn Declaration 16
Solid fuels 113, 117-22, 146, 168, 174, 178
South Africa 160, 161
South Korea 160
Spain 139, 140
Special Fund 17
Standard Oil of California 4
Standard Oil of New Jersey 4
Statoil 18
Strategies, new 74-7
Structural changes 73
Surcharges 96
Swaziland 161
Sweden 139, 140
Switzerland 20
Synthetic fuels 132
Synthetic Fuels Corporation 132, 133
Syria 162

Taif Conference 17
Taiwan 160
Tanker capacity 60
Tariffs 105
Taxation 130-3
Technological change 102
Tehran Agreement 14, 15
Tehran pricing system 14

Texaco 4
Thailand 160, 161
Third World. See Developing Countries; North-South dialogue
Three Mile Island 99, 124, 125, 136
Tidal energy 117
Trade barriers 20, 105, 107
Trans-Arabian Peninsula Crude and NGL parallel pipelines 60
Transnational oil companies 4-5, 11-3, 16, 19, 22, 41, 56, 73, 78, 166
Transportation 52, 60, 98, 135
Trinidad and Tobago 161, 162
Tripoli Agreement (1971) 14
Tunisia 162

UNCTAD 83
Underground storage 60
United Arab Emirates 152, 154
United Kingdom 4, 6, 97, 141, 143
United Nations (UN) 12, 49, 53, 68, 83, 87, 89
United States of America (USA) 4, 13, 15, 19, 41, 46, 71, 87, 88, 96-8, 102, 104, 116, 123-38, 141, 174
  coal 124, 126, 130, 136
  confrontation scenario 137
  co-operation scenario 135-6
  energy conservation 124, 130
  energy economy 126
  energy future 134
  energy goals 130
  energy imports 126
  energy legislation 128
  energy plan 128-9
  energy policy 126-34
  energy problem 128
  energy problem summarized 123
  energy resources 123
  energy supply and demand 124-6
  Environment Regulations 127
  future energy consumption 134
  health and safety controls 127
  latest developments 133-4
  national energy programme 123
  neutral scenario 136-7
  oil imports 126, 130, 134
Uranium reserves resources 123, 140, 153, 161, 171, 177, 181
USSR 7, 98, 104, 109, 110, 163, 170-5
  basic problems 172-3
  coal 171, 172
  confrontation scenario 174
  co-operation scenario 173-4
  energy consumption 174
  energy problems summarized 170
  energy resources 170-1
  energy supply and demand 171-2

neutral scenario 174
oil consumption 173, 174

Venezuela 4, 11, 17, 153, 154
Vietnam War 19
Vittachi, Varinda Tarzie 87

Washington Energy Conference 41, 47
Wasteful practices 52
West Africa 4
West Germany 20, 46, 96, 139, 141, 143
Western Europe 5, 46, 71, 88, 98, 104, 139-46, 171, 172
  coal 139, 140, 142, 143
  confrontation scenario 146
  co-operation scenario 144-5
  energy conservation 140, 142, 143
  energy consumption 140, 141, 145
  energy policy 141-3
  energy problem summarized 139
  energy resources 139-40
  energy supply and demand 140-1
  neutral scenario 145
  oil consumption 146
  oil imports 139
Wind energy 117
'Windfall profits' tax 133
World
  coal 117, 118
  confrontation scenario 109-10, 118
  co-operation scenario 108, 116-7
  energy balance 109, 110
  energy demand 91-2, 94, 117
  neutral scenario 108, 117-8
World Alternative Energy Strategies of 1977 116
World Bank 49, 83-5, 102, 163, 165, 166
World Energy Conference 79

Yamani, His Excellency Shaikh Ahmed Zaki 29
Yugoslovia 160, 161, 171